INTERCONNECT TECHNOLOGY AND DESIGN FOR GIGASCALE INTEGRATION

INTERCONNECT TECHNOLOGY AND DESIGN FOR GIGASCALE INTEGRATION

Edited by

JEFF DAVIS
Georgia Institute of Technology

JAMES D. MEINDL
Microelectronics Research Center, Georgia Institute of Technology

Kluwer Academic Publishers
Boston/Dordrecht/London

Distributors for North, Central and South America:
Kluwer Academic Publishers
101 Philip Drive
Assinippi Park
Norwell, Massachusetts 02061 USA
Telephone (781) 871-6600
Fax (781) 871-6528
E-Mail <kluwer@wkap.com>

Distributors for all other countries:
Kluwer Academic Publishers Group
Post Office Box 322
3300 AH Dordrecht, THE NETHERLANDS
Telephone 31 78 6576 000
Fax 31 78 6576 474
E-Mail <orderdept@wkap.nl>

Electronic Services <http://www.wkap.nl>

Library of Congress Cataloging-in-Publication

Title: Interconnect Technology and Design for Gigascale Integration
Author: Edited by Jeff Davis and James D. Meindl
ISBN: 1-4020-7606-1

Printed on acid-free paper.

Printed in the United States of America

Contents

Preface

The International Technology Roadmap for Semiconductors (ITRS) projects that by 2011 over one billion transistors will be integrated into a single monolithic die. The wiring system of this billion-transistor die will deliver power to each transistor, provide a low-skew synchronizing clock to latches and dynamic circuits, and distribute data and control signals throughout the chip. The resulting design and modeling complexity of this GSI multilevel interconnect network is enormous such that over 10^{17} coupling inductances and capacitances throughout a nine-to-ten-level metal stack must be managed. This book will address the limits and opportunities for GSI interconnect design and technology in the twenty-first century.

This book is the cumulative effort from academic researchers at Georgia Tech, MIT, and Stanford, as well as from industry researchers at IBM T.J. Watson Research Center, LSI Logic, and SUN microsystems. The material found in this book is unique in that it spans IC interconnect topics ranging from IBM's revolutionary copper process to an in depth exploration into interconnect-aware computer architectures. This broad swath of topics presented by leaders in the research field is intended to provide a comprehensive perspective on interconnect technology and design issues so that the reader will understand the implications of the semiconductor industry's next substantial milestone – gigascale integration.

A summary of the chapters provided by each author is listed below.

Chap. 1: Interconnect Opportunities for GSI

Throughout the past four decades semiconductor technology has advanced at exponential rates in both productivity and performance. In recent years multilevel interconnect networks have become the primary limit on the productivity, performance, energy dissipation, and signal integrity of GSI. Consequently, a broad spectrum of novel solutions to the multi-faceted interconnect problem must be explored. In this chapter, we review recent salient results of this exploration. Based on prediction of the complete stochastic signal interconnect length distribution of a megacell, optimal reverse scaling of each pair of wiring levels provides a prime opportunity to minimize cell area, clock period, power dissipation, or number of wiring levels. Using a heterogeneous version of Rent's rule, a design methodology for the global signal, clock, and power/ground distribution networks for a system-on-a-chip has been derived. Wiring area, bandwidth, and signal integrity are the prime constraints on the design of the networks. Three dimensional integration offers the opportunity to reduce the length of the longest global interconnects in a distribution by as much as 75%. Wafer level batch fabrication of chip input/output interconnects and chip scale packages provides new opportunities such as I/O bandwidth enhancement, simultaneous switching noise reduction, and lower cost of packaging and testing. Microphotonic interconnects have long term potential to improve latency, bandwidth, power dissipation, and crosstalk.

Chap. 2: Cu BEOL Interconnects for Silicon CMOS Logic Technology
In August of 1997, IBM announced its schedule for what would be the first implementation of Cu interconnect technology on integrated circuit (IC) chips. This has been one of the most significant revolutions in BEOL technology since the invention of the integrated circuit. In order to implement Cu as the wiring metal in IC chips, significant advances had to be made in several aspects of BEOL technology. These include a detailed understanding of the microstructure of Cu and its dependence on deposition method and process conditions, its influence on the electromigration behavior of these interconnects, dual damascene integration and all the associated patterning and integration challenges posed by this technique, and the invention of an effective liner scheme that would prevent the diffusion of Cu into the insulator. In this chapter, all these aspects of Cu interconnect technology will be reviewed, beginning with a brief look at the evolution of the BEOL interconnect through the years leading up to key innovations such as dual damascene processing and chemical mechanical polishing which have enabled the introduction of copper interconnects.

Chap. 3: Interconnect Parasitic Extraction of R, C, and L
Interconnect parasitic resistance, capacitance and inductance extraction is critical for determining delay, signal integrity and power consumption in deep sub-micron IC technology. After briefly reviewing electromagnetics, fundamental RC extraction algorithms as well as practical extraction techniques are discussed in this chapter. With circuit clock and data frequencies operating in the multi-gigahertz range and faster signal rise and fall times, parasitic inductance of the wires can no longer be ignored. Modeling of on-chip inductance is then presented for chips with power/ground wires and grids that emulate those used in practical circuits. The models and analytical formulae capture 3-D geometry, multi-conductor environment and substrate effects.

Chap. 4: Distributed RC and RLC Transient Models
This chapter outlines the solutions to the partial differential equations (PDEs) that describe the transient voltages along global and semi-global VLSI interconnects that can be modeled with distributed RC or RLC networks. The exact transient solutions for various regular GSI interconnect structures are simplified such that closed-form expressions for the time delay, overshoot, and crosstalk are derived. The implications of non-ideal return paths are discussed for current and future GSI designs.

Chap. 5: Power, Clock, and Global Signal Distribution Techniques
The objective of this chapter is to provide a technique for the design and optimization of global interconnect networks in an integrated architecture for global signal, clock, and power-supply networks. To enhance the understanding of limits associated with the interconnect architecture, salient sets of models for global signal, clock, and power-supply distribution networks are required to describe the interconnect characteristic in future generations of technology. These new interconnect models are utilized to create an integrated architecture for global interconnects in a GSoC.

Chap. 6: Stochastic Multilevel Interconnect Modeling & Optimization
It has been shown that optimized logic networks have certain collective properties that can be described with Rent's Rule. Using Rent's Rule as a basis, a wire length distribution is rigorously derived, and from this distribution, basic models for dynamic power dissipation, critical path models, and chip area are derived. Finally, a unique application of a stochastic wiring distribution and a cycle time model is the optimization of a multilevel wiring network to produce complex digital systems with optimized clock frequency, die size, or power dissipation. The distinguishing feature of the repeater insertion methodology in these multilevel interconnect architectures is that it simultaneously solves two distinct problems - optimal wire sizing and wiring layer assignment – using independent constraints on maximum repeater area for efficient design space exploration to optimize the area, power, clock frequency and metal levels of a GSI logic megacell.

Chap. 7: Interconnect-Centric Computer Architectures
Designing "interconnect-aware" architectures requires a study of algorithm, architecture, and technology to localize computation and communication. The multimedia application domain provides new opportunities to define an efficient computation substrate for embedded systems that scale well with poor wires. These architectures have simpler, distributed processing elements for fine-grain processing. This chapter presents system models that capture interconnect demand and uses these models to explore the potential of different

architectures and technology. Needed architectural research employs these approaches to localize computation and communication in an efficient computing substrate. Projected technology capabilities from the National Technology Roadmap for Semiconductors (NTRS) are used to make realistic design choices and to offer a view of performance in future designs. Key performance metrics such as chip area, cycle-time, and power dissipation are used for comparison. The main tasks to study the relationship between architecture and wire demands are the definition and implementation of system models that ties architecture description to wiring models. The validity of the models is verified with comparisons against actual chip data. Architectural explorations can then be performed to provide insights into the performance of different designs.

Chap. 8: Chip-to-Module Interconnect
The role of the chip-to-module connection technology is to provide vital services and communications to the IC and inflict a minimum degradation in performance and minimum yield loss at the lowest possible cost. The methods of chip-to-module interconnection for GSI are largely evolutionary in nature because they depend on complimentary advances in substrates (i.e. printed circuit board) and ICs. In this chapter, the goals and requirements for an effective chip-to-module connection technology will be presented. This will be developed by first providing a brief historical review of substrate and connection methodologies which have served the industry. The review will provide a background as to the current infrastructure. Then, the approaches to compliant chip-to-module interconnections will be presented along with future challenges.

Chap. 9: 3-D IC DSM Interconnect Performance Modeling & Analysis
This chapter analyzes the limitations of existing interconnect technologies and design methodologies and presents a novel 3-dimensional (3-D) chip design strategy that exploits the vertical dimension to alleviate the interconnect related problems and to facilitate SoC applications. A comprehensive analytical treatment of these 3-D ICs has been presented and it has been shown that by simply dividing a planar chip into separate blocks, each occupying a separate physical level interconnected by short and vertical inter-layer interconnects (VILICs), significant improvement in performance and reduction in wire-limited chip area can be achieved, without using any other circuit or design innovations. A scheme to optimize the interconnect distribution among different interconnect tiers is presented and the effect of transferring the repeaters to upper Si layers has been quantified in this analysis for a two-layer 3-D chip. Furthermore, one of the major concerns in 3-D ICs arising due to increased power density has been analyzed and an analytical model has been presented to estimate the temperatures of the different active layers. It is demonstrated that

advancement in heat sinking technology will be necessary in order to extract maximum performance from these chips. Implications of 3-D device architecture on several circuit designs and CAD methodologies and tools have also been discussed with special attention to SoC design strategies. Finally, some of the promising technologies for manufacturing 3-D ICs have been outlined.

Chap. 10: Silicon Microphotonics
A rebuilding of the world's information infrastructure is taking place to give instantaneous availability of data, voice and video. This revolution of the Information Age is being gated more by the introduction of new materials and components, than by the design of systems, software and networks. Electrons transmitted through metal wires have an information carrying capacity limited by the resistance and capacitance of the cable and the terminating electronic circuits. Photons transmitted through fiber are capacity limited only by the dispersion of the medium. Each network node that requires transduction from photonics to electronics limits the performance and affordability of the network. The key frontier is the large scale integration and manufacturing of photonic components to enable the distribution of high bit rate optical streams to the individual information appliance. Microphotonics is the platform for large scale, planar integration of optical signal processing capability.

The editors would like to the thank the authors for their time and patience in the completion of this book. We feel that the coordination and the time invested by everyone involved has produce a book of high quality and broad interest to both the industry and academic readers.

Jeff Davis and James Meindl

School of Electrical and Computer Engineering
Georgia Institute of Technology
Atlanta, GA

Chapter 1

INTERCONNECT OPPORTUNITIES FOR GIGASCALE INTEGRATION (GSI)

J.D. Meindl, J.A. Davis, P. Zarkesh-Ha*, C. Patel**, K.P. Martin, and P.A. Kohl

Georgia Institute of Technology

**LSI Logic Corporation*

***IBM T.J.Watson*

1.1 Introduction

Throughout the past four decades semiconductor technology has advanced at exponential rates in both productivity and performance. Three generic strategies have guided these advances: 1) scaling down minimum feature size, 2) increasing die size, and 3) enhancing packing efficiency (defined as the number of transistors or length of interconnect per minimum feature square of silicon area). Scaling of transistors reduces their cost, intrinsic switching delay, and energy dissipation per binary transition. Scaling of interconnects serves to reduce cost but increases latency in absolute value and energy dissipation relative to that of transistors. These increases result from relatively larger average interconnect lengths (measured in gate pitches) and larger die sizes for successive generations. Therefore, interconnects have become the primary limit on both the performance and the energy dissipation of GSI.

Following this brief introduction, Section 1.2 quantifies the key facets of the interconnect problem. The principal generic opportunities to resolve it including new materials and processes, scaling, and novel architectures are reviewed in Section 1.3 with emphasis on scaling. Reverse scaling of multilevel interconnect networks is based upon prediction of stochastic signal wiring distributions to achieve minimum area, power dissipation, clock period or number of metal levels. A methodology to derive an integrated architecture for global signal, power, and clock distribution networks for a system-on-a-chip is reviewed in Section 1.4. Sections 1.5, 1.6, and 1.7 explore three unconventional approaches to alleviating the on-chip interconnect problem. These are novel three-dimensional structures, high-density input/output interconnect enhancements, and compatible microphotonic interconnects, respectively. A brief conclusion is provided in Section 1.8.

1.2 The Interconnect Problem

What is the quintessential purpose of an interconnect? In a single word it is communication. To be more complete, it is communication between distant points with small latency. A lucid illustration that displays this key purpose is a graphic whose vertical axis is reciprocal interconnect length squared and whose horizontal axis is interconnect latency[1]. Using logarithmic scales on both axes, a diagonal line is a locus of constant distributed resistance- capacitance product, the principal figure of merit of the large majority of interconnects used for GSI. As illustrated in Figure 1.1, reducing the distributed resistance-capacitance product moves the diagonal locus toward the lower left corner of the display therefore providing smaller latency for a given interconnect length. However, during the past four decades interconnect scaling has increased the distributed resistance- capacitance product thus moving toward the upper right corner of the display and therefore demanding larger latency for a given interconnect length as illustrated in Figure 1.2(a). In stark contrast as illustrated in Figure 1.2(b), scaling of transistors reduces the power-delay product or switching energy of a binary transition therefore moving toward the lower left corner of the power-delay plane to reduce simultaneously both average power transfer and delay.

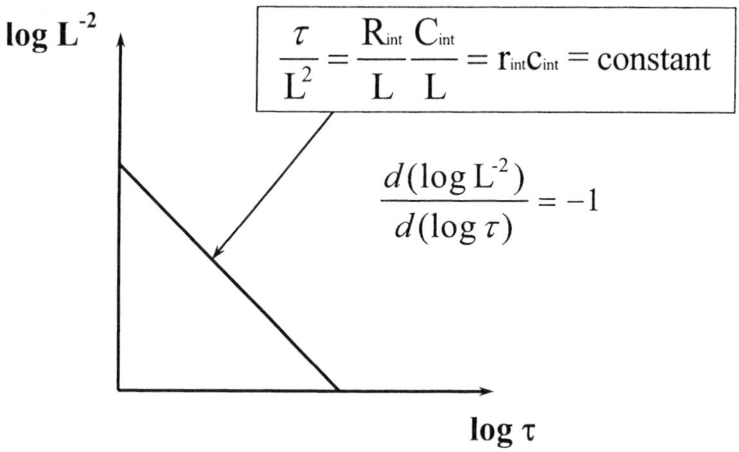

Figure 1.1. Interconnect reciprocal length squared ($1/L^2$) vs. latency (τ) with log scales illustrating diagonal lines as loci of constant distributed resistance-capacitance product ($r_{int}c_{int}$).

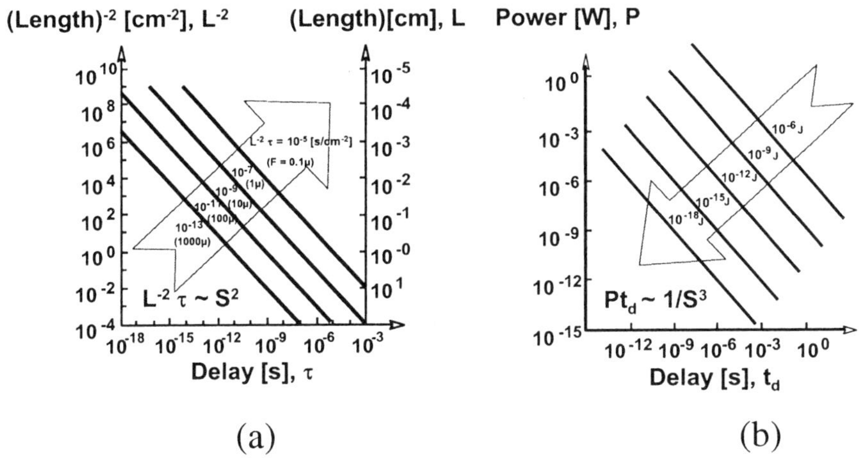

Figure 1.2. (a) Interconnect ($1/L^2$) vs. (τ) and (b) transistor average power transfer (P) vs. transition delay (t_d) illustrating diametrically opposite results of scaling down minimum feature sizes.

Table 1.1. MOSFET and interconnect latency for 1.0 μm, 100 nm, and 35 nm technology generation [2]

Technology Generation	MOSFET Switching Delay ($t_d = CV/I$)	"RC" Response Time ($L_{int} = 1$ mm)	Time-of-flight ($L_{int} = 1$ mm)
1.0 μm (Al, SiO_2)	$\approx$ 20 ps	$\approx$ 1 ps	$\approx$ 6.6 ps
100 nm ($Cu, \kappa = 2.0$)	$\approx$ 5 ps	$\approx$ 30 ps	$\approx$ 4.6 ps
35 nm ($Cu, \kappa = 2.0$)	$\approx$ 2.5 ps	$\approx$ 250 ps	$\approx$ 4.6 ps

In order to quantify the exploding disparity between the latency of interconnects and transistors consider the comparisons illustrated in Table 1.1. For the one micron generation technology of the late 1980's, the "CV/I" or intrinsic switching delay of a MOSFET [3] before it is loaded with parasitic or wiring capacitance is approximately 20 ps. However, for the same generation, the total resistance-capacitance product or RC delay of a "benchmark" 1.0 mm long interconnect is about 1.0 ps. In comparison, for the 100 nm generation projected for early production in 2005, the CV/I delay of a MOSFET decreases to 5 ps while the RC latency of a 1.0 mm long wire increases to 30 ps. The relevant observation is that as semiconductor technology is advancing from the 1.0 μm to the 100 nm generation, the RC delay or response time of a benchmark 1.0 mm long interconnect is devolving from 20 times faster to six times slower than transistor intrinsic switching delay. Furthermore, the 1999 ITRS projection for 35 nm technology in 2014 suggests a 2.5 ps transistor delay and a 250 ps RC latency for a 1.0 mm long interconnect [2]. For completeness, the time-of-flight (ToF) of a 1.0 mm long interconnect is included in Table 1.1. As indicated, ToF delay is independent of scaling but does depend on the value of the relative permittivity of the interconnect dielectric.

In order to underscore the formidable challenge presented by interconnects to continued performance improvements for GSI, it is noteworthy that the numerical values for RC delay cited in Table 1.1 represent simple best case calculations. For example, the results do not account for the adverse results of surface scattering, high frequency skin effect, liner thickness for copper interconnects or temperature rises in a multilevel wiring network.

Beyond latency , interconnects present an energy dissipation problem illustrated in Table 1.2 that also limits the performance of GSI as a consequence of practical constraints on the heat removal capacity of the package of a gigascale chip or the energy storage capacity of its portable power source. Again comparing technology generations, it is evident that the energy dissipation associated with a binary transition of a minimum geometry MOSFET versus a 1.0 mm long interconnect is 33%, five times, and thirty times larger for the interconnect for the 1.0 μm, 100 nm, and 35 nm technology generations, respectively. These gross imbalances clearly indicate that the power dissipation problem of gigascale chips is essentially an interconnect problem.

Table 1.2. ITRS projections for switching delay, switching energy, clock frequency, total chip current drain, maximum number of wiring levels, maximum total wire length per chip and chip pad count for 1.0 μm, 100 nm, and 35 nm technology generations [2].

	Technology Generations		
ITRS Parameter	1.0 μm	100 nm	35 nm
MOSFET Switching Delay	$\approx$ 20 ps	$\approx$ 5 ps	$\approx$ 2.5 ps
Interconnect "RC" Response Time ($L_{int} = 1$ mm)	$\approx$ 1 ps	$\approx$ 30 ps	$\approx$ 250 ps
MOSFET Switching Energy	$\approx$ 300 ps	$\approx$ 2 fJ	$\approx$ 0.1 fJ
Interconnect Switching Energy	$\approx$ 400 ps	$\approx$ 10 fJ	$\approx$ 3 fJ
Clock frequency	$\approx$ 30 MHz	$\approx$ 2-3.5 GHz	$\approx$ 3.6-13.5 GHz
Supply Current ($V_{dd} = 5.0, 1.0, 0.5$ V)	$\approx$ 2.5 A	$\approx$ 150 A	$\approx$ 360 A
Max. number of wiring levels	3	8-9	10
Max. total wirelength per chip	$\approx$ 100 m	$\approx$ 5000 m	( )
Chip pad count	$\approx$ 200	$\approx$ 3000-4000	$\approx$ 4000-4400

The preceding discussion of latency and energy dissipation problems presented by interconnects is concerned with signal wiring. Historical records and ITRS projections [2] of clock frequencies for high performance microprocessors summarized in Table 1.2 indicate 30 MHz, 3.0 GHz, and 13 GHz as the nominal clock frequencies for the 1.0 μm, 100 nm, and 35 nm technology generations, respectively. These rapidly escalating requirements place enormous new demands on the interconnects that implement clock distribution networks of gigascale chips. Bandwidth, power dissipation, skew, and jitter represent increasingly formidable issues.

Although gigascale signal and clock distribution network problems are daunting, power distribution may well match them in difficulty. As noted in Table 1.2, estimated maximum chip current drain is 2.5 A, 150 A, and 360 A for the 1.0 μm, 100 nm, and 35 nm technology generations, respectively. Concurrently, power supply voltage scales down from 5.0 V, to 1.0 V, to 0.5 V for the corresponding generations. These aggressive expectations for high current, low voltage power distribution impose utterly unprecedented demands on interconnect networks.

Finally, the targets for number of wiring levels, maximum total interconnect length, and number of bonding pads or input/output interconnects per chip cited in Table 1.2 add significantly to expectations for future interconnect capabilities. In short, the highly demanding requirements that are projected for on-chip wiring compel comprehensive research over the most extensive and multidimensional solution space that can be conceived.

1.3 Reverse Scaling

Approximate expressions for the latency (τ) of a single isolated interconnect that is RC limited with an ideal return path are given by

$$\tau_{90\%} \approx r_{int}c_{int}L^2 + 2.3R_{tr}c_{int}L + 2.3C_L(r_{int}L + R_{tr}), \tag{1.1}$$

$$\tau_{90\%} \approx r_{int}c_{int}L^2 + 2.3R_{tr}c_{int}L \quad \text{for} \quad C_L << c_{int}L \tag{1.2}$$

and

$$\tau_{90\%} \approx r_{int}c_{int}L^2 \quad \text{for } C_L << c_{int}L \quad \text{and } R_{tr} << r_{int}L, \tag{1.3}$$

where r_{int} and c_{int} are the interconnect resistance and capacitance per unit length, respectively, R_{tr} is the source resistance, C_L is the load capacitance, and L is interconnect length. The latency of a low resistance interconnect that is resistance, inductance, and capacitance or RLC limited is given by

$$\tau_{90\%} \approx ToF = L/[c_0/(\epsilon_r)^{1/2}] \tag{1.4}$$

where

$$\frac{R_{\text{int}}}{Z_o} \leq 2\ln\left[\frac{4Z_o}{R_{tr} + Z_o}\right], \qquad R_{tr} < 3Z_o \qquad \text{and} \qquad C_L << c_{\text{int}}L \qquad (1.5)$$

are required for ToF response. In (1.5), Z_o is the characteristic impedance and $R_{int} = r_{int}L$ is the total resistance of the interconnect; c_o is the velocity of light in free space and ϵ_r is the relative permittivity of the interconnect insulator. Since RC limited performance is far more common than ToF limitations, the RC case is considered in this section.

The simple relationship given by (1.3) serves as the basis for reviewing the principal generic opportunities for solving the key latency problem. The latency of an RC limited interconnect can be expressed as the product of three factors as indicated in (1.6).

$$\tau = [\rho\varepsilon]\left[\frac{1}{\text{HT}}\right]\left[\text{L}^2\right] \qquad (1.6)$$

The resistivity-permittivity factor $[\rho\epsilon]$ identifies opportunities to reduce latency through new materials and processes such as the replacement of aluminum with double damascene copper [4]. The $[1/HT]$ factor, where H defines metal height and T defines insulator thickness, represents device and circuit level [1] opportunities to reduce latency through reverse scaling. Finally, L defines the length of an interconnect and the $[L^2]$ factor represents system level [5] opportunities to improve latency through the use of new microarchitectures that serve to "keep interconnects short." Solutions to the latency problem must be pursued at each of the levels – material and process, device, circuit, and system [1] represented in (1.6). The scope of this section is confined to device, circuit, and system level opportunities to reduce latency though reverse scaling. In comparison to alternatives such as new materials and processes as well as novel architectures, the compelling advantages of reverse scaling are: 1) minimal time to implementation, 2) low cost of implementation, 3) low risk, and 4) high payoff.

The key to optimal reverse scaling is the capability to predict the complete stochastic interconnect density distribution for a projected next generation product. Consider the case of a macrocell consisting of a random logic network of N microcells or logic gates. As illustrated in Figure 1.3, the macrocell can be modeled as a square array of logic gates. Rent's Rule ($R = kN^p$) [6] and the principle of conservation of interconnects are applied recursively to the macrocell as indicated in Figure 1.3. A closed form expression for the complete stochastic signal wiring distribution resulting from this process is given by [7]

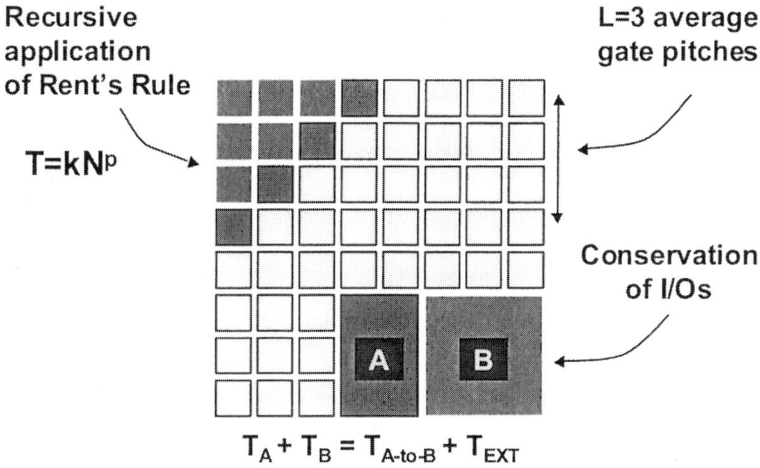

Figure 1.3. Diagram of a macrocell consisting of a random logic network of N microcells or individual logic gates configured in a square array. Rent's Rule and the principle of conservation of interconnects are applied recursively to derive the complete stochastic interconnect length distribution of the random logic network.

Region I: $1 \leq L < \sqrt{N}$

$$f(L) = \Gamma \frac{\alpha k}{2} \left(\frac{L^3}{3} - 2\sqrt{N} L^2 + 2NL \right) L^{2p-4} \tag{1.7}$$

Region II: $\sqrt{N} \leq L \leq 2\sqrt{N}$

$$f(L) = \Gamma \frac{\alpha k}{6} \left(2\sqrt{N} - L \right)^3 L^{2p-4} \tag{1.8}$$

where

$$\Gamma = \frac{2N\left(1 - N^{p-1}\right)}{\left(-N^p \frac{1+2p-2^{2p-1}}{p(2p-1)(p-1)(2p-3)} - \frac{1}{6p} + \frac{2\sqrt{N}}{2p-1} - \frac{N}{p-1} \right)}. \tag{1.9}$$

Equation (1.7) applies to shorter interconnects and (1.8) to longer interconnects in the distribution. These expressions reveal the dependence of interconnect density ($f(L)$ in units of number of interconnects of length L per gate pitch) versus interconnect length L in gate pitches. The dependencies on interconnect length L, number of gates in the network N, Rent's coefficient k, and Rent's exponent, p are evident. As demonstrated in Figure 1.4, this stochastic wiring distribution is found to be in close agreement with experimental data characterizing commercial products [7]. The key to obtaining close agreement between predicted and actual wiring distributions for a new product is to derive appropriate values of Rent's coefficient k and exponent p using data from previous generations of a product family. These two empirical parameters appear to have *genetic* characteristics.

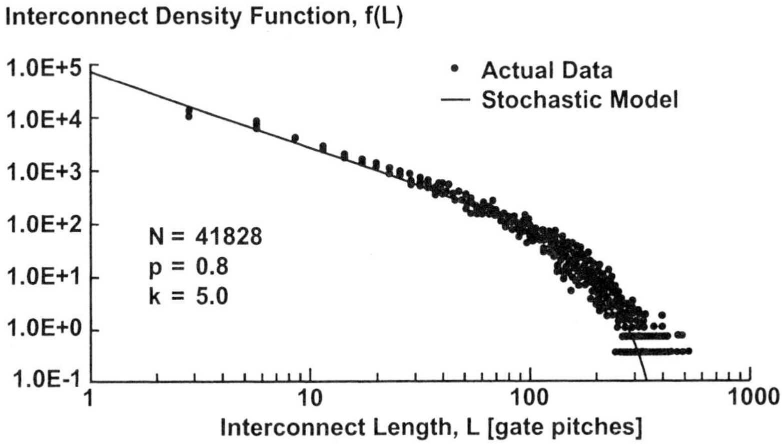

Figure 1.4. Interconnect density distribution. The vertical axis represents the density of interconnects of length L in units of number of interconnects per gate pitch and the horizontal axis represents interconnect length in gate pitches. A gate pitch is the center-to-center spacing of the gates in the square array of Figure 1.3. Actual data is taken from a commercial microprocessor [7].

An optimal architecture for a multilevel interconnect network that minimizes macrocell area, power dissipation, clock cycle time , or number of wiring levels can be derived using the stochastic interconnect distribution given by (1.7-1.9). A derivation for minimum macrocell area begins with the formulation of a wiring area "supply and demand" equation (1.10) [8]

$$2e_w A_m = \chi p_n \sqrt{\frac{A_m}{N}} \int_{L_{n-1}}^{L_n} L f(L) dL \tag{1.10}$$

$$p_n = 2\sqrt{\frac{1.1\rho\varepsilon_r\varepsilon_0 6.2 f_c}{\beta}}\sqrt{\frac{A_m}{N}} L_n \tag{1.11}$$

$$p_n = 2.5\frac{2f_c}{\beta}\sqrt{6.2\rho\varepsilon_r\varepsilon_o R_o C_o}\sqrt{\frac{A_m}{N}} L_n \tag{1.12}$$

The available area for an orthogonal pair of wiring levels can be expressed as $2e_w A_m$ where e_w is a wiring efficiency factor that must be estimated from previous designs and A_m is the area of the macrocell. The required area is defined by the right hand side of (1.10) where $\chi < 1$ converts point- to-point wire length to net length [7]. (Net length is the total length of wiring that connects the output terminal of a driver gate to the inputs of its load gates.) The factor p_n is wire pitch , the square root factor is gate pitch (in cm), and the integral represents the total length of wire in gate pitches between its upper (L_n) and lower (L_{n-1}) length limits. Based on a distributed RC network model, (1.11) imposes a latency requirement on the longest interconnect (of length L_n) on a given pair of wiring levels. The required latency is expressed by β/f_c where $\beta < 1$ and $1/f_c$ is the clock period. In essence, the first and second equations are solved simultaneously for the minimum pitch p_n and maximum corresponding wire length L_n for each pair of wiring levels until L_n equals the maximum required wire length of the macrocell on its top pair of wiring levels. Equations (1.10) and (1.11), are solved simultaneously if repeaters are not used while (1.10) and (1.12) apply if optimal repeaters are used [9]. The parameters R_0 and C_0 represent the output resistance and input capacitance , respectively, of a minimum geometry MOSFET used as the basis for the repeater circuits [10].

An example of minimization of macrocell area is illustrated in Figure 1.5. A random logic network consisting of 12.4 million gates implemented with 100 nm generation technology using eight levels ($n = 8$) of copper interconnects and operating at a clock frequency f_c = 578 MHz is considered. Two alternative wiring network architectures are compared. The first architecture shown on the left is restricted to two and only two different cross-sectional dimensions (or two tiers) for eight levels of wiring. It requires two levels of 100 nm and six levels of 540 nm wiring as well as a macrocell area A_m = 2.34 cm^2 to interconnect the macrocell. The second architecture shown on the right is optimized to use three tiers of wiring in order to minimize cell area. It therefore requires four levels of 100 nm wiring, two levels of 150 nm wiring, and two levels of 300 nm wiring as well as a macrocell area A_m = 0.70 cm^2. The decisive macrocell area advantage of the three tier architecture is achieved using the methodology defined in (1.10), (1.11), and (1.12), whose central feature is demand prediction based upon a complete stochastic wiring distribution, $f(L)$ [8], [9].

A second and currently more realistic example of an optimal multilevel network architecture is illustrated in Figure 1.6. In this case the macrocell consists of an 11.3 million gate random logic network implemented with 100 nm technology using eight levels of copper wiring ($n = 8$) and operating at a clock frequency of 1.56 GHz. If the pitch is chosen *a priori* to double for every pair of levels, the resulting architecture consists of two levels each of 100, 200,

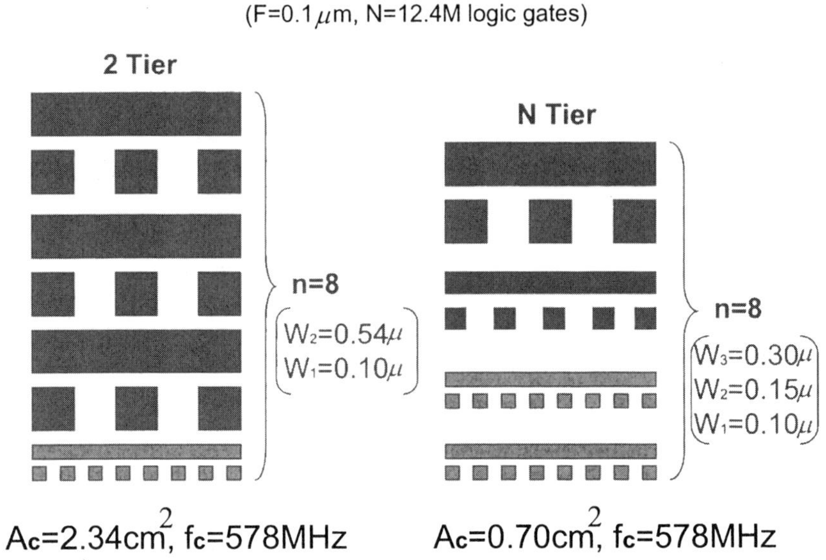

Figure 1.5. Comparison of wiring limited macrocell areas A_m required for a two tier versus an optimal three tier multilevel interconnect network architecture.

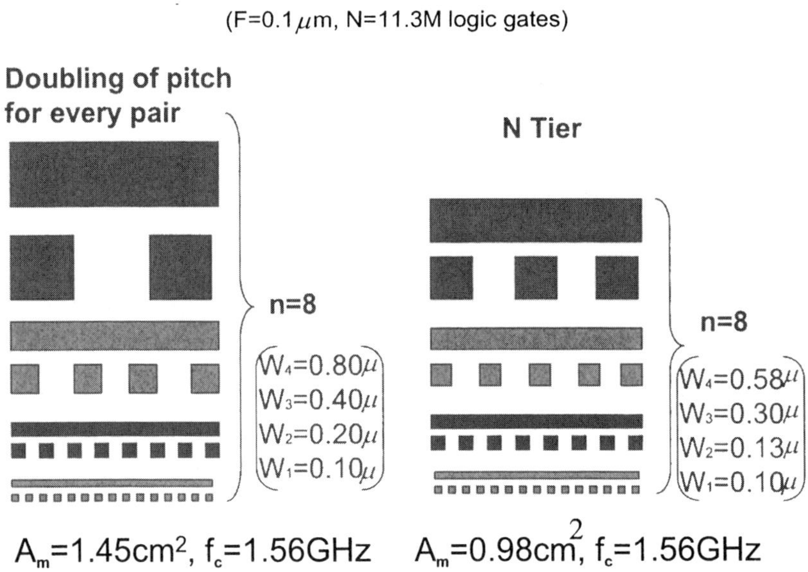

Figure 1.6. Comparison of non-optimized and optimized four tier architectures.

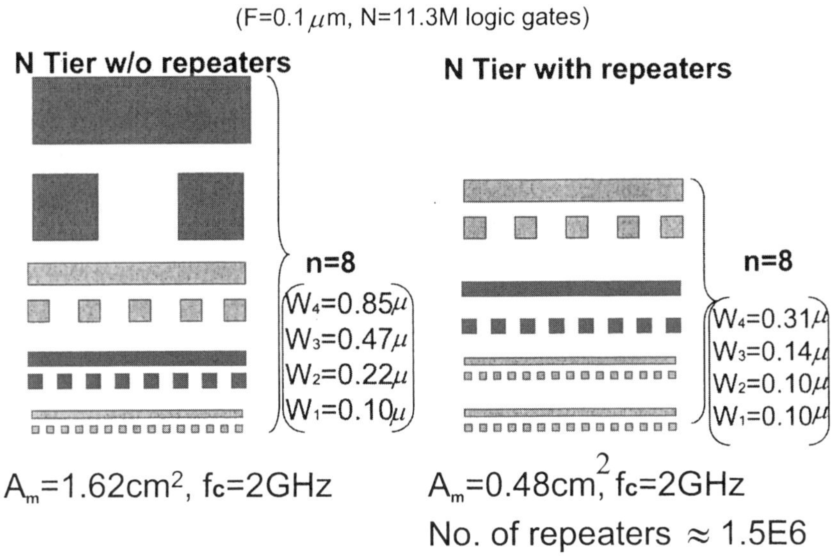

Figure 1.7. Comparison of optimized four tier architectures without and with repeaters.

400, and 800 nm wiring, which require a 1.45 cm^2 area. In contrast, using the methodology prescribed by (1.10) and (1.11), the optimal wire level pair dimensions are 100, 130, 300, and 580 nm, which yield a macrocell area of 0.98 cm^2 or approximately a 32% reduction. If 1.5×10^6 repeaters are used, macrocell clock frequency can be increased to f_c = 2.0 GHz and area reduced to 0.48 cm^2 as illustrated in Figure 1.7.

As indicated by (1.10) , determination of the area available for signal wiring on an orthogonal pair of levels requires estimation of the wiring efficiency factor e_w based on results of previous designs. As the number of wiring levels and the number of repeater circuits increase, via blockage tends to reduce wiring efficiency. The impact of via blockage can be estimated by calculation of a via blockage factor

$$B_V = A_V / A_m \tag{1.13}$$

where A_V is the area blocked by vias on a given level of wiring and A_m is the macrocell or chip area. As illustrated in Figure 1.8(a), terminal vias (that connect a particular interconnect net to a transistor) cause a "ripple effect" that reduces the number of wiring tracks available in a given area. In contrast, turn vias (that connect two wiring levels) do not cause via blockage. To elucidate with a simple example illustrated in Figure 1.8(a), $B_V = 0$ for the five uninterrupted wiring tracks on the left without terminal vias and $B_V = 0.2$ for the four wiring tracks on the right where 20% of the available wiring area is blocked by terminal vias, (three of which are shown in Figure 1.8(a) [11]. Assuming a uniform distribution of terminal vias as illustrated in Figure 1.8(b), a general expression can be derived for B_V in terms of the geometry of the wiring layout [11]:

$$B_V = \sqrt{\frac{N_V(2W+s\lambda)^2}{A_m}} \tag{1.14}$$

where N_V is the total number of terminal vias for a particular metal level on a chip and W, s, and λ are defined in Figure 1.8(b). The number of terminal vias, N_V, for a given wiring level is determined by the total number of interconnects on wiring levels above the given wiring level using the methodology defined by (1.7), (1.8), and (1.9). From (1.13) and (1.14), the via blockage factors for the eight wiring levels used in two macrocells (with F = 100 nm and N = 12.4 million gates) similar to those described in Figure 1.5 are illustrated in Figure 1.9 [11].

Figure 1.9 reveals two striking features of via blockage due to signal interconnects as predicted by the new model. First, via blockage is more problematic for relatively small area macrocells due to their greater interconnect density. More importantly, via blockage is severe only on the first level of wiring where 15-30% of the total wiring area of a representative macrocell can be blocked. The via blockage estimate based on a previous model [12] is also illustrated in Figure 1.9.

1.4 System-on-Chip

The previous section deals with reverse scaling of signal wiring for a macrocell that may be modeled as a largely homogeneous block of microcells. A second commonly encountered situation is a system-on-a-chip consisting of a number of heterogeneous megacells such as control logic networks, cache memory arrays, arithmetic logic units, register files etc. Each of these megacells can be characterized by a peculiar equivalent number of gates N_{Gi}, Rent's coefficient K_i, and Rent's exponent P_i [13]. The question to be addressed is the following : How can the global signal, power, and clock distribution networks for the heterogeneous SoC be designed to: i) fit all of the global wiring into the top two metal levels, ii) meet the required system clock frequency, and iii) limit the crosstalk noise to a specified maximum value? An initial response to this question follows.

The methodology begins by engaging a recently derived heterogeneous version of Rent's rule [13]. For the heterogeneous system-on-a-chip illustrated in Figure 1.10, this expanded version is defined by

$$T_{eq} = K_{eq} N^{P_{eq}} \tag{1.15}$$

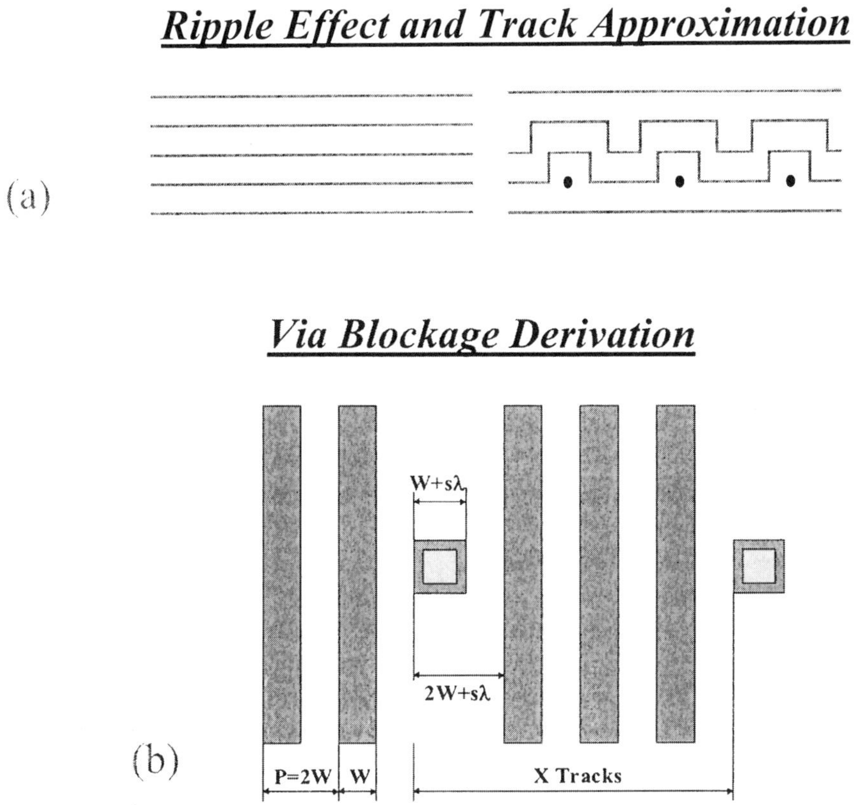

Figure 1.8. Sketch of wiring layouts used for derivation of via blockage factor $B_V = A_V/A_m$ where A_V is the area of a wiring level blocked by vias, A_m is the total macrocell or chip area and N_V is the number of terminal vias piercing a wiring level as indicated by the ripple effect and track approximation [11].

where

$$K_{eq} = \left(\prod_{i=1}^{n} K_i^{N_{Gi}}\right)^{\frac{1}{N_{Geq}}}, \tag{1.16}$$

$$P_{eq} = \frac{\sum\limits_{i=1}^{n} P_i N_{Gi}}{N_{G_{eq}}} \tag{1.17}$$

and

$$N_{G_{eq}} - \sum_{i=1}^{n} N_{Gi}. \tag{1.18}$$

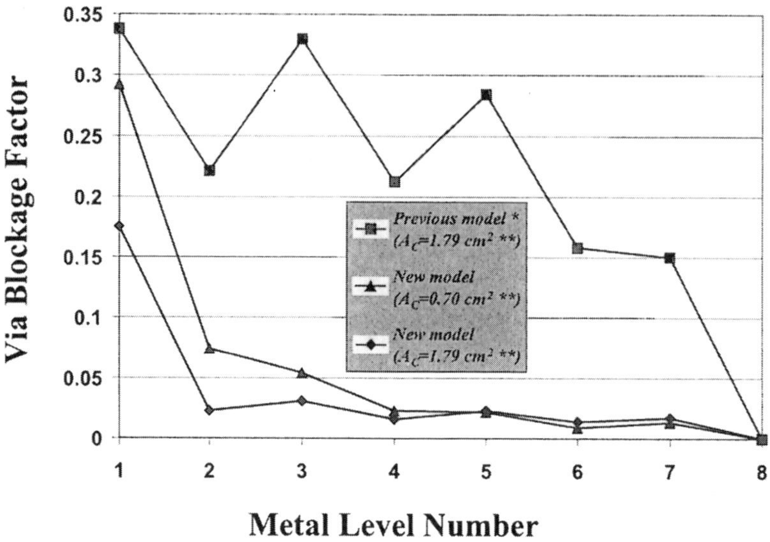

Figure 1.9. Via blockage factor versus metal level number for two interconnect networks (with minimum feature size F = 100 nm and number of gates N = 12.4 million) similar to those described in Figure 1.5 [11]. The previous model is described in [12].

In this power law relationship (1.15) Rent's coefficient K_{eq} is expressed as a weighted geometric average (1.16) and Rent's exponent P_{eq} as a weighted arithmetic average (1.17). Heterogeneous Rent's rule is used to derive three probability density distributions as summarized in Figure 1.11 [13]. The first is a net fan-out (FO) distribution that defines the number of nets $N_{Net}(m)$ versus the number of net terminals $m = FO + 1$ where N_m is the total number of megacells in the SoC. The second is a net bounding area distribution that describes the number of nets versus average net bounding area for nets with a specific number of terminals m. The average bounding area dimension of a square net $a(m)$ is shown in Figure 1.11 where η_p is an empirical placement efficiency factor that is estimated based on previous designs [13]. The third distribution is an average net length distribution that describes the number of nets versus average net length for nets with a specific number of terminals m. An expression for the average value of net length $L_{av}(m)$ is given in Figure 1.11. These three distributions are combined to derive the total global signal wiring requirement L_{tot} as shown in Figure 1.11 [13].

Figure 1.12 summarizes this new methodology and compares model predictions with data from a commercial product. The graph in Figure 1.12 plots number of interconnect nets per mm or net density versus average interconnect

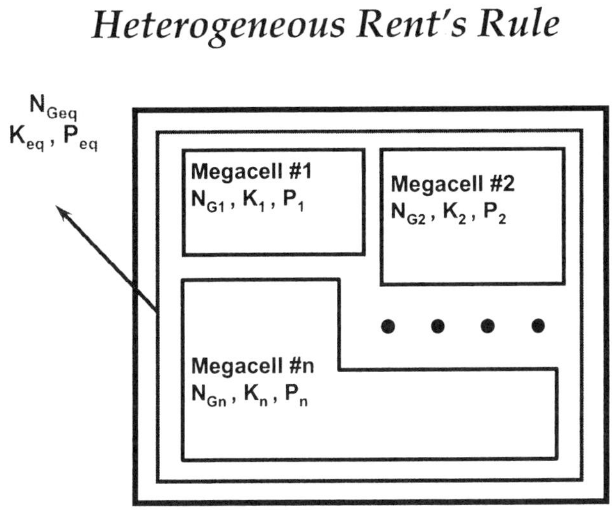

Figure 1.10. SoC layout used for definition of heterogeneous Rent's rule that applies to a heterogeneous set of megacells #1 through #n comprising a system-on-a-chip [13].

net length in mm. The first dashed locus describes the density of nets with a fan out of one; the second describes nets with a fan out of two; the third a fan out of three etc. The solid locus is the total interconnect net density distribution in number of nets per mm versus average interconnect length as calculated using the new model. The open circles represent data describing a commercial microprocessor consisting of 20 heterogeneous megacells [13].

In essence, the summation in Figure 1.11 defines the total length of global signal wiring required for a heterogeneous SoC. The next wiring resource requirement that must be defined relates to power distribution. Figure 1.13 presents the results of modeling the required area for power distribution A_{power}, for the cases of peripheral bonding pads or an area array of bonding pads. For peripheral bonding pads it is assumed that an equipotential ring surrounds the chip as illustrated in Figure 1.13(a). For area array bonding pads illustrated in Figure 1.13(b), it is assumed that V_{dd} is the potential of each bonding pad and that the current drain at each orthogonal intersection of the power grid lines is constant. A_{SoC} is the total SoC area [13]. In Figure 1.13, $\delta = \Delta V_{dd}/V_{dd}$ is the normalized voltage drop from a bonding pad to the most distant via at the intersection of an orthogonal pair of power grid lines, V_{dd} is supply voltage, H is metal height, P_{tot} is total chip power dissipation and ρW is metal resistivity. Note that A_{Power} for area array bonding pads can be reduced effectively by increasing the number of bonding pads n_{pad}.

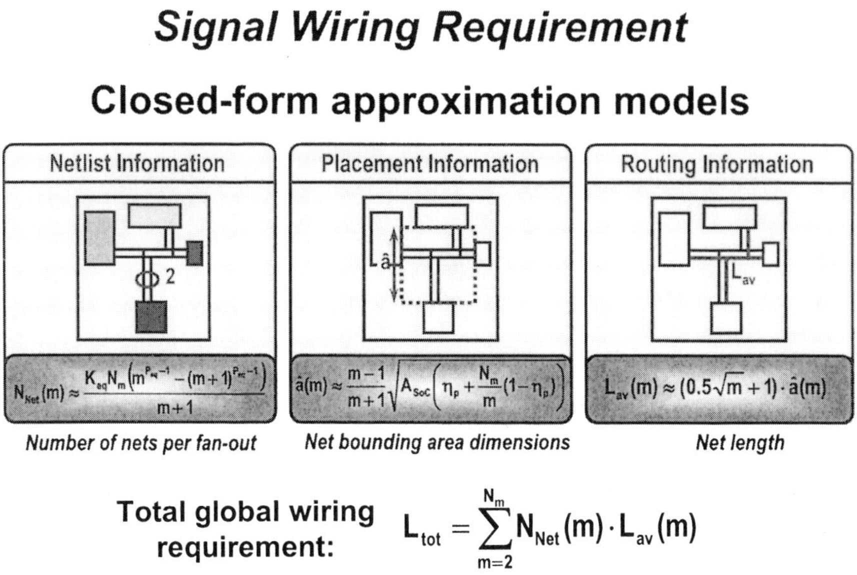

Figure 1.11. Netlength distribution (# of nets $N_{Net}(m)$ versus number of net terminals m), average net bounding area $a(m)$, average net length $L_{av}(m)$, and total global wire length requirement L_{tot} [13].

The most critical clock distribution network requirement that must be met is imposed by the bandwidth necessary for rapid transitions of the clock waveform. It is assumed that global clock distribution is implemented with a balanced H-array. This array is modeled as a distributed RC network whose maximum length extends from the chip clock input pad to a terminal buffer/repeater of the global H-array. The approximate value of this maximum length is the dimension of the chip edge l. Figure 1.14 defines the clock frequency limit f_{Clock} as a function of chip area A_{SoC} = 12 [13].

The final performance requirement that is imposed on the global wiring network is a crosstalk noise limit. A model used for an approximate calculation of global crosstalk noise is illustrated in Figure 1.15. In this representation, a global signal line or victim is assumed to be surrounded by two near and two far attackers. Simultaneous in-phase switching of the four attackers causes crosstalk noise on the victim due to coupling of both mutual capacitance and mutual inductance. A high quality nearby return path is assumed to be available. Treating the five coupled lines as distributed RLC networks, a set of partial differential equations quantifies the problem [14], [15], [16]. Some results of a solution to this set of equations are illustrated in Figure 1.16, which plots the ratio of crosstalk-to-binary signal voltage swing versus time [16]. Compar-

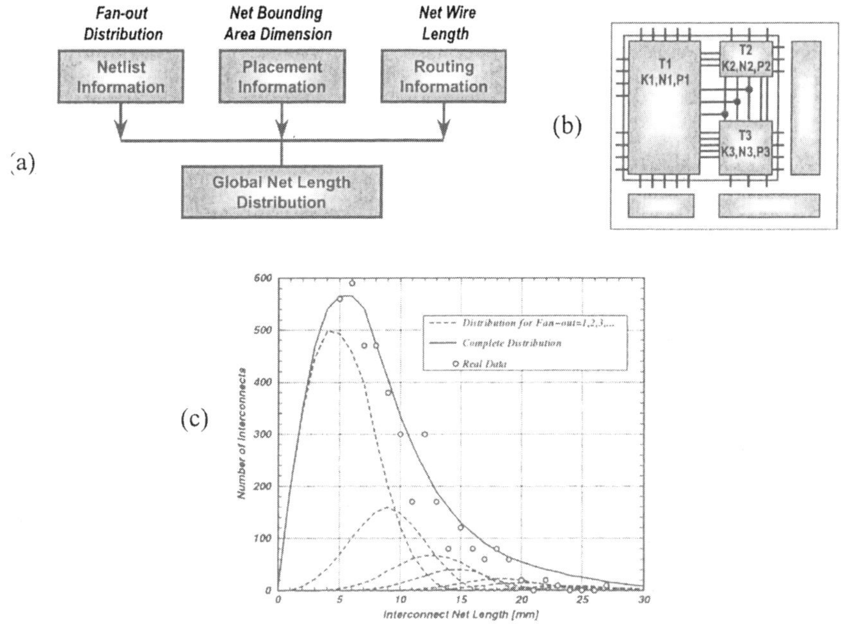

Figure 1.12. Summary diagram of derivation of interconnect net density distribution (number of interconnects per mm versus average interconnect net length in mm for nets with a specific number of terminals m) and comparison of model predictions (solid curve) versus actual data (open circles) [13].

ing the three and five line loci, it is evident that in the presence of a nearby high quality return path, the near attackers shield the victim from the far attackers. Therefore using the three-line model, simplified expressions for peak crosstalk voltage derived from the solutions of the set of partial differential equations are:

$$\frac{V_n}{V_{dd}} \cong \frac{1}{2}\frac{c_{mutual}}{c_{line} + c_{mutual}} \tag{1.19}$$

for distributed RC models [17] and

$$\frac{V_n}{V_{dd}} \cong \frac{\pi}{2}\frac{1}{2}\frac{c_{mutual}}{c_{line} + c_{mutual}} \tag{1.20}$$

for distributed RLC models [14], [15], [16] where c_{mutual} is the line-to-line distributed capacitance and c_{line} is the line-to-return path distributed capacitance .

A summary of the complete set of three compact models that define the primary global interconnect design requirements is given in Figure 1.17. The

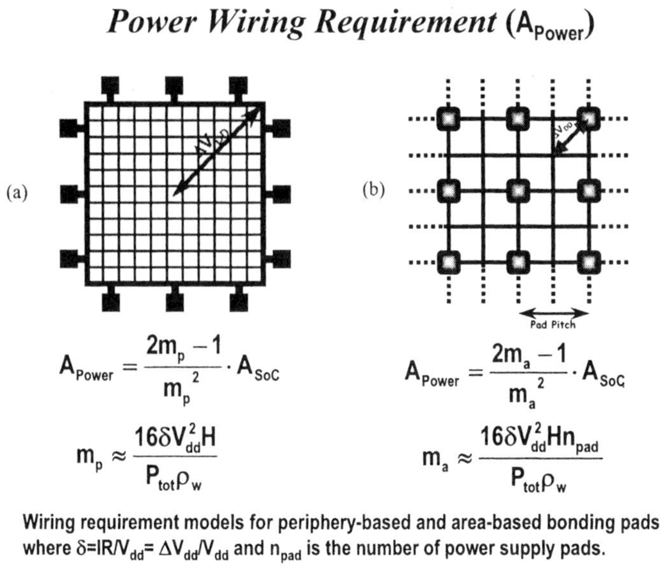

Figure 1.13. Power distribution network area requirement A_{Power} as a function of chip or macrocell area ASoC where V_{dd} is supply voltage, H is metal height, P_{tot} is total chip power dissipation and ρ_w is wire resistivity for peripheral and area array bonding pads.

models are expressed in terms of the physical parameters of the two global wiring levels illustrated in Figure 1.18. In the wiring resource requirement, the three terms represent the signal and power wiring areas, as well as unused area, respectively. The second and third models describe the clock wiring bandwidth and signal wiring crosstalk noise limit, respectively. In Figure 1.19, the three models are applied to a particular SoC consisting of 20 heterogeneous megacells containing a total of approximately six million transistors [13]. In the global interconnect design plane, the vertical axis represents interconnect thickness H and the horizontal axis interconnect width W. The allowable design region that satisfies all primary global wiring requirements is the zone bounded by the resource, bandwidth, and noise limit loci. For example, an interconnect width $W \approx 2.4\mu$m and height $H \approx 2.0\mu$m satisfies the prime design constraints with a minimum pitch. Projections of the allowable design regions for several future generations of technology are illustrated in Figure 1.20. Here it is evident that the amount of compression of the allowable design region becomes unacceptable and additional flexibility such as expansion of the number of global wiring levels appears to become necessary. In summary, the methodology presented in this section enables early projections of key physical parameters of a global interconnect network that simultaneously

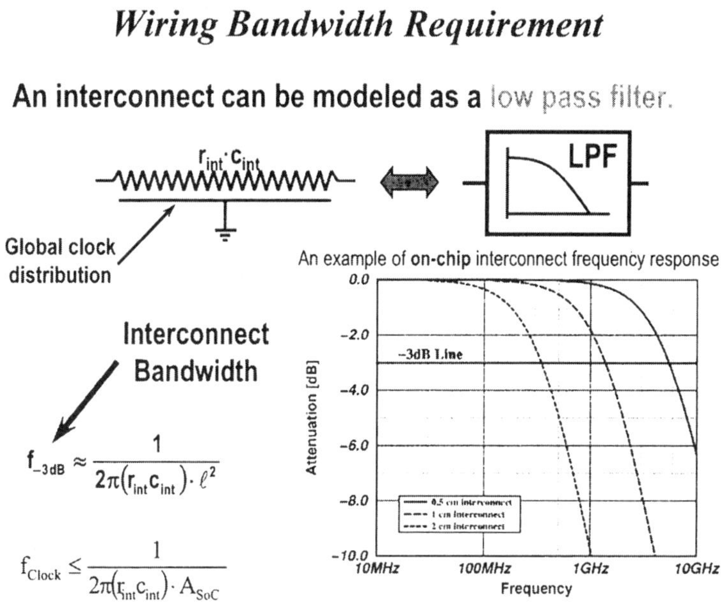

Figure 1.14. Summary of global H-array clock distribution network bandwidth requirement f_{-3dB} and limit on clock frequency f_{Clock} for a distributed RC network model of the H-array.

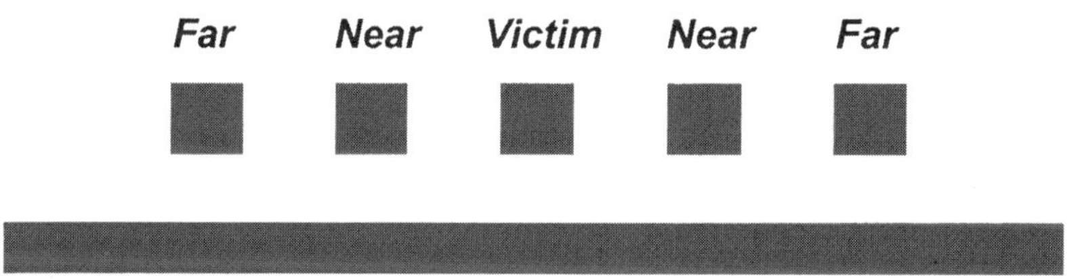

Figure 1.15. Model used for calculation of crosstalk noise induced in a global signal interconnect assuming a victim surrounded by near and far attackers in the presence of a nearby high quality return path.

satisfies the primary requirements of a SoC for signal, power, and clock distribution. The compact physical models that serve to implement the methodology offer a convenient opportunity to establish a quantitative guide to detailed design of a SoC. Therefore, the methodology may serve as a useful precursor to final design. Enhancements of this methodology that include, for example, the

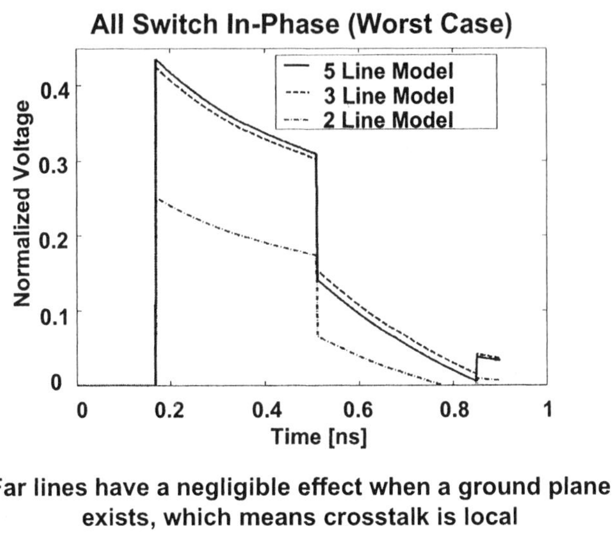

Figure 1.16. Normalized crosstalk voltage induced on a global victim interconnect by one, two, and four attacker interconnects.

Equation Description	Simplified Expressions for Global Wiring Requirements in Terms of w, s, H, and T_{ox}
Wiring resource requirement	$(w+s)\sum_{m=2}^{N_m} N_{Net}(m)\cdot L_{av}(m)+\frac{2m_p-1}{m_p^2}A_{SoC}+0.5\left(1-\frac{1}{m_p}\right)^2 A_{SoC}\le A_{SoC}$ where: $N_{Net}(m)\approx\frac{K_{eq}N_m\left(m^{P_{eq}-1}-(m+1)^{P_{eq}-1}\right)}{m+1}$ $L_{av}(m)\approx(0.5\sqrt{m}+1)\frac{m-1}{m+1}\sqrt{A_{SoC}\left(\eta_p+\frac{N_m}{m}(1-\eta_p)\right)}$, $m_p=\frac{16\delta V_{dd}^2 H}{P_{tot}\rho_w}$
Wiring bandwidth requirement	$f_c\le\frac{1}{4\pi\rho_w\varepsilon_o\varepsilon_r\left(1/HT_{ox}+1/ws\right)A_{SoC}}$
Wiring noise limit	$\frac{\pi}{4}\frac{1/ws}{\left(1/HT_{ox}+1/ws\right)}\le\%Noise$

Figure 1.17. Summary of complete set of requirements to be imposed on global signal, power, and clock distribution networks expressed in terms of the geometry of the two global wiring levels.

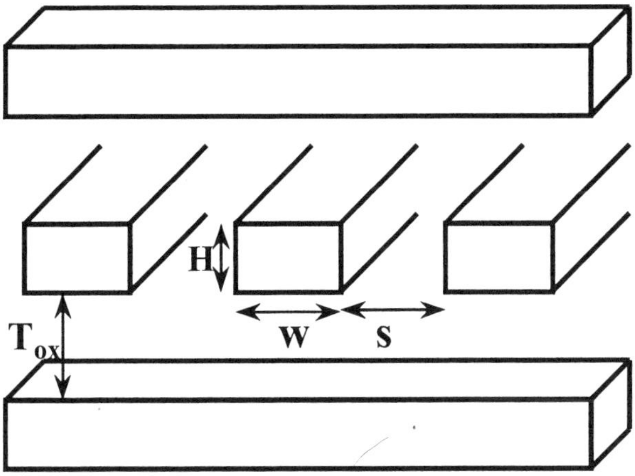

Figure 1.18. Physical representation of orthogonal interconnect system

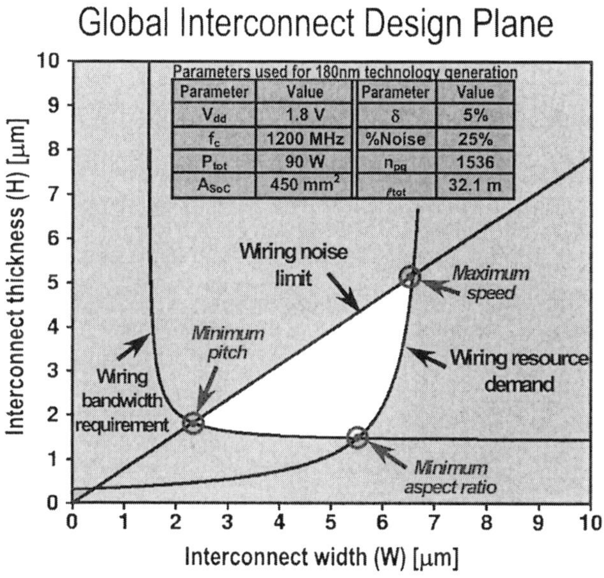

Figure 1.19. Global interconnect design plane plotting interconnect thickness H versus width W for the interconnect requirements summarized in Figure 1.17 as applied to a SoC consisting of 20 heterogeneous megacells including a total of approximately six million transistors [13].

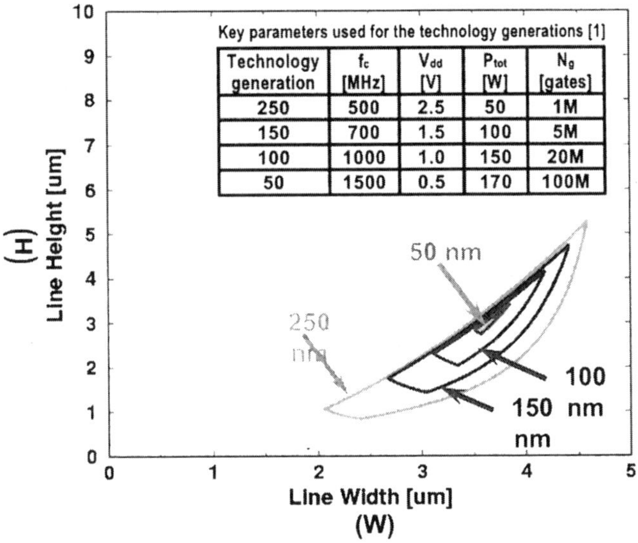

Figure 1.20. Global interconnect design plane projections illustrating compression of allowable design region and consequently the necessity for greater flexibility such as expansion of the number of wiring levels.

effects of clock skew, non-ideal return paths, and simultaneous switching noise are needed.

1.5 Three-Dimensional Integration

To achieve three-dimensional (3-D) integration in semiconductor technology requires the capability to stack multiple strata each containing both transistors and multilevel interconnect networks as discussed in preceding sections. This is a formidable challenge that is unlikely to be engaged seriously absent a convincing case for substantive benefits. Therefore, what are the primary benefits that can be projected for 3-D integration? It appears that the singular generic advantage of 3-D integration is a substantial reduction in length of the longest global interconnects used in a SoC.

Several rigorous derivations of stochastic interconnect distributions for 3-D random logic networks [18], [19], [20] based upon the 2D distribution discussed in Section 1.3 [7], [8] have been reported. Using the analytic models derived in [20], the stochastic interconnect distributions for a 4.0 million gate random logic network implemented with 1, 4 and 16 strata are illustrated in Figure 1.21. Note that for simplicity these distributions assume that the interstratal pitch $r = 1$, which strictly imposes the condition that the interstratal pitch equals the intrastratal logic gate pitch. The loci of Figure 1.21 clearly indicate that multiple strata or 3-D integration exerts very little impact on the density of local interconnects but that it has a profound effect on the length of the longest interconnects of the logic network. This observation is illustrated with greater clarity in Figure 1.22. The right vertical axis indicates a length of

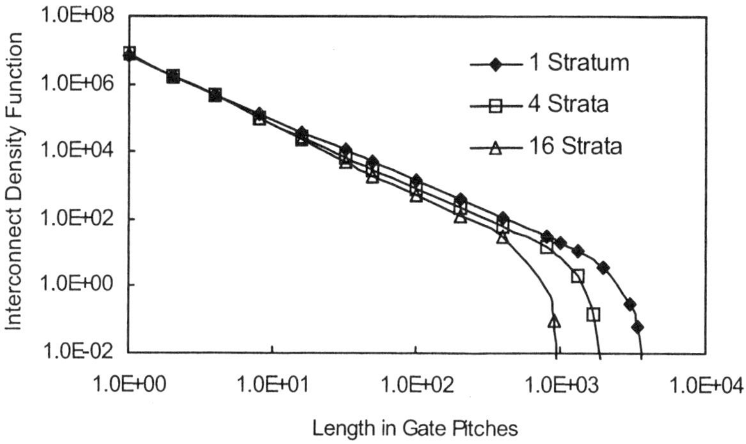

Figure 1.21. Interconnect density per gate pitch versus length in gate pitches for a random logic network of approximately four million gates implemented on one stratum, four strata, and sixteen strata.

approximately 4000 gate pitches for a corner-to-corner interconnect in a singe stratum implementation, 2000 gate pitches for a four strata implementation, and 1000 gate pitches for a 16 strata implementation. For time-of-flight limited global interconnects this could result in a 4:1 reduction of latency and the possibility of approximately a fourfold increase in global clock frequency – for the expense of a 16 strata implementation of the system.

A key simplifying assumption limiting the projections illustrated in Figures 1.21 & 1.22 is that the interstratal pitch equals the intrastratal gate pitch or $r = 1$. Setting aside this assumption, a generic 3-D wiring distribution for a 4.0 million gate random logic network whose interstratal pitch is treated as an independent variable has been rigorously derived [20]. Figure 1.23 illustrates a key result of this new derivation for interstratal pitches $r = 1$ and $r = 50$. The two distributions are quite similar for short local and long global interconnect lengths. The only region in which the two loci deviate is the mid-range of interconnect lengths where interconnect length and stratal pitch are roughly equal. Consequently, it appears that interstratal separation distance is not a critical parameter in determining 3-D wiring distributions. The generic benefit of substantial reductions in length of the longest global interconnects in

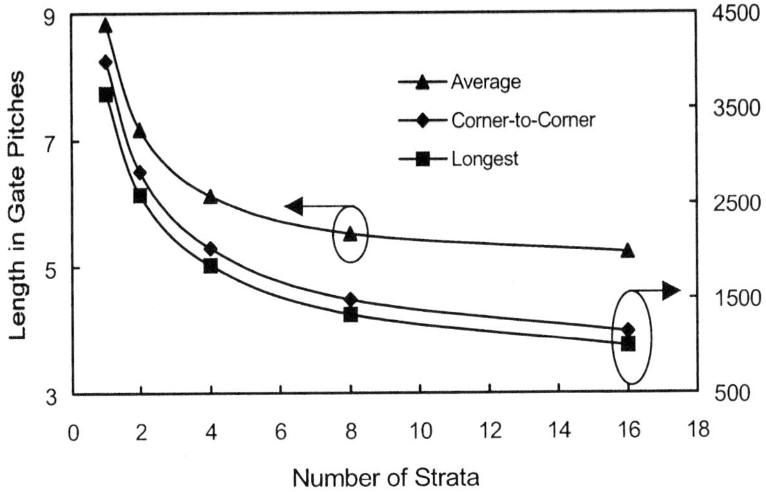

Figure 1.22. Interconnect length versus number of strata for average, corner-to-corner, and longest interconnects of the random logic network of Figure 1.21.

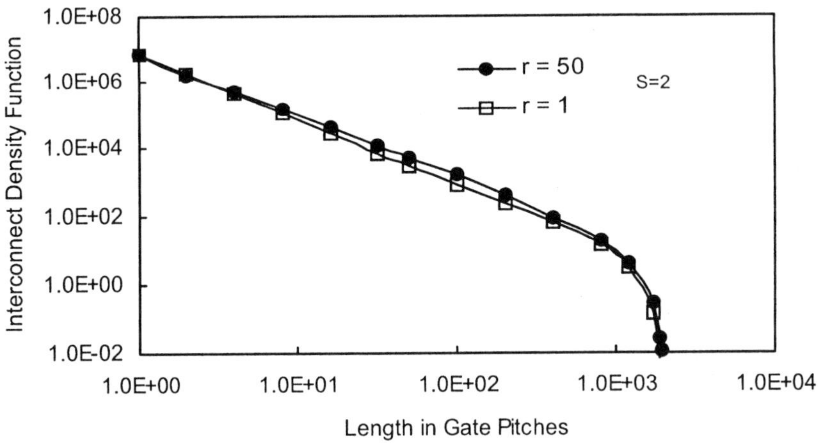

Figure 1.23. Interconnect density per gate pitch versus length in gate pitches for the random logic network of Figure 1.21 implemented on two strata for interstratal pitches $r = 1$ and $r = 50$ intrastratal gate pitches.

a distribution resulting from 3-D integration is an inherent advantage of 3-D wiring. A concomitant inherent disadvantage of 3-D structures is heat removal [21]. Beyond these general issues, the attraction of 3-D integration for specific applications may be dominated by the peculiar features of the application itself. For example, two-dimensional sensor arrays that require direct access to each sensor cell for immediate signal preprocessing are interesting prospects for 3-D integration [22]. More broadly, the capacity to explore opportunities for extraordinary performance enhancements through 3-D integration would benefit from generic advances in capabilities to fabricate 3-D structures.

1.6 Input/Output Interconnect Enhancements

The intent of input/output interconnect enhancements is to improve the cost, size, reliability, and performance of a gigascale SoC. Historically, bonding wires have been the dominant approach to chip input/output (I/O) interconnects [23]. IBM pioneered the introduction of solder-bump I/O interconnects using flip-chip technology with a thin layer of glass passivation sealing the chip encapsulated in silicone gel, which prevented the formation of continuous water films [24], [25]. A particular novel technology that is currently under investigation for I/O enhancements is described as Sea of Leads (SoL) [26]. This technology proposes the use of wafer-level-batch- fabrication of compliant polymer packages, ultra high density ($> 104/cm^2$) x-y-z flexible metal leads, and solder-like bumps attached to the lead tips as illustrated in Figure 1.24. A short sequence of full wafer SoL batch fabrication processes constituting a "tail-end-of-the-line" (TEOL) are envisaged to follow conventional back-end-of-the-line (BEOL) wafer processing. The further intent of SoL technology is to complete all final electrical testing and burn-in operations prior to wafer dicing that yields known good packaged die ready for immediate shipment to customers. The flexible leads are designed to provide sufficient x-y-z compliance to accommodate typical differences in the thermal coefficients of expansion between a silicon chip and the substrate to which it is attached. The need for epoxy underfill is thereby precluded and the possibility of convenient detachment of a chip from a substrate module is enabled.

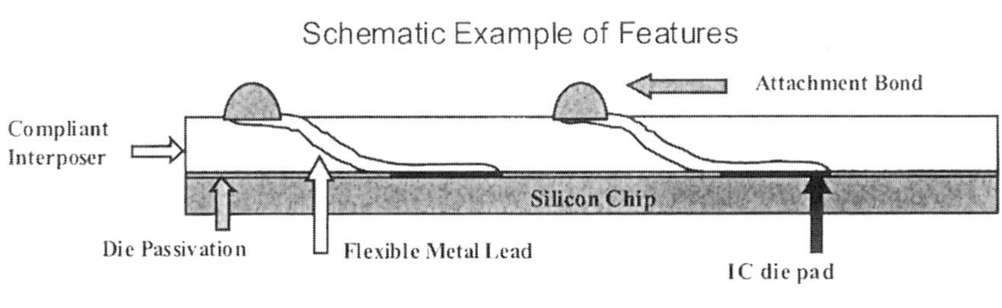

Figure 1.24. Schematic representation of salient objectives of Sea of Leads input/output interconnect technology. All silicon dice remain unseparated in a wafer until all compliant interposer packages, flexible metal input/output leads, and bonding material on the lead tips are batch fabricated. Then all dice undergo full electrical testing and burn-in prior to dicing the wafer to yield known good packaged die ready for shipment.

Compressible, Very High Density I/O

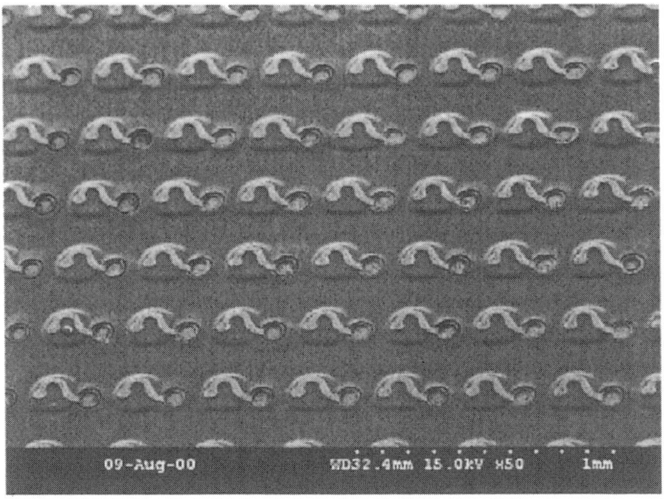

Figure 1.25. Photomicrograph of a Sea of Leads.

Concurrent fabrication of packages and leads of all chips on a wafer extends the historically potent economies of wafer level batch processing to the relatively costly die-by-die assembly, bonding, packaging , testing, and burn-in operations [27], [28]. Moreover, the size of the SoL package is the minimum for a chip-scale-package (CSP). Significant reliability improvement may result from avoidance of epoxy underfill often needed to relieve stress on relatively rigid solder ball connections between chip and substrate. Figure 1.25 is a photomicrograph of a Sea of Leads. The circular pattern is the via linking a die bonding pad with the lead itself, which is the "question-mark-shaped" copper pattern. This peculiar shape is designed to provide a high degree of x-y axis flexibility and thus accommodate chip-substrate thermal expansion differences. The somewhat rounded region beneath the copper lead defines the boundaries of a polymer interposer air cavity that is introduced to enhance z-axis compliance. This compliance is added in order to provide convenient and reliable temporary electrical contacts between an array of electrical test probes and the leads of the dice under test, especially when the probe tips are not in a precisely planar arrangement. A photomicrograph of the cross-section of an air cavity is shown in Figure 1.26. An SEM of a 1x1 cm^2 die with a SoL density of 12,000 per cm^2 is shown in Figure 1.27. The leads are oriented along the contours of expansion of the die to provide a higher degree of compliance proceeding radially outward from the center to the edge of the die.

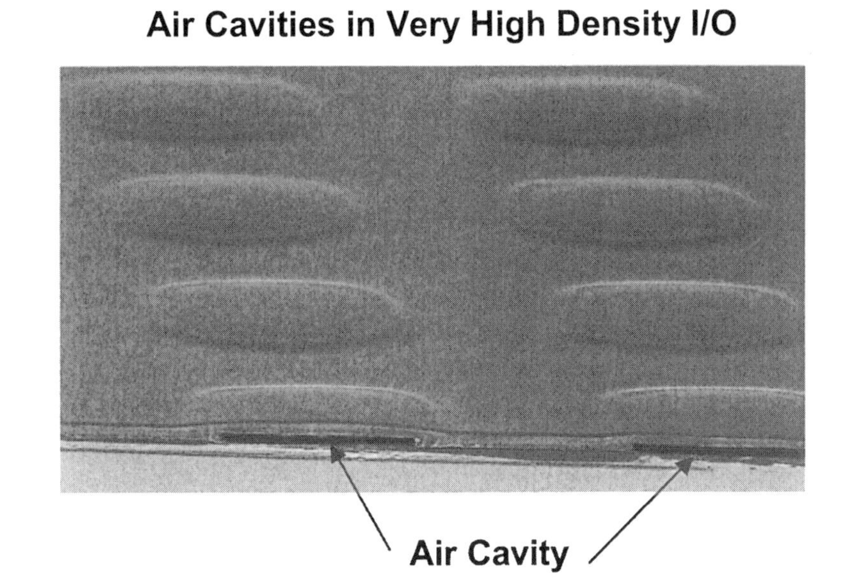

Figure 1.26. Photomicrograph of the cross section of a SoL air cavity.

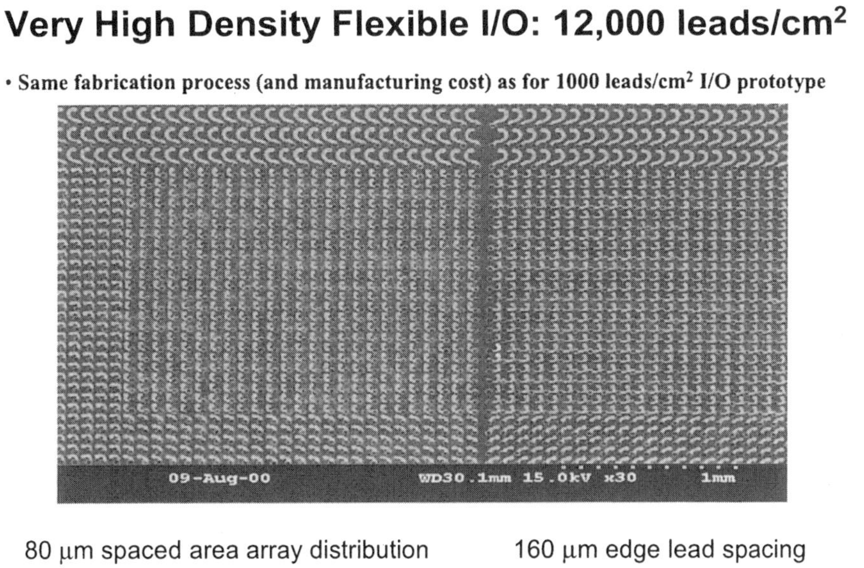

Figure 1.27. SEM of a SoL with a density of 12,000 per cm^2.

Key performance enhancements that appear to be in the offing for SoL technology include the following [26], [29]: i) substantially increased input/output bandwidth for a chip resulting from the significantly larger (e.g. $\approx 10\times$) number of signal leads that are available; ii) "time-of-flight" global signal interconnect latency for a chip resulting from exiting and then reentering the die using external on-module wiring or "exterconnects" to implement very low loss time-of-flight internal global wiring links for the chip; iii) reduced global clock skew due to use of time-of- flight exterconnects to implement global clock trees; iv) reduced global clock power dissipation through recycling the energy of reflected clock pulses distributed through low loss exterconnects [30]; v) suppression of far-attacker crosstalk noise on global signal interconnects due to use of exterconnects with nearby high quality return paths provided by module power and ground planes; vi) suppression of simultaneous switching noise (SSN) and reduced parasitic IR voltage drop in the power/ground distribution networks resulting from the significantly larger (e.g. $\approx 10\times$) number of power and ground leads that are available; and vii) improved isolation and reduced interference in mixed signal systems resulting from use of separate power/ground input/output leads for analog and digital signals.

Additional opportunities that are available through SoL include the capacity to satisfy the voracious appetite of 3-D integration for I/O capacity and the potential for compatibility of electrical, RF wireless, and photonic I/O interconnects.

In short, SoL can be described as a "disruptive" technology because the intent is to use batch-fabricated ultra-high density input/output leads to improve the cost, size, reliability, and performance of a SoC [26], [29].

1.7 Photonic Interconnects

An exposition of interconnect opportunities for GSI would not be complete without consideration of photonic or optical interconnects [31], [32], [33], [34]. In order to be competitive with electrical interconnects for GSI, photonics must provide low power, high speed, small size, low cost photon emitters, detectors, and conductors or waveguides – that are compatible with CMOS technology. Consequently, this section focuses on compatible photonics or photonic technologies with the long range potential to satisfy the extremely stringent and peculiar demands of GSI.

The most challenging objective for CMOS compatible photonic interconnects is an efficient room temperature silicon light emitter. A novel silicon diode that exploits dislocation loops to introduce a local strain field that modifies the band structure to confine carriers near the junction and therefore enhance light emission was recently demonstrated [35].

Short of high quality silicon photoemitters, a most interesting approach to compatible photonics is based upon heteroepitaxial deposition on Si of SiGe, followed by Ge, followed by GaAs, and finally A1GaAs [32], [33]. The close lattice constant match of Ge and GaAs provides a basis for growing high quality single crystal layers of GaAs. This heteroepitaxial approach to compatible photonics has the potential to provide III-V compound semiconductor lasers, Ge detectors, and poly or monocrystalline Si waveguides. Figure 1.28 illustrates the current-voltage curves of heteroepitaxial SiGe and GaAs diodes on a Si substrate [33]. Figure 1.29 displays photomicrographs of a right angle bend and a spilt in of a polycrystalline Si waveguide [34]. Transmission loss is less than 0.5 dB in the bend and 1.0 dB in the split. Waveguide width is 0.5 μm, which is comparable to dimensions of upper level metal interconnects. These recent advances are encouraging demonstrations of the long range promise of compatible microphotonic interconnects.

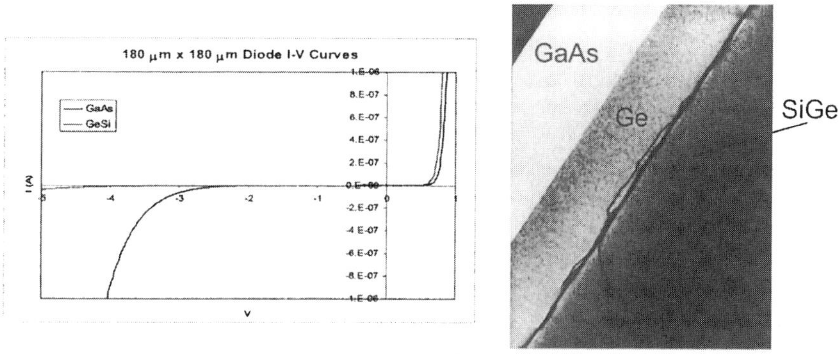

TDD < 1x10⁶ cm⁻²

Figure 1.28. I-V curves for heteroepitaxial SiGe and GaAs diodes on a Si substrate. [Courtesy of Prof. Gene Fitzgerald, MIT.]

It has long been proposed that the most likely point of entry of photonic interconnects into silicon integrated electronics is in clock distribution [36], [37]. Recently, a polymer waveguide network with volume grating output couplers embedded in a printed wiring board (PWB) was proposed to transfer photons from a printed wiring board to one or more silicon photodetectors fabricated in a CMOS chip [38]. This approach to optical clock distribution does not utilize on-chip photon emitters and enables a planar package configuration.

1.8 Conclusions

Interconnect latency is now the primary performance issue for GSI and the problem only promises to become more serious for future generations of tech-

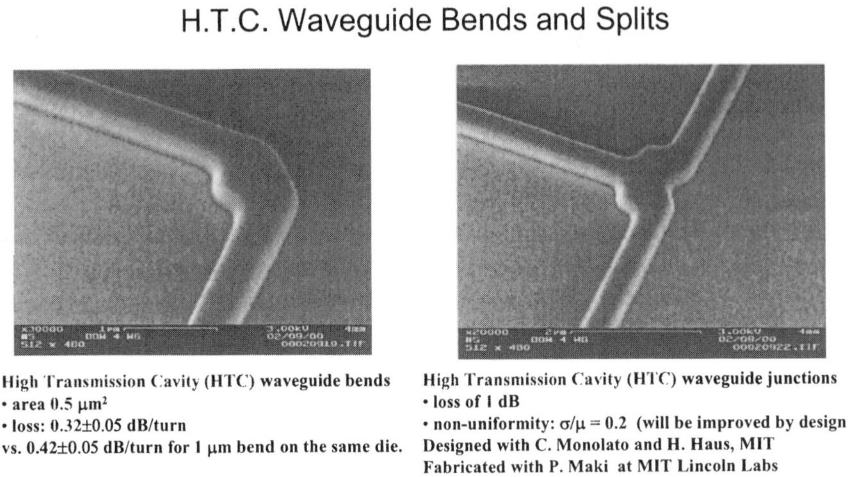

Figure 1.29. Photomicrograph of polycrystalline Si waveguide bends and junctions. [Courtesy of Prof. Lionel Kimerling, MIT.]

nology. Opportunities to address the problem range, for example, from carbon nanotube conductors that may enable ultra-high speed ballistic transport [39] to new single chip, distributed shared memory, cellular arrays of microprocessors [40], [41] that serve to keep interconnects short. The second interconnect problem that is not broadly recognized as such is energy dissipation. The keys to solution of this problem are short interconnects – and transistors with the smallest possible subthreshold swing and therefore the smallest possible binary signal swing. Crosstalk and simultaneous switching noise represent a third interconnect problem – signal integrity – which is difficult to describe using compact physical models.

For virtually any family of gigascale chips, the key to optimal reverse scaling of multilevel signal interconnect networks is prediction of the complete stochastic wiring distribution of a next generation product. More general signal integrity models that can be incorporated into reverse scaling methodologies are needed.

The task of conjointly optimizing the architecture of the global signal, clock, and power/ground distribution networks of a system-on-a-chip consisting of a set of heterogeneous megacells is demanding. A first attempt to address this task comprehensively engages a new stochastic model for global signal wiring, a new model for global power/ground wiring area, a global clock bandwidth requirement, and a crosstalk noise requirement. Enhancements of current methodologies that include, for example, the effects of clock skew, nonideal return paths , and simultaneous switching noise are needed.

The generic benefit of substantial reductions (e.g. $> 50\%$) in length of the longest global interconnects in a distribution is an inherent advantage of 3-D integration. The capacity to explore novel opportunities for extraordinary performance enhancements through 3-D integration would benefit from generic advances in capabilities to fabricate 3-D structures.

In order to maintain historic rates of advance of monolithic semiconductor technology, more attention to ancillary features and particularly to input/output interconnects is unavoidable. Sea of Leads represents an early effort to more intimately couple the chip itself to its environment and then to exploit concomitant new opportunities. Key projected performance enhancements include substantially increased input/output bandwidth, reduced global signal interconnect latency, reduced global clock skew, reduced global clock power dissipation, greater suppression of simultaneous switching noise, and improved signal integrity in mixed signal systems. More broadly, Sea of Leads represents an effort to extend the quintessential feature of semiconductor technology – wafer level batch fabrication of several hundred chips – to the traditional die-by-die packaging and testing domains.

To be become widely used in GSI, photonics must provide low power, high speed, small size, low cost photon emitters, detectors, and conductors or waveguides – that are compatible with CMOS technology [42]. Recent advances in heteroepitaxial deposition on Si of SiGe, followed by Ge, followed by GaAs to demonstrate light emitting and detecting diodes as well as Si waveguides are promising.

Acknowledgments

The intellectual contributions of Azad Naeemi, Raguraman Venkatesan, Muhannad Bakir, Hiren Thacker, Qiang Chen, James Joyner and Tony Mule of the Georgia Institute of Technology Microelectronics Research Center to this paper are gratefully acknowledged. In addition, the authors wish to express their appreciation to DARPA, contract F33615-97-C-1132, MARCO, contract MDA 972-99-1-002, and the SRC, contract 448:048 for their generous support.

References

[1] J.D. Meindl, "Low power microelectronics: Retrospect and prospect," *Proc. IEEE*, vol. 83, pp. 619–635, Apr. 1995.

[2] Semiconductor Research Association, San Jose, CA, *International Technology Roadmap for Semiconductors*, 1999.

[3] M. Bohr *et al.*, "A high performance 0.25 micron logic technology optimized for 1.8v operation," in *Proceedings of IEDM*, (San Francisco, CA), pp. 847–850, 1996.

[4] T.N. Theis, "The future of interconnection technology," *IBM J. Res. Develop.*, vol. 44, pp. 379–390, May 2000.

[5] J. Hennessy *et al.*, " Cache-coherent distributed shared memory: perspectives in its development and future challenges," *Proc. IEEE*, vol. 87, pp. 418–429, Mar. 1999.

[6] B. Landman and R. Russo, "On a pin versus block relationship for partition of logic paths," *IEEE Trans. on Computing*, vol. C-20, pp. 1469–1479, Dec. 1971.

[7] J.A. Davis *et al.*, "A stochastic wire-length distribution for gigascale integration (GSI) - part I: Derivation and validation," *Trans. Electron Devices*, vol. 45, pp. 580–589, Mar. 1998.

[8] J.A. Davis *et al.*, "A stochastic wire-length distribution for gigascale integration (GSI) - part II : Applications to clock frequency, power dissipation and chip size estimation," *Trans. Electron Devices*, vol. 45, pp. 590–597, Mar. 1998.

[9] R. Venkatesan *et al.*, "Optimal *n-tier* multilevel interconnect architectures for gigascale integration (GSI)," *Trans. Very Large Scale Integration (VLSI) Systems*, vol. 9, pp. 899–912, Dec. 2001.

[10] H.B. Bakoglu and J.D. Meindl, "Optimal interconnection circuits for VLSI," *IEEE Trans. Elect. Devices*, vol. 32, May 1985.

[11] Q. Chen *et al.*, "A compact physical via-blockage model," *IEEE Trans. VLSI Systems*, vol. 8, pp. 689–692, Dec. 2000.

[12] G.A. Sai-Halasz, "Performance trends in high-end processors," *Proc. IEEE*, vol. 83, pp. 18–34, Jan. 1995.

[13] P. Zarkesh-Ha *et al.*, "Prediction of net-length distribution for global interconnects in a heterogeneous system-on-a-chip," *IEEE Trans. VLSI Systems*, pp. 649–659, Dec. 2000.

[14] J.A. Davis and J.D. Meindl, "Compact distributed RLC interconnect models - part i : Single line transient, time delay and overshoot expressions," *Trans. Electron Devices*, vol. 47, pp. 2068–2077, Nov. 2000.

[15] J.A. Davis and J.D. Meindl, "Compact distributed RLC interconnect models - part ii : Coupled line transient expressions and peak crosstalk in multilevel interconnect networks," *Trans. Electron Devices*, vol. 47, pp. 2078–2087, Nov. 2000.

[16] A. Naeemi *et al.*, "Analytical models for coupled distributed RLC lines with ideal and non-ideal return paths," in *Proceedings of IEDM*, (San Francisco, CA), pp. 689-692, 2001.

[17] T. Sakurai, "Closed form expressions for interconnection delay, coupling and crosstalk in VLSI's," *IEEE Trans. Electron Devices*, vol. 40, pp. 118–124, Jan. 1993.

[18] A. Rahman *et al.*, "Wire-length distribution of three-dimensional integrated circuits," in *Proceedings of IITC*, (San Francisco, CA), pp. 233–235, 1999.

[19] S.J. Souri and K.C. Saraswat, "Interconnect performance modeling for 3d integrated circuits with multiple silicon layers," in *Proceedings of IITC*, (San Francisco, CA), pp. 24–26, 1999.

[20] J.W. Joyner *et al.*, "Impact of three dimensional architectures on homogeneous digital circuits," *Trans. Very Large Scale Integration (VLSI) Systems*, vol. 9, pp. 922–928, Dec. 2001.

[21] H.B. Bakoglu, *Circuits, interconnections and packaging for VLSI - Chapter 3.* Reading, MA: Addison-Wesley, 1990.

[22] J. Burns *et al.*, "Three-dimensional integrated circuits for low-power, high bandwidth systems-on-a-chip," in *Proceedings of ISSCC*, (San Francisco, CA), pp. 268–269, 2001.

[23] B.L. Gehman, " Bonding wire microelectronic interconnections," *IEEE Trans. on Comp.*, vol. CHMT-3(3), p. 375, Sept. 1980.

[24] E.M. Davis *et al.*, "Solid logic technology: Versatile high performance microelectronics," *IBM J. Res. Develop.*, vol. 8, p. 102, 1964.

[25] P.A. Totta and R.P. Sopher, "SLT device metallurgy and its monolithic extension," *IBM J.Res. Develop.*, vol. 5, pp. 226–238, May 1969.

[26] A. Naeemi *et al.*, " Sea of leads: a disruptive paradigm for a system-on-a-chip (SoC)," in *Proceedings of ISSCC*, (San Francisco, CA), pp. 280–281, 2001.

[27] C.S. Patel *et al.*, "Low cost high density compliant wafer level package," in *Proceedings of High-Density Interconnect and Systems Packaging Conference*, (Denver, CO), pp. 262–268, 2000.

[28] C.S. Patel *et al.*, " Cost analysis of compliant wafer level packages," in *Proceedings of ECTC*, (Las Vegas, NV), pp. 268–269, 2000.

[29] A. Naeemi *et al.*, "Performance improvements using on-board wires for on-chip interconnects," in *Proceedings of EPEP*, pp. 325–328, 2000.

[30] P. Zarkesh-Ha *et al.*, "Stochastic net length distribution for global interconnects in a heterogeneous system-on-a-chip," in *Proceedings of Symposium on VLSI Technology*, (Honolulu, HI), pp. 44–45, 1998.

[31] A. Krishnamoorthy and D. Miller, "Scaling optoelectronic-VLSI circuits into the 21st century: a technology roadmap," *IEEE J. Quantum Electronics*, vol. 2, pp. 55–76, Apr. 1996.

[32] L.M. Giovane *et al.*, "Materials for monolithic silicon microphotonics," in *Proceedings of Materials and Devices for Silicon-based Optoelectronics Symposium*, (Warrendale, PA), pp. 45–56, 1998.

[33] E. Fitzgerald and L. Kimerling, " Silicon-based microphotonics and integrated optoelectronics," *MRS Bulletin*, vol. 23, Apr. 1998.

[34] L.C. Kimerling, " Silicon microphotonics," *Appl. Surf. Sci.*, vol. 159-160, pp. 8–13, June 2000.

[35] W.L. Ng *et al.*, "An efficient room-temperature silicon-based light-emitting diode," *Nature*, vol. 410, pp. 192–194, Mar. 2001.

[36] J.W. Goodman *et al.*, " Optical interconnections for vlsi systems," *Proc. IEEE*, vol. 72, pp. 850–866, July 1984.

[37] S.K. Tewsbury and L.A. Hornak, " Optical clock distribution in electronic systems," *J. VLSI Signal Proc.*, vol. 16, pp. 225–246, June 1997.

[38] A.V. Mule *et al.*, "Input coupling and guided wave distribution schemes for broad-band intrachip guided wave optical clock distribution networks using volume grating coupler technology," in *Proceedings of IITC*, (San Francisco, CA), pp. 128–130, 2001.

[39] C. Zhou *et al.*, " Electrical measurements of individual semiconducting single-walled carbon nanotubes of various diameters," *Applied Physics Letters*, vol. 76, pp. 1597–1599, Mar. 2000.

[40] V. Mulutinovic and P. Stenstrom, "Special issue on distributed shared memory systems," *Proc. IEEE*, vol. 87, pp. 399–403, Mar. 1999.

[41] W.J. Dally and J.W. Poulton, *Digital Systems Engineering*. Cambridge, U.K., NY, USA: Cambridge Univ. Press, 1998.

[42] D.A.B. Miller, "Rationale and challenges for optical interconnects to electronic chips," *Proc. IEEE*, vol. 88, pp. 728–748, June 2000.

Chapter 2

COPPER BEOL INTERCONNECTS FOR SILICON CMOS LOGIC TECHNOLOGY

Satya V. Nitta, Sampath Purushothaman, James G. Ryan*, Daniel C. Edelstein, Panayotis Andricacos, Chao-Kun Hu, Thomas M. Shaw, Robert Rosenberg, and James R. Lloyd

IBM T.J. Watson Research Center, Yorktown Heights, NY 10598

* *IBM Microelectronics Division, Hopewell Junction, NY 12533*

2.1 Introduction

The semiconductor industry has been at the forefront of the technological revolution that has changed the way we live over the last three decades. The increase in performance of the integrated circuit (IC) chip has largely been due to the decreasing dimensions on the IC chip, leading to an increase in speed of the transistor. The transistors in an IC chip need to be connected to the outside world and the first level of hierarchy in this connection is the so called " back end of the line (BEOL) interconnect". A typical BEOL interconnect consists of a metallic wire that is surrounded by an insulating cladding called the interlayer dielectric. Over the last few technology generations, it has become increasingly evident that shrinking device dimensions alone will not continue to achieve the increases in IC chip performance that the semiconductor industry needs. This is because the delay associated with sending signals through the BEOL interconnect begins to dominate the overall delay in the IC chip as seen in Figure 2.1 [1]. The first step towards reducing this delay is to find a suitable metal that has a lower resistivity than aluminum (Al) to replace it as the wiring metal. Of all the elements in the periodic table, the most suitable candidate to replace Al as the wiring metal is copper (Cu). However, replacing Al with Cu as the metal of choice in BEOL interconnects is far from trivial.

In August of 1997, IBM announced its schedule for what would be the first implementation of Cu interconnect technology on integrated circuit (IC) chips [2]. This has been one of the most significant revolutions in BEOL technology since the invention of the integrated circuit. Figure 2.2 shows a micrograph illustrating the complexity of the wiring pattern for the 32-bit RISC

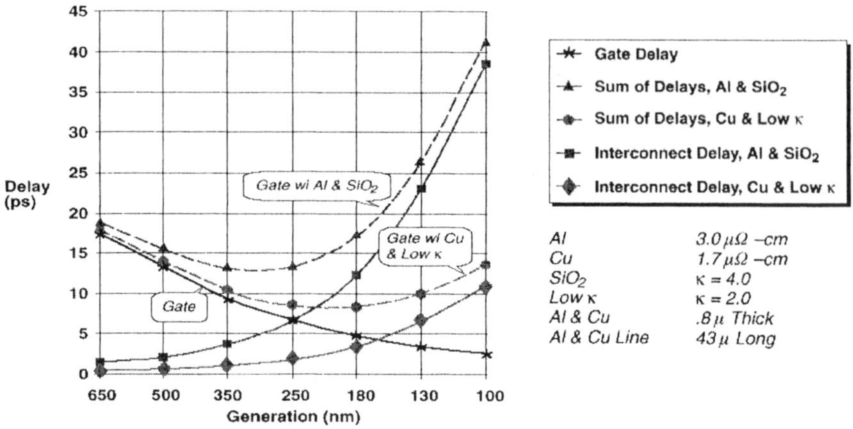

Figure 2.1. Delay as a function of semiconductor technology generation. Interconnect delay dominates device delay for deep submicron device generations [1].

CPU announced [2]. In order to implement Cu as the wiring metal in IC chips, significant advances had to be made in several aspects of BEOL technology. These include a detailed understanding of the microstructure of Cu and its dependence on deposition method and process conditions; its influence on the electromigration behavior of these interconnects; dual damascene integration [3] and all the associated patterning and integration challenges posed by this technique; and the invention of an effective liner scheme that would prevent the diffusion of Cu into the insulator. In this chapter, all these aspects of Cu interconnect technology will be reviewed, beginning with a brief look at the evolution of the BEOL interconnect through the years leading up to key innovations such as dual damascene processing and chemical mechanical polishing which have enabled the introduction of copper interconnects.

2.2 BEOL Evolution

Figure 2.3 shows the evolution of minimum first-wiring-level-contacted pitch for IBM logic and memory semiconductor products as a function of the year of introduction into manufacturing. Several notable sections of the curves highlight technology changes. The 1975-1978 reduction in pitch for logic products was a result of the change from wet-etched Al based wiring to the use of the lift-off patterning method [4]. The prevailing metal patterning technology in the early 1970s was wet subtractive etching but this technique limited the ability to achieve tighter pitch because etching bias and tolerance were high for Al and even worse for the Al(Cu) alloys used. Furthermore, wet etching undercut structures and formed holes in the metal at any recursive topography

Figure 2.2. Multilevel interconnect with copper wiring for 32-bit RISC CPU (Courtesy of T. Way, IBM Microelectronics Division, Burlington, VT.

on the wafer surface. These factors provided the motivation to migrate to the lift-off approach for metal patterning. This method was first implemented in manufacturing in 1975 for self aligned metal oxide semiconductor (SAMOS) metal-gated DRAMs.

The lift-off approach allowed significant improvements in contacted metal pitch compared to that achievable with metal etching [4]. However, as the metal pitch was tightened with successive generations, topography and reflectivity issues reduced the acceptable lift-off process window. The migration from lift-off to metal reactive-ion etching (RIE) occurred in the mid eighties along with the introduction of tungsten (W) studs .

Metal polishing at via levels was the first microelectronics use of the metal patterning technique known as "damascene" (now in wide use in chip manufacturing). This technique involves etching trenches or vias in an insulating layer which is then filled with metal. Chemical Mechanical Polishing (CMP) is then used to remove the extraneous material (overburden) until the metal remains only in the trenches and vias. The damascene technique was initially used for aluminum and tungsten wiring but its use has now been extended to copper wiring. The advantages of damascene processing include fewer process steps, easier lithographic alignment, improved planarity, and elimination of metal etching or RIE. A form of the damascene process called the "dual

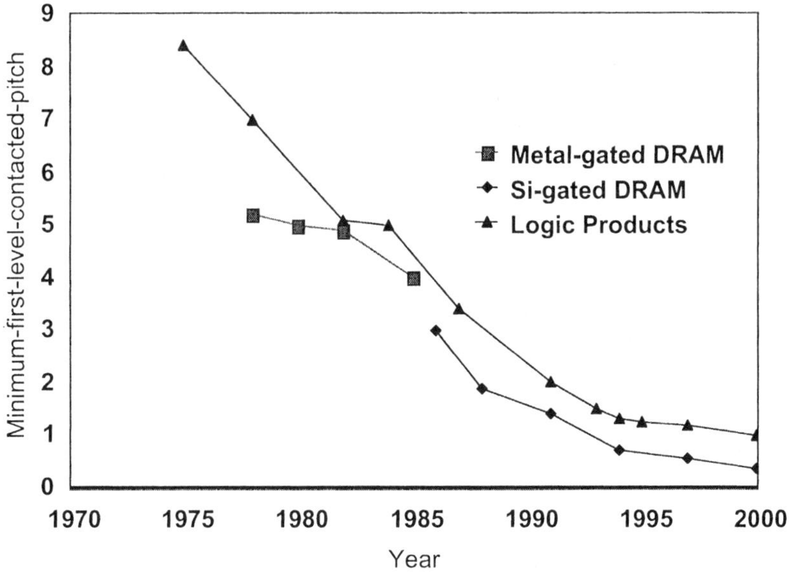

Figure 2.3. Minimum first-wiring-level contacted pitch as a function of year of introduction into manufacturing at IBM [4]

damascene" [3] process has now gained favor for fabricating Cu based interconnects. A typical dual damascene process flow is shown in Figure 2.4 below. In the dual damascene process, an insulator, which will eventually house the metal wire and the via, is deposited and lithographically patterned such that the line and via patterns can be sequentially transferred into it. The patterns are transferred using reactive ion etch(RIE). Following this, the trench and via holes are lined with a barrier film and Cu seed layer for metal deposition. The Cu is then deposited using electroplating. This process ends up overfilling the features. The excess metal is then polished off by CMP to leave the Cu behind in the vias and trenches. Several methods of practicing dual damascene processing are known and new variations to the basic theme are regularly discovered. Regardless of which dual damascene variation is practiced, the costs associated with manufacturing interconnects with the dual damascene process are always less than those associated with the single damascene process or metal RIE process.

2.3 The Case for Copper

The advantages of Cu wiring include significantly lower resistance (R), the ability to pass higher amounts of current densities through the interconnect, and increased scalability relative to Al(Cu) wiring [5]. These benefits have

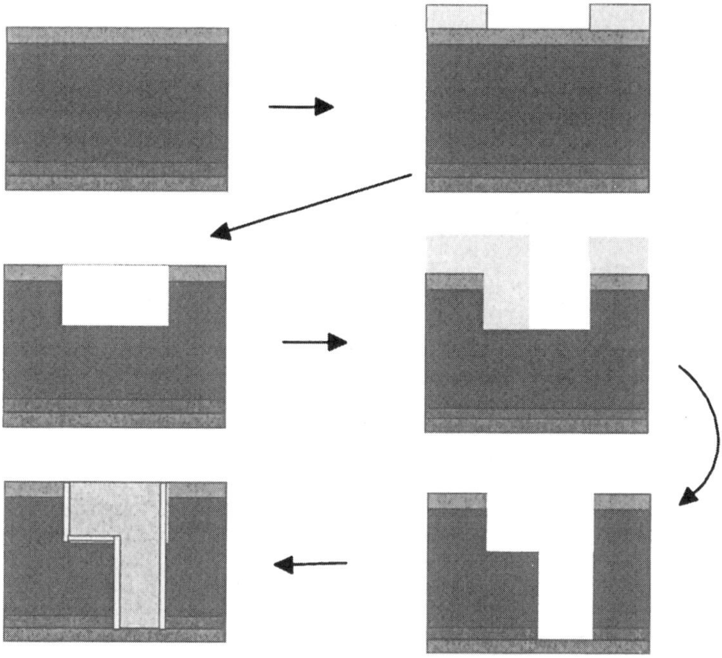

Figure 2.4. Typical line first dual damascene process flow

enabled the scaling of pitch and thickness, which in turn has enabled the fabrication of high density lower wiring levels which have reduced capacitance (C) and crosstalk, as well as global wiring levels which have low RC.

The performance impact of wiring can be related to the delay associated with charging a capacitive load through a small resistive device as well as the RC delay of the wire itself. Figure 2.5 shows a hierarchical wiring scheme which is driven by the varying needs of the wiring (low capacitance, low RC and low resistance) [6]. As can be seen from the figure, the lower wiring levels are at the minimum possible pitch and thickness to minimize capacitance and crosstalk and maximize wiring density for most of the circuits where short lines are used. At the same time, the upper levels are scaled uniformly (both vertically and horizontally) to maintain a constant capacitance-per-unit length while reducing resistance by the square of the scaling factor. Thus, low-RC wires are available for medium and long distance signal paths. This hierarchical scheme helps avoid the resistance catastrophe that would otherwise occur with scaled ULSI wires in high performance CPU's [7, 8]. The dual damascene process is especially amenable to such a hierarchical wiring scheme because of the high aspect ratios involved.

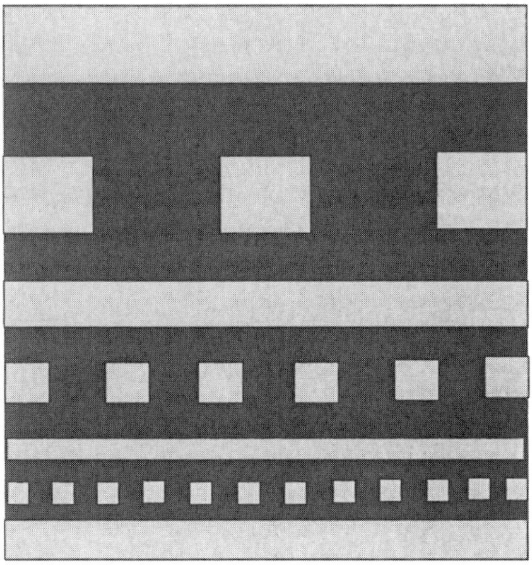

Figure 2.5. Schematic of a hierarchical wiring scheme with thin wires at the bottom closer to the device level and fat wires at the top closer to the package level to optimize performance [5]

An additional advantage of using Cu as the wiring metal is that it is possible to obtain near bulk resistivity values for Cu wires in submicron interconnects. This was not the case with Al wiring where the need to alloy the Al led to much higher resistivities. It is estimated [9] that there is about a 45% reduction in effective resistivity (which includes the resistance of the liner) of the wire with Cu as compared to the Ti/Al(Cu)/Ti/TiN lines. Hu et. al., [10] show that this advantage is maintained at least down to 0.1 μm line widths at 4.5:1 aspect ratios. An added benefit of this resistance advantage is that the Cu lines can be scaled down in thickness leading to a reduction in capacitance and crosstalk between adjacent lines. As can be seen in Figure 2.6, scaling the Cu thickness down to the point where the resistance equals that of a similar Al(Cu) line reduces the capacitance by approximately 20% and crosstalk by 15%. This reduction is equivalent to that obtained if a switch in the insulator is made from SiO_2 to a low-k material with a dielectric constant of 3. This prevents the need for an immediate replacement of SiO_2 with a low-k material as the insulator.

Finally, as will be discussed in the section on electromigration, the electromigration lifetime with Cu wires is at least two orders of magnitude longer at stress temperature than for the corresponding Ti/Al(Cu)/Ti lines. This enables the use of higher current densities and extends the use of Cu wiring for future technology generations which will have tighter pitches.

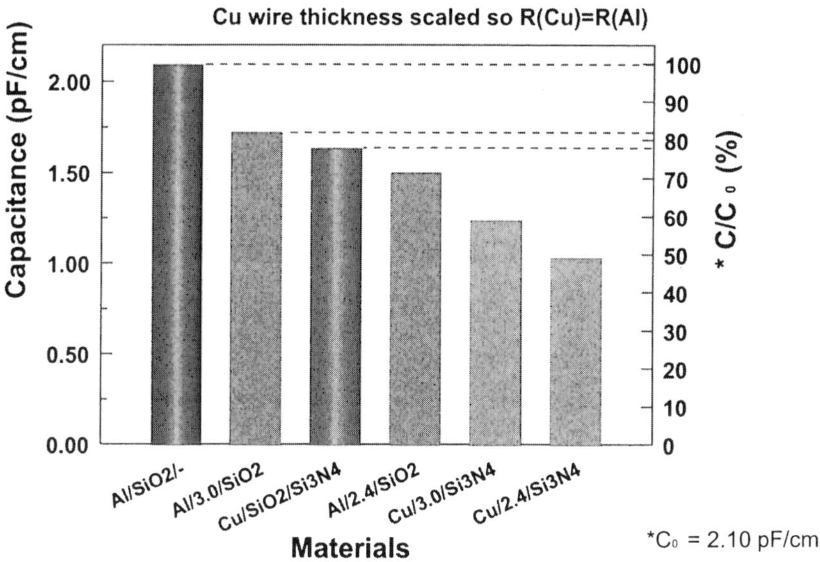

Figure 2.6. 3D capacitances with different metal/dielectric pairs [5].

2.4 Electroplating of Cu

Electroplating of Cu is one of the chief processes that has enabled the use of Cu wires in BEOL interconnects. While other processes such as electroless plating, laser-enhanced plating, jet plating, CVD and PVD [11] could also be used to deposit Cu into submicron vias and trenches, it has become clear that electroplating offers the most significant reliability improvements as well as cost of ownership reductions [12].

2.4.1 Integration of Electroplating in Device Fabrication

In order for Cu to be deposited on the surface of a wafer by electroplating, it is first necessary to cover the surface with a seed layer. The function of this layer is to act as a nucleation site for copper and to conduct the current from a contact at the wafer edge to all points on the wafer where a deposit is desired. The requirement of a seed layer has led to a variety of approaches for the integration of plating [12]. Two such approaches are illustrated in Figure 2.7.

The first of these is a process called "Through-mask plating" which uses a masking material on top of the seed layer such that electroplating occurs only on those areas of the seed layer that are not covered by the mask. The masking material and the surrounding seed layer are subsequently removed. Through-

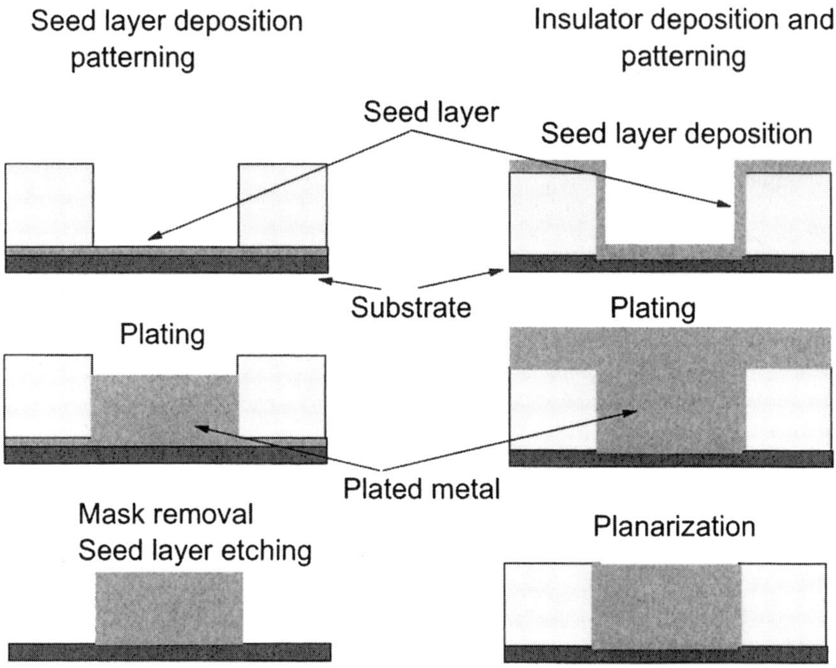

Figure 2.7. Two different approaches to integrating metal with electroplating as the deposition technique [12].

mask plating has been implemented in the fabrication of thin-film recording heads [13, 14] and C4 interconnections [15].

The second is damascene plating, which involves deposition of the seed layer over the insulator. Excess metal must be removed by a planarization step such as chemical-mechanical polishing (CMP). As discussed earlier, electroplating is ideally suited for the fabrication of interconnect structures, since it allows inlaying of metal simultaneously in dual damascene via holes and overlying line trenches [16] by a process called dual damascene. Further, it allows for the deposition of a barrier layer between the seed layer and the insulator. The barrier prevents interaction between the metal and the insulator [17].

In order for plating to be used as the deposition process of choice in dual damascene integration, it needs to fill trenches and vias completely, without any voids or seams which generally lead to low electromigration lifetimes due to the fact that they provide a fast path for the diffusion of Cu atoms. In the following section, it will be shown that, through appropriate control of the evolution of the profile of the plated film, it is possible to obtain void-free and seamless deposits by electroplating .

2.4.2 Profile Evolution in Damascene Copper Plating

Figure 2.8 shows the three possible ways for the profile of plated copper to evolve during the deposition. As the name implies, in conformal plating, a deposit of equal thickness occurs at all points of a wafer. This leads to the creation of a seam, or, if the shape of the feature is re-entrant, a void. Subconformal plating results when substantial depletion of the cupric ion in the plating solution inside the feature leads to significant concentration overpotentials which, in turn, cause the current to flow preferentially to more accessible locations outside the feature. Additionally, if the feature depth is large (say in excess of 50 μm), the ohmic drop in the electrolyte may cause nonuniformity in the distribution of the current in favor of external feature locations. Subconformal plating leads to the formation of a void even in straight-walled features [12].

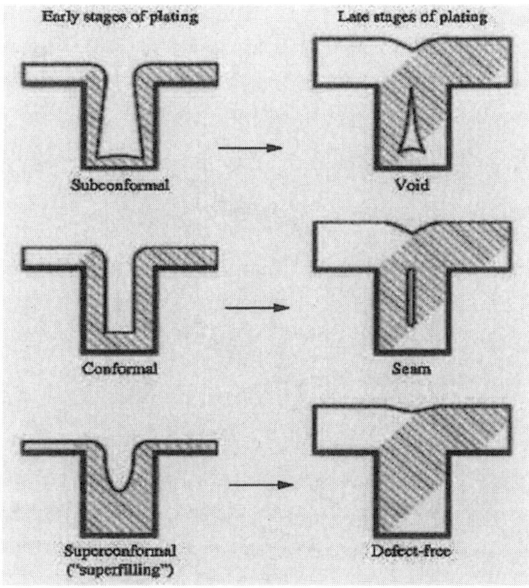

Figure 2.8. Evolution of the plating profile with three different deposition conditions that exhibit varying degrees of conformality. Superfilling leads to the best results [12]

For defect-free filling, a higher deposition rate in the bottom than on the sides of the feature is desired. This mode of plating, called superfilling [12] as shown in Figure 2.8, can be accomplished by the use of additives. Superfilling produces void-free and seamless deposits inside trenches and vias at dimensions commonly used in BEOL interconnect processing. This helps eliminate fast diffusion paths for Cu electromigration which would otherwise exist with subconformal or conformal plating [12]. Another phenomenon that contributes to good Cu reliability is the low temperature self annealing behavior of additive based electroplated Cu [18 - 20]. Andricacos et al., [12] have demonstrated this effect and have also shown that, through the use of such (proprietary) additives during plating, it is possible to control the profile. As shown in Figure 2.9, plating was interrupted before the feature was completely filled with Cu in order to assess the shape of the profile of the deposited metal. In case (a), it can be seen that superfilling occurs whereas in case (b), a nearly conformal deposit is obtained. Superfilling also involves rounding of the corners of the trench bottom. As can be seen, rounding is observed in both cases. Andricacos et. al., [12] have also developed a mathematical model of superfilling that is based on differential inhibition by diffusion-controlled additives. They reason that since the additive is diffusion-controlled, shape- induced concentration-field effects drive a very wide range of additive fluxes over the microprofile. These include extremely low fluxes in deep interior corners, low fluxes at the bottom center, moderate fluxes at sidewalls, and high fluxes at shoulders. The continuous variation of inhibition with additive flux over a very wide flux range enables the strong position dependence of the deposition rate, especially the differentiation between bottom and sidewall that promotes void-free and seam-free filling.

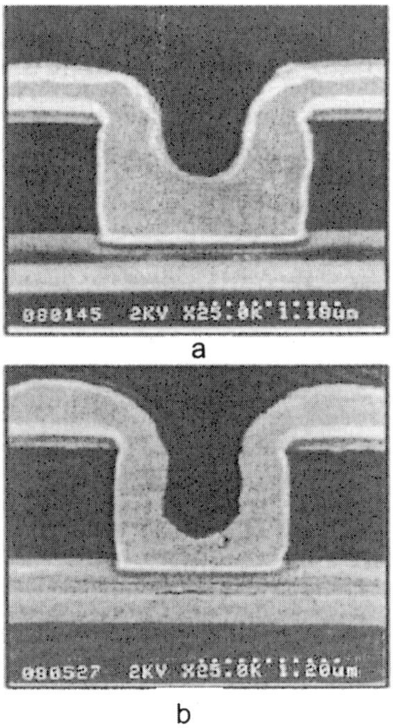

Figure 2.9. (a) Example of superfilling (b) Example of a conformal profile [12]

Considerable insight into the mechanism of superfilling has been gained recently as a result of various experimental observations that simple models such as the one described above cannot rationalize e.g. [21]. Real Cu plating baths contain several types of additives called accelerators, suppressors, and levelers. Individual additives can dominate the superfilling phenomenon either alone or in combination with others. It is possible that instead of the leveler-induced 'differential inhibition', an accelerator-induced 'differential acceleration' is at play [22-25]. The accelerator can generate an accelerating reaction intermediate whose concentration in the features is higher causing enhanced rates of Cu deposition inside features, hence superfilling. It appears that accelerator-dominated baths offer better extendibility; hence understanding their operation in detail is very important for the continued relevance of the Cu plating process in the interconnect application [26].

2.4.3 A Liner Scheme for Cu Interconnects

One of the key differences between the Al and Cu interconnects is the inability of copper to form a self passivating oxide layer like aluminum. As a result, the integration and use of copper in BEOL interconnects requires that the free surfaces and interfaces associated with copper be passivated by suitable barrier layers. In the damascene structures there are two distinct interfaces between copper and the surrounding dielectric that need to be passivated. First, the bottom interface that separates the copper from the bottom and side walls of the damascene trench and via structure and second, the top interface that separates the CMP planarized copper line top from the next level intermetal dielectric (IMD). The requirements for the barriers, and hence the materials used, are distinct for these two interfaces. In this section, the first of the two interfaces and the rationale behind the selection of appropriate materials to be used as barriers there will be discussed.

The barrier material employed in the bottom and sidewall interface is usually referred to as the liner to reflect the fact that it is expected to conformally "line" this interface. The liner is expected to serve multiple functions in the structure. First, it is required to act as a barrier against copper diffusion during multilevel interconnect build (typically 400 to 450C for several hours). This is essential because most of the dielectrics used as IMDs either become electrically leaky or exhibit significant degradation of electrical breakdown when contaminated with copper. Additionally, the liner should act as a barrier to the ingress and oxidation of copper by oxygen, water vapor or other oxidizing species that the structure might be exposed to during processing and actual use. This is especially important for future copper BEOL interconnects employing low dielectric constant (k) and ultra low-k IMD films. Third, it has to serve as an adhesion layer between copper and the IMD. This is important to ensure that the copper damascene fill does not delaminate from the trenches and vias during CMP and thermal cycling. Fourth, it is preferable that the liner has good lateral (in plane) electrical conductivity so that it can serve as the redundant metal layer in the interconnect wires. This attribute enables the liner to serve as a current supporting layer in the case of voiding and loss of connection in the copper lines as a result of thermal stresses or electromigration. As mentioned earlier and as will be discussed in greater detail subsequently, copper interconnects can exhibit superior electromigration resistance, but are not immune to failure by such phenomena and as such can benefit from the redundant conductive liner. Resistivity on the order of 40-50 $\mu\Omega$-cm is highly desirable in this regard. A further requirement which is more pertinent to the selection of the deposition process employed to form the liner, is that the liner conform and cover the bottom and sidewalls in the width and aspect ratios of patterns expected in the BEOL interconnects.

Many candidate materials including Cr, Ti, Ta, nitrides of Ti, W and Ta, silicon nitrides of these metals, and metal/metal nitride bilayers have been explored for the liner application over the years. Ta based films have emerged as the most suitable choice based on their ability to meet the above mentioned criteria. The effectiveness of Ta as a copper diffusion barrier was reported first by Hu et al. [27] who showed that the films in the β-phase crystalline form were good barriers to copper diffusion. An essential requirement for achieving good barrier performance was later found to be the control of oxygen level in the Ta films since copper diffusion through Ta grain boundaries was sensitive to this factor [28]. In one of the earliest published studies on the dual damascene integration of Cu interconnects, β-Ta was used as the liner to prevent Cu diffusion into the polyimide IMD due to its superior adhesion to the dielectric [29]. However, when Silicon Dioxide was chosen as the IMD for use in advanced processors, a change from the β-Ta liner was necessary due to the inadequate adhesion of Ta or Ta_2N to silicon dioxide [30,31]. Additionally, it was found that TaN which can be formed under appropriate reactive sputtering conditions has an excellent adhesion to SiO_2 [32]. However, the adhesion between the copper seed layer required for plate up and the TaN film is quite poor necessitating a second adhesion layer between the two [32]. Adhesion at this interface is also critical since it is believed that apart from mechanical reliability, good electromigration resistance is also achieved as a result [33]. Ta has been found to be the best choice for this second adhesion layer because it affords good wetting of Cu [34] and atomic scale mixing [35] without alloying into Cu. Alloying of Cu is undesirable because the resistivity of Cu increases dramatically even with small amounts of solute elements [36] thus partially negating the benefit of replacing Al with Cu. The bilayer structure of TaN/Ta has been successfully integrated with Cu/SiO_2 multilevel dual damascene interconnects by taking advantage of the unique properties of the individual layers.

TaN exhibits excellent adhesion to SiO_2 while Ta provides the transition layer between TaN and the Cu seed layer. Ta can also be easily deposited in the same pump down sequence with TaN by just changing the sputtering gas chemistry thus facilitating a convenient process with ease of manufacturability. It has been found that sputter deposited Ta provides better step coverage than lighter barrier metals such as Ti due to the greater directionality of the heavier species, Ta, and its ability to recoil around inside trench features [37]. This is attractive from the perspective of conformally coating high aspect ratio structures with minimum barrier layer thickness to maximize the volume of copper in the lines.

Other serendipitous benefits of the use of the bilayer stack have also been discovered further making its use very attractive [32]. It has been found that the TaN composition best optimized for adhesion and low resistivity yields a new crystalline phase of TaN (hexagonal close packed versus the more commonly known face centered cubic version of TaN) [38]. This TaN phase nucleates α-Ta which has a bcc structure and has a resistivity of about 15-30 micro ohm-cm [39, 40]. This value is about ten times lower than that of the β-Ta which tends to form when Ta is deposited directly on insulators like SiO_2 [32]. Thus a judicious combination of hcp TaN and the α-Ta in a bilayer structure allows the formation of a liner stack that has an overall effective liner resistivity less than 40 micro ohm-cm, which is preferred in BEOL wiring structures. This level of liner resistivity also enables the liner to perform adequately as a redundant conductor in case of copper electromigration fails.

The diffusion barrier efficacy of the liners with TaN is also excellent. Edelstein et al. [32, 41] have reported that 25 nm thick TaN films act as an effective barrier between Si and Cu up to temperatures as high as 800 ^{o}C. They attribute this barrier performance to the nearly amorphous and/or microcrystalline structure with stuffed grain boundaries. Structures with the bilayer TaN/Ta liner are expected to provide even better barrier performance when used in actual products, due to the presence of the α-Ta over layer. Thus the bilayer liner enables optimum adhesion, good diffusion barrier efficacy and conductor redundancy in copper interconnects. Enhancements such as ionized sputter deposition have further improved the manufacturability of the sputtered TaN/Ta liner stack. Single wafer cluster tools that can deposit the liner and the Cu seed layer efficiently in a single tool platform have been developed and have since become the workhorse systems for manufacturing plants the world over.

For future technology generations, it is expected that the liner choice will have to be revisited as inter metal dielectrics migrate to low k and ultra-low k films, feature sizes get smaller and aspect ratios become more aggressive. Decreasing line widths require that the liner thickness scale concomitantly to maintain a reasonable copper cross sectional area. This drives a more stringent need for diffusion barrier coverage at lower liner thickness. It is necessary to ensure that future dual damascene features be reliably and conformally coated at such liner thickness levels. Such requirements would most likely drive the liner choice towards films deposited by CVD or ALD processes and new metallurgies and stacks would likely be developed. The concept of near zero thickness or self-forming liners may indeed be needed for the copper interconnects of the next decade.

2.5 Reliability of Cu Interconnects

Several reliability questions pertaining to copper BEOL interconnects had to be investigated and appropriate solutions identified before the successful implementation of chips with copper wiring. The most prominent among these issues is that of electromigration failures in copper interconnects. The underlying mechanisms that govern electromigration, the interplay between them and the copper microstructure had to be also developed. Additionally a methodology used to analyze and predict acceptable use conditions based on test data had to be developed. A brief overview of this important reliability aspect will be presented below and the reader is referred to more recent publications for details [42-44]. In addition, a quick survey of two other reliability issues, namely, stress migration and voiding, and time dependent dielectric break down will also be presented.

2.5.1 Electromigration

Electromigration in Cu and Al is a phenomenon wherein the movement of electrons through the metal interconnects under the applied electric field causes the drift of atoms in the direction of the electron flow. The electron motion is often referred to as an electron wind that sweeps the atoms with it causing mass depletion (and hence void formation) in the cathode end and mass accumulation at the anode end of the interconnect. Electromigration in aluminum interconnects was first discovered in the mid-1960's and mitigated by the alloying of Al with Cu and the introduction of redundant current carrying liner layers [42]. Electromigration still remains a key limitation on the current carrying capacity of interconnects even with the use of Cu wiring and continues to be a field of active research and development.

The effects of various factors on the electromigration behavior can be understood through stress testing of representative structures with known metal microstructures, and analysis of failed sites and the correlation of this information with the test data using the framework provided by equation (2.1) below. In this equation, electromigration induced atom flux, J_e , is given by Nv_d where N is the density of atoms in the interconnect material and v_d is the atom drift velocity. The drift velocity in turn can be expressed as a product of the intrinsic atomic mobility M and the electromigration driving force F_e through the Nernst-Einstein equation,

$$v_d = F_e \cdot M = Z^*_{eff}\, e\, E\, \frac{D_{eff}}{kT} = Z^*_{eff}\, e\, j\, \rho\, \frac{D_{eff}}{kT} \tag{2.1}$$

where e is the electron charge, Z^*_{eff} is the apparent effective charge number, E is the electric field responsible for the current density j, D_{eff} is the effective diffusivity of the diffusing atom through the interconnect line, ρ is the specific resistivity of the interconnect material, k is the Boltzmann constant and T is the temperature in degrees Kelvin. Equation (2.1) can be used to understand the relation between the microstructure of the interconnects and the failure modes that occur due to it. Extrapolation of the test data can also be performed to determine acceptable use conditions to yield a value of current density supportable by a given interconnect design at the nominal use temperature.

Structures typically employed for electromigration testing are shown in a schematic in Figure 2.10 [44]. These test sites are comprised of long lines (about 400 μm long) and contain multiple metal and via levels designated as M0/CA/M1/V1/M2, M1/V1/M2/V2/M3 or M0/CA/M1 as shown in Figures 2.10(a), 2.10(b) and 2.10(c) respectively. M0 represents the first local interconnect level. CA is the contact level between M0 and M1 on the chip and is made of W. M1, M2 and M3 represent the intermediate interconnect levels and V1and V2 the via levels made of Cu. Single damascene processing is used to fabricate the W levels and the M1 Cu level while dual damascene processing is used to fabricate the M2/V1 and M3/V2 combinations. These details are pertinent to the flow of Cu during the electromigration stressing and the associated failure modes as will be briefly discussed below. The lines and vias in the test structures are fabricated using the same fabrication methods and tools employed for the product wafers so that meaningful information relevant for product structures can be gleaned. The dimensions of the lines and vias are selected to include typical values expected in the particular technology generation of product and additionally some wider lines that serve as model structures for evaluating specific issues associated with factors such as flux divergence and the relative contributions of grain boundary and interface diffusion paths to the electromigration atom flux. The structures are annealed prior to test to simulate the thermal annealing the product structure will experience during build so that the microstructure of the interconnect materials would be comparable to what would be expected in actual product structures.

Electromigration testing is usually carried out at accelerated test conditions in the 250 to 350°C test temperature range in an inert or reducing ambient. Typical current densities used in the test are such that they lead to electromigration failures in a few hundred hours. The resistance of test lines is monitored as a function of time at test conditions and failure time is usually defined as the time at which the resistance increases by a certain percentage from the

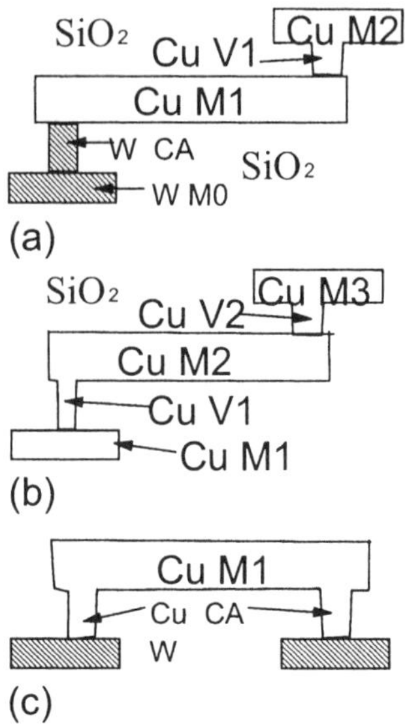

Figure 2.10. Electromigration test structures [44]. See text for details.

initial value (usually 4 to 10%). Typical plots of resistance versus time curves for samples of the type shown in Figure 2.10(a) [44] with electron flow from Cu via to Cu line are shown in Figure 2.11. A range of data showing short times to failure as well as seemingly failure resistant behavior is evident from this example. The data from a large collection of such tests are then replotted as cumulative percentage failure versus log failure time (τ) as shown in the sample Figure 2.12 [44]. The data in this particular plot represent a set of samples with the structure shown in Figure 2.10(b). A mean value of life time that corresponds to 50% cumulative fails, designated as t_{50}, and an associated standard deviation σ are extracted based on a single or multiple log normal distribution fit of the data. Cross sectional analysis of failed sites is usually carried out by focus ion beam (FIB) sectioning followed by SEM or TEM characterization. By analyzing the location and morphology of the failed sites and their correlation to the length of time to failure, useful information on the underlying mechanism and atom migration pathways associated with specific failure modes can be deduced. This information is also used to validate the particular methodology used for fitting the data (single versus multiple log normal distributions of failures) and extrapolating the same to use conditions.

Electromigration induced failures arise from Cu atom flux divergence which in turn is caused by the discontinuity in the metal types and the availability of diffusion paths in the direction of electron flow. Discontinuity in metal

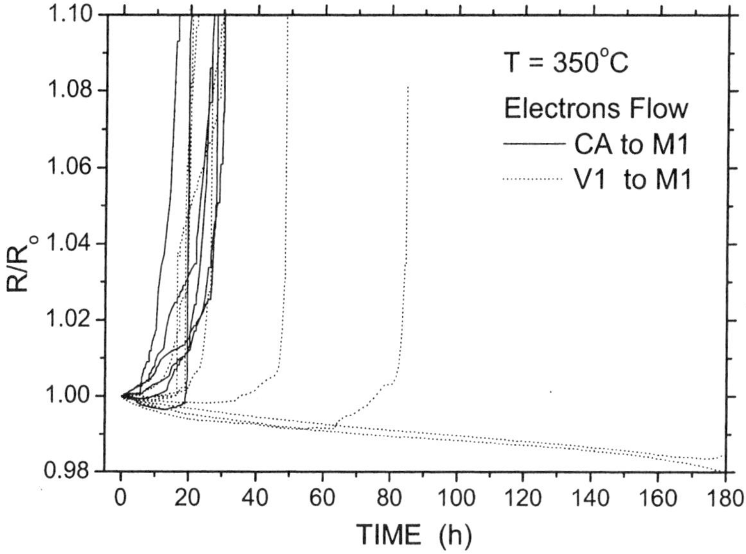

Figure 2.11. Typical experimental data showing line resistance versus test time at 350°C for samples with test structure shown in Figure 2.10(a) [44].

types occurs in the W to Cu interfaces and the Cu to liner interfaces in the structures depicted in Figure 2.10. These interfaces are blocking interfaces in that diffusion of copper atoms through them is practically negligible. This in turn can lead to a depletion void or a pile up of atoms in the vicinity of such a blocking interface. Figure 2.13 depicts an SEM cross section of one such case where a void is formed above the W at a W to Cu interface where the electron flow is from the W stud to the Cu line. More recent tests with dual damascene line structures and analysis have shown that the data in the case where such blocking boundaries are present at the cathode end of the line can be represented as a combination of two log normal distributions [43]. A typical data plot from such a case with a structure similar to the one in Figure 2.10(c) is shown in Figure 2.14. The bimodal nature is related to two distinct failure modes identified by FIB cross sectional analysis. The first one in which void formation occurs at the interface between the MC level W and the CA level Cu and the associated failure time is short. Figure 2.15(a) is FIB cross sectional image of such a failure site. The second mode is one where the voids form at the line ends only or at the line end and at the CA/MC interface, with a longer failure time. Typical FIB sections of such failure sites are shown in Figure 2.15(b) and Figure 2.15(c) respectively. The electromigration activation energy values determined by Arrhenius analysis of the failure time data for the two

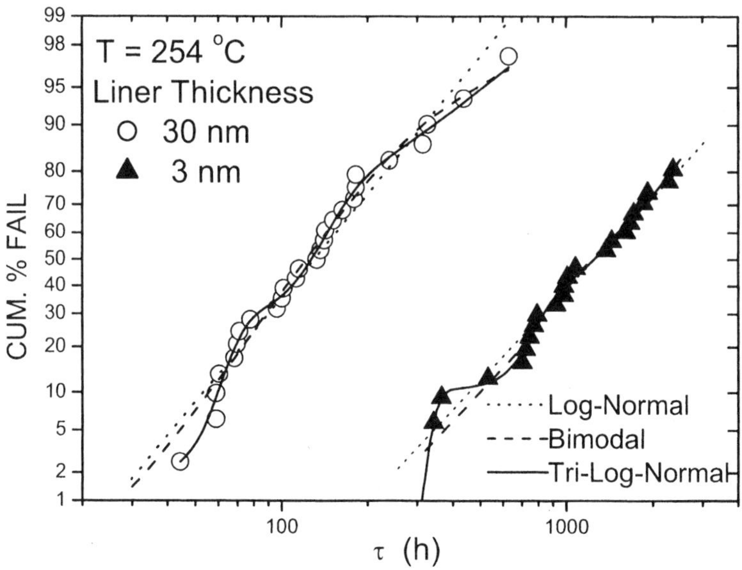

Figure 2.12. Cumulative percent failure vs. log time for samples with test structure Figure 2.10(b) with thin and thick liner and hence different Cu blocking efficacy. Lines are least square fits of single, bimodal and trimodal log-normal fits to the data [44].

failure modes are similar and are about 1 eV [43] indicating that the diffusion pathways are likely to be the same in both cases. The difference in time to failure is therefore related to the volume of Cu that needs to be transported before a significant resistance change is detected. The bimodal analysis based on such failure mode determination allows the extraction of meaningful t_{50} and σ values from experimental data in such interconnect schemes.

More complex behavior is possible in the case of structures such as the one in Figure 2.10(b) where the interfaces between vias and lines are made of liner layers and their permeability could vary depending on the aspect ratio of the structures and the type of liner deposition process conditions employed. A trimodal fit of data was necessary to account for the data and three different failure modes were also observed corresponding to the three regimes of failure times as shown in Figure 2.12. The ability of copper atoms to diffuse through some of the more permeable interfaces can reduce flux divergence at these interfaces and hence increase the time to form voids. In some cases voids form at line ends which are removed from the active via/line interfaces and hence do not cause an electrical open in the test duration. Such characteristics can be exploited to enhance the electromigration life time of copper interconnect structures provided the appropriate process and integration techniques to con-

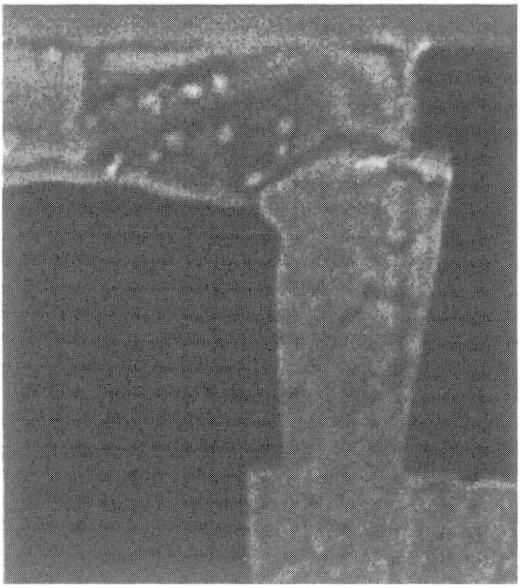

Figure 2.13. SEM micrograph of a void at the cathode end of a Cu line connected to a blocking W via [42].

trollably produce these structures can be developed [45, 46]. The reader is referred to the paper by Hu et al [44] for a more extensive discussion of the nuances involved in these more complex cases.

The other important aspect that governs the kinetics of the electromigration void growth is the diffusion path way for the atomic flux. Since the bulk diffusivity of Cu is very small at the test temperatures employed, the most likely candidates for the diffusion pathway are the Cu/liner interface, Cu/dielectric barrier cap interface and the Cu grain boundaries [42]. Plated Cu interconnects of submicron widths nominally exhibit a bamboo-like microstructure after the normally expected thermal anneals in the multilevel interconnect processing [42]. Such a microstructure is characterized by grain boundaries that are predominantly perpendicular to the line length and width. Grain boundaries of such an orientation do not contribute to Cu atom flux along the length of the line. This leaves one of the other two interfaces as possible diffusion pathways for Cu flux. It is believed that the Cu/dielectric barrier interface is more likely to be the main contributor for copper atom migration at the typical test temperatures and those expected during the useful service of electronic chips. The fact that all the different failure modes alluded to in the preceding discussion exhibit nominally the same activation energy lends support to the hypothesis that a single diffusion pathway is active under these conditions.

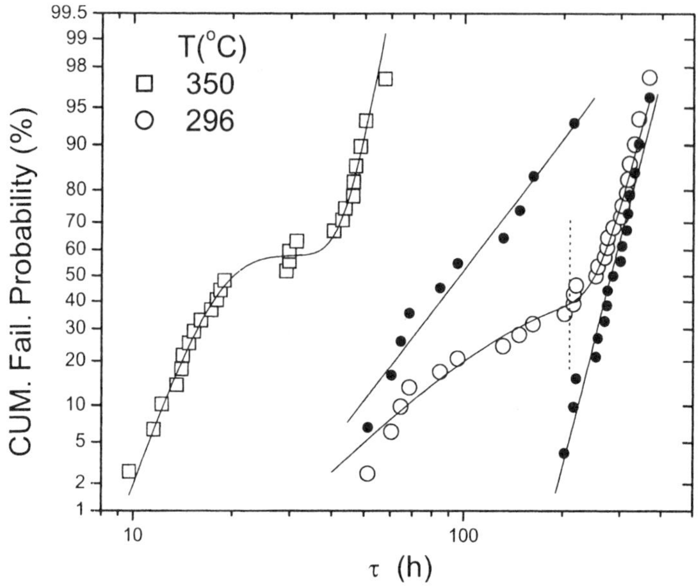

Figure 2.14. Cumulative percent failure vs. time on a log-normal scale illustrating the separation of the data into two distinct failure distributions [43].

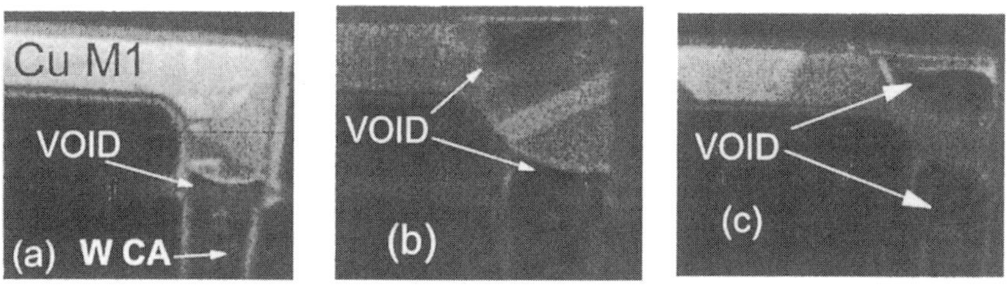

Figure 2.15. FIB cross section images of void formation illustrating different failure modes [43]. See text for details.

Cu lines which are several microns wide do not possess a clear bamboo structure but rather have a polycrystalline or a mixed bamboo-like and polycrystalline microstructure [42]. Such lines do show indications that more than one diffusion path way is active in the electromigration failure. A more detailed analysis where D_{eff} is expressed in terms of the diffusivity of copper through the different pathways [42] needs to be invoked in such a case. The drift velocity underlying the migration process can be modeled as a superposition of the two pathways, namely, the free Cu surface and the copper grain boundaries over such a line width range [42]. For fine interconnects that are of importance to very large scale integrated chips, where the Cu/dielectric barrier interface dominates as the diffusion pathway this analysis predicts that the lifetime is independent of line widths at a constant line height. However, the current supportable by these fine lines actually decreases because line heights typically scale with line widths due to processing and performance reasons. Methods to mitigate electromigration rates by slowing down the diffusion kinetics for Cu at the appropriate transport interface would therefore be a key area of future research.

2.5.2 Stress Induced Voiding (SIV)

Another reliability issue associated with submicron interconnects is the phenomenon known as stress induced voiding (SIV). This phenomenon was first observed in Al interconnects [47] but its quantitative characterization has been elusive [48, 49]. In a simplistic sense stress voiding can be described as the formation of voids in metal interconnects as a result of diffusion and creep processes driven by tensile stresses developed in the metal features constrained by the inter-metal dielectric. The stresses and stress gradients have their origin in the thermal strains arising from differences in the coefficient of thermal expansion (CTE) between the metal and the dielectric, which in turn leads to local stresses as the metal is constrained from accommodating these stresses by the fairly rigid IMD. When heated to moderately high temperatures (150 to 250°C), aluminum interconnects with bamboo structure exhibit void formation in grain boundary sites which grow to become wedges that can lead to line opens [50] (metal still under tension at this range). The problem usually shows a temperature dependent rate that is very low near a stress-free temperature, peaks at an intermediate critical temperature (180 -250°C) and tapers off again at higher temperatures due to metal expansion driving the metal into compressive stress relaxation. Although there is a general consensus on the plausible mechanisms underlying SIV, distinguishing specific contributions from different mechanisms has been difficult due to the variability in void morphology, line and via geometry and in the properties of the metallization and the IMD [51]. The SIV problem has been tackled in Al interconnects by the use of SIV resistant redundant metal layers (refractory materials such as $TiAl_3$) so that SIV does not lead to line opens.

In the case of copper interconnects, one might be tempted to surmise that stress voiding ought to be a lesser concern because the thermal activation energies of comparable diffusion pathways are higher in Cu than Al while the thermomechanical driving force is similar [51]. However, SIV is indeed observed in structures containing small vias located on top of wide lines when the Cu interconnects are constrained by the IMD and the dielectric caps. The problem is not observed in vias connected to submicron lines. This problem occurs when the electroplated Cu is not fully annealed to a stable microstructure before the post-CMP dielectric cap is applied on top of the lines. It is known that Cu interconnects formed by electroplating possess a very fine and nonuniform grain structure in the as-plated condition. Such a structure is characterized by an abnormal grain growth at room temperature which is deemed to be responsible for the resistance transient that results in the decrease of the line resistance with time [52]. It is also believed that such a Cu interconnect structure can have a high supersaturation of vacancies when it is not properly annealed prior to being constrained by the post-CMP cap layer [51]. In a recent analysis, Ogawa et al. [51] postulate that SIV void growth can occur readily in such a scenario by utilizing the already available excess of vacancies and the various diffusion pathways such as grain and interface boundaries. The associated activation energy for such pathways will be quite low (about 0.74 eV) as seen in the SIV data. One could conclude that it is very critical to ensure that the electroplated copper interconnect lines be fully annealed to a stable grain structure before any caps are deposited to avoid this problem. Interestingly, other microstructural aspects such as the crystallographic texture of the copper films, could also play a role in the SIV process [53] once the larger issue of grain growth is addressed. However, the successful implementation, large scale production and use of chips with submicron copper interconnects with silicon oxide and fluorinated silicon oxide IMD indicate that the SIV problem can be mitigated with the appropriate post-plating process regimen.

2.5.3 Time Dependent Dielectric Breakdown (TDDB)

As mentioned earlier, one key feature of the aluminum interconnects is the ability of aluminum to form a chemically inert and passivating oxide layer that precludes any concern of metal penetration into the inter-metal dielectric. In contrast, copper does not form such an oxide barrier on its surface and, unless prevented by dielectric or metallic barriers, will diffuse into the IMD rather readily and rapidly. While the initial concern about this issue was poisoning of the devices, a more serious concern is the penetration of copper into the IMD and the attendant formation of trap levels in the band gap of the IMD. This in turn can lead to leakage conductivity in the dielectric. When a sufficient level of trap density is achieved in the IMD percolation effects can lead to dielectric breakdown over time in the presence of an applied electric field. This time dependent dielectric break down (TDDB) can occur at moderate test temperatures (about 150°C to 300°C). In fact TDDB is quite sensitive to small quantities of Cu in the IMD and has been used as a means to evaluate the efficiency of copper diffusion barriers [54]. TDDB stress is usually performed at about 1 to a few MV/cm in the temperature range mentioned above [55, 56] and failure is defined as time to exceed a preset leakage current density level (usually of the order of 0.1 to 1 mA/cm^2). Results are plotted as log-normal plots of cumulative percent failed versus time in a manner similar to the electromigration data.

A recent study [55] reported TDDB behavior of blanket PECVD oxide films deposited from the commonly used tetraethylorthosilicate (TEOS) precursor. By testing intentionally copper doped TEOS based oxide films, the authors showed that there appears to be a threshold of about 10^{12} atoms/cm^2 below which TDDB failure was not detected. At the upper end of concentration (about 10^{15} atoms/cm^2) time to failure shows a saturation effect most likely due to a copper solubility limit in the oxide. Activation energy and field acceleration factors can be derived from such data to extrapolate failure times at expected use conditions in chips.

Different IMD films and even the same IMD film deposited by different methods can be expected to show different levels of TDDB sensitivity. In copper damascene structures, in addition to the copper "doping" effect described above, TDDB behavior can also be influenced by the surface integrity of the IMD after the typical process exposures it encounters during the interconnect build. Noguchi et al. [56] report that plasma preclean treatments commonly employed to clean the copper line surfaces after CMP but prior to a silicon nitride barrier deposition can affect the TDDB behavior of the oxide IMD. They report that ammonia plasma clean resulted in a better TDDB behavior compared to a nitrogen plasma treatment or a hydrogen ambient anneal [56]. They attribute this in part to the reduction of the CuO on the lines and its conversion to form CuN which bonds well to the nitride cap layer. They also postulate that H radicals in the ammonia plasma passivate the dangling bonds on the oxide surface improving its TDDB. Wet chemical treatments postulated to remove surface defects and damage layers resulting from CMP are also shown to improve the TDDB [56].

As the copper-silicon oxide based interconnect fabrication process evolves and matures, reliability issues such as the cleanliness of the copper line surface, efficacy of the metallic and dielectric barriers and the quality of the IMD surface after CMP will be addressed and the associated failure propensity will be lessened. However, as one moves into the era of Cu-low k and Cu-ultra low k IMD structures [1], one has to revisit these issues in light of the new process exposures and integration schemes employed to fabricate these interconnects and the sensitivity of the new IMD materials to the same. One aspect of low k and ultra low k dielectrics is that many of them derive their low k values at least in part through the presence of molecular free volume and/or intentionally created porosity. This opens up the possibility of ingress of ambient gaseous species (oxygen, water vapor) into the interconnect structure during interconnect build and during actual chip use. This added environmental effect has to be factored in future reliability assessments for these systems.

2.6 Processing of Cu Interconnects

2.6.1 Tooling Changes in the Transition to Cu

While the transition from the Al(Cu) subtractive etch processing to Cu dual damascene processing was indeed a revolution, it was achieved with minimal disruption to the tooling requirements in the BEOL section of chip fabrication plants. To achieve this transition while using the existing fabrication infrastructure, only one new tool, an automated Cu electroplating system was needed. This transition also ensured that some of the existing tools which hosted processes such as metal RIE , dep-etch SiO_2 etc were now obsolete. However, while the disruption to the tooling infrastructure was minimal, a large effort was required to develop new processes on existing tools to achieve high yield of Cu interconnects in SiO_2 dielectric. For example, this was the first time a multi-level build was performed using dual damascene integration and consequently, a tremendous number of lithography and reactive ion etch challenges had to be overcome for successful integration. Some tooling issues such as the modification of some of the dielectric deposition processes from a gapfill application to planar interlevel dielectric application needed to be resolved as well. Similarly, as discussed in the section on liners above, new barrier deposition processes needed to be developed. Additionally, a large development effort was required to develop processes to polish Cu and liner to leave behind planar interconnects. Finally, it was important to understand the integration challenges posed by all these new processes. An intimate understanding of the interplay between these processes and their effect on the final microstructure of the Cu in the interconnect as well as the interfaces formed between Cu and the surrounding liner and dielectric materials was needed in order to achieve high yield and good reliability of Cu interconnects [57].

2.6.2 Manufacturability Considerations

For a process flow to be manufacturable in large volumes at low cost, it is essential for all the unit processes to be optimized. Further, it is important that the integration of these processes produces chips with high yield and good reliability. For example, as shown in Figure 2.9, it can be demonstrated that through appropriate optimization of the process chemistry, it is possible to plate individual features completely without voids or seams. However, for this process to be considered a manufacturable process, this has to be replicated on a full wafer, and repeated from wafer to wafer and lot to lot. Figures 2.16 - 2.18 show the data on blanket wafers for Cu plating obtained from wafer marathons and production runs using a proprietary plating process. Figure 2.16 shows that it is possible to deposit $2\mu m$ films on a thin seed layer with about 1.0% (1σ) non-uniformity across the wafer which is well within the acceptable limits. This can be achieved despite the fact that it is generally difficult to deposit thin films with good uniformity via electroplating. Figure 2.17 shows the average thickness non-uniformity obtained over a 5000 wafer run for 1.3 μm thick films. As can be seen, the average non-uniformity is 1.72% which is also well within acceptable limits. The thickness itself can be controlled quite well from wafer to wafer as can be seen from Figure 2.18 which shows a variation in thickness of 0.65% during a 17000 wafer marathon. The calculated Cu resistivity from the mean sheet resistance presented in Figure 2.18 and from measuring the thickness of the plated films, is 1.79 $\mu\Omega$-cm. This value is the same as the value obtained from integrated Cu interconnect resistances [2] and confirms that the principal advantage of Cu, namely its low resistivity, is preserved by using the electroplating process. The reproducibility exhibited by the process on blanket wafers as demonstrated by Figures 2.16 - 2.18 leads to excellent thickness control and reproducibility on actual product lots. As can be seen from Figure 2.19, the lot to lot thickness variations over a period of many months is well within the specified limits. This is obtained partly due to a very stable bath chemistry whose concentration is tightly controlled [57].

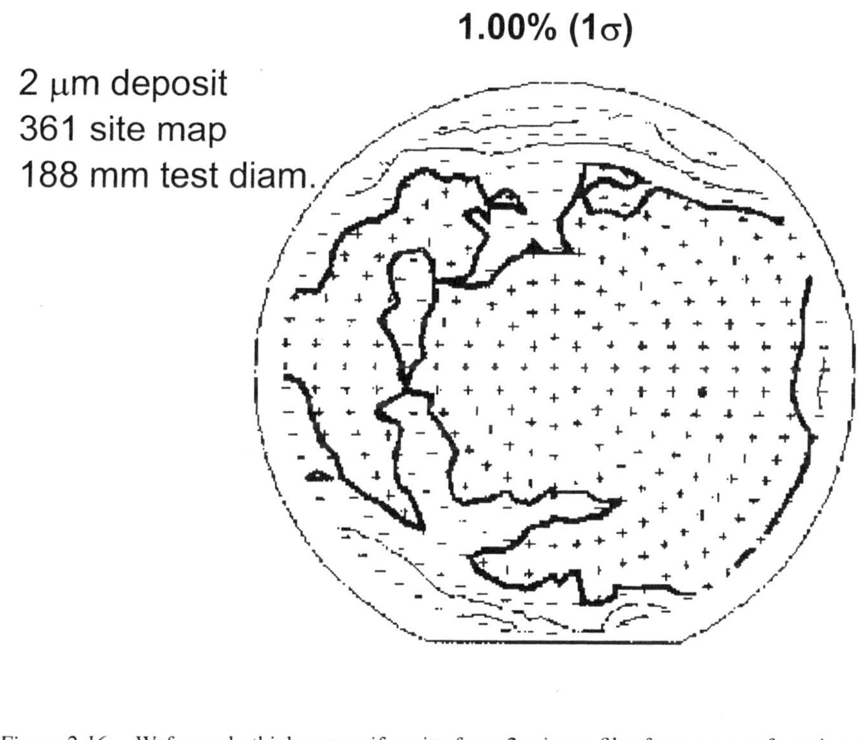

Figure 2.16. Wafer scale thickness uniformity for a 2 micron film from a manufacturing Cu electroplater [57].

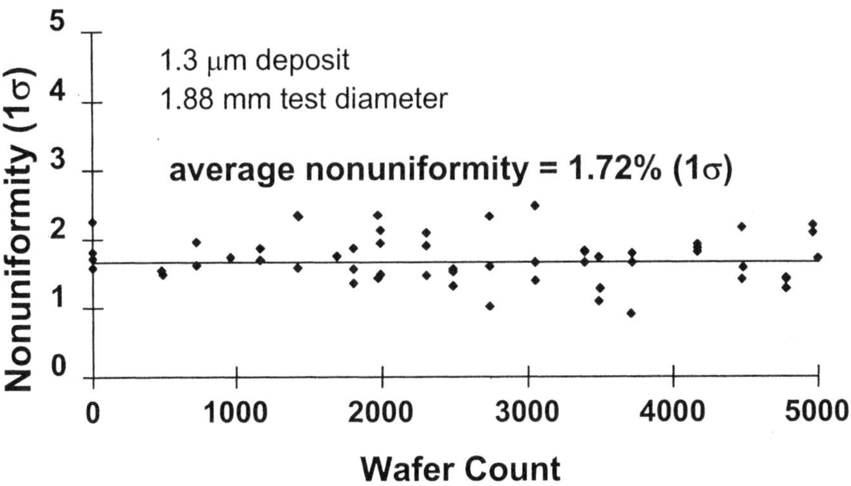

Figure 2.17. Wafer-scale thickness uniformity over 5000 wafers for a 1.3 micron deposit [57].

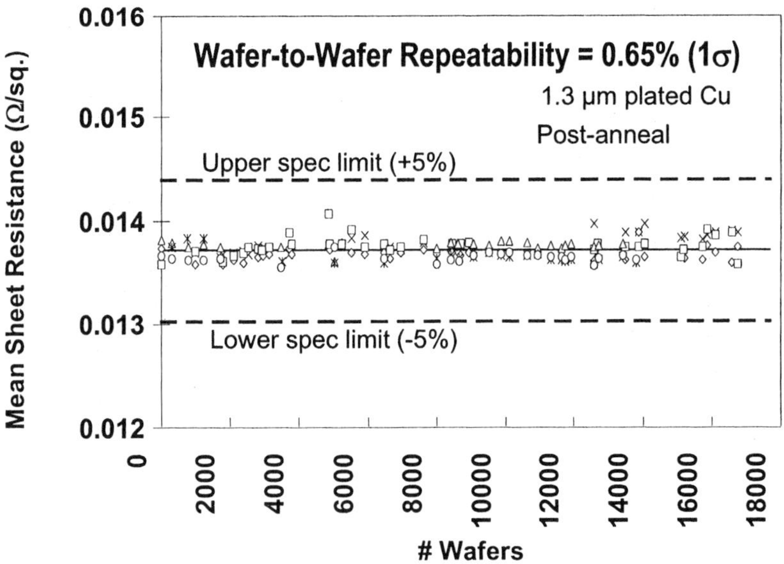

Figure 2.18. Wafer-wafer mean thickness repeatability for a 17000 wafer run [57].

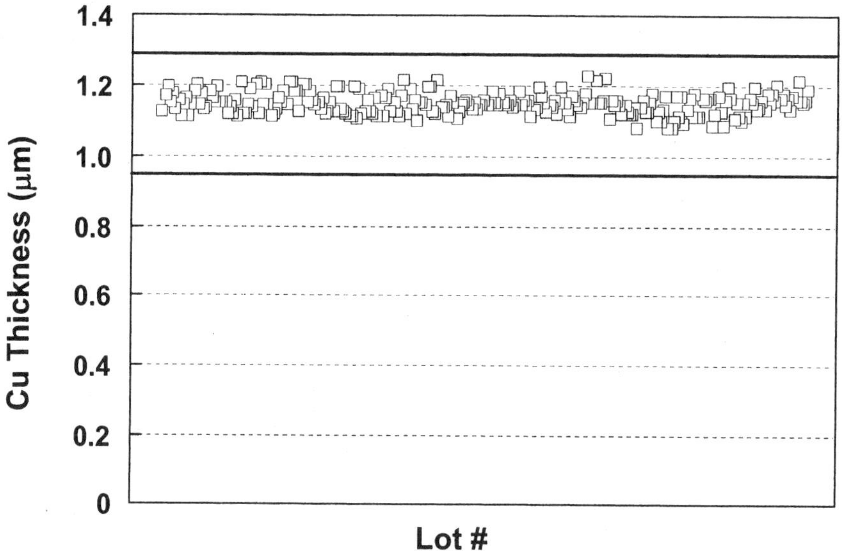

Figure 2.19. Wafer-wafer mean thickness repeatability for production microprocessor lots [57].

As the previous paragraph illustrates, it is possible to obtain a manufacturable process by optimizing the chemistry as well as the tooling infrastructure for a particular unit process. However, in order to obtain high yield and good reliability, it is necessary to optimize all the other unit processes such as Cu CMP, lithography, liner deposition and cleaning processes.

As demonstrated in reference [57], single vias and via chains (two unlanded vias and one line segment per link) can be fabricated with low contact resistance (< 0.3 Ω/ via and < 0.6 Ω/link over many lots). This again demonstrates that the principal advantage of Cu, namely, that of low resistivity, can be preserved during process integration. This data also confirms that Cu vias have less than half the resistance as compared to Al (Cu)/W vias. Low via resistances contribute significantly to the performance and yield of multilevel interconnect structures.

Further, as has been demonstrated by Stamper et al, [58] geometric considerations such as via size, via slope, via overlay, trench aspect ratio etc., also play a role in improving the reliability of Cu interconnects. They find that increased aspect ratios prior to the PVD Cu seed deposition degraded reliability and that increasing the via contact area and slope increased wiring reliability. They also report that dual damascene integration has improved wiring reliability as compared to single damascene integration. Evaluation of the interconnect profiles with cross-sectional SEM analysis as well as analysis of parametric test results during integration are useful tools to analyze the maturity of a process flow.

Ultimately, the yield and reliability of the chip at the final test is the deciding metric for the maturity and manufacturability of a process. Reference [57] shows the yield data during a typical production ramp of a large, high performance, 6 metal level microprocessor. It can be seen that during the initial stages, the yield obtained is below the target level as standard production problems are encountered. However, as these problems are fixed, the yield increases to a point that is well above the target yield at which point the process becomes manufacturable.

In the early stages of Cu interconnect development, it was widely anticipated that contamination of silicon devices with Cu could occur due either to practices employed during fabrication or diffusion of Cu into the device region. Diffusion of Cu into the device regions could lead to catastrophic results due to the degradation of device properties since Cu forms a deep level in the band gap of Si which contributes to excess generation currents during reverse bias and excess recombination currents during forward bias of a pn junction [59]. However, through a careful choice of diffusion barrier materials as well as an appropriate control of the fabrication environment, it is possible to prevent this contamination. Wachnik, et al. [60] have performed extensive evaluations of the ability of Cu to diffuse into device regions on a variety of test structures including some structures that were deliberately defective. They report that they find no degradation in device behavior attributable to contamination due to the presence of Cu in the device.

Edelstein et al. [2] have also investigated the potential for Cu contamination of MOS devices at high temperatures (140 - 290°C) and under a wide range of bias stresses (-2 to -50 V) on devices with large area exposures to Cu wires. After exhaustive testing, they conclude that they observe no fails in the devices due to Cu poisoning either.

2.7 Summary

The shift to Cu as the interconnect metal has been one of the most challenging undertakings that the semiconductor industry has encountered. Several key innovations were needed to achieve this advance. Among these was the ability to plate Cu films with no voids or seams in aspect ratios typical to BEOL interconnect processing. Another key innovation was the development of the TaN/Ta based liner scheme to prevent Cu diffusion into the surrounding dielectric. Further, it was important to understand the evolution of the profile of Cu films during electroplating, and the role of the microstructure of Cu during reliability stressing. Other key enabling processes that needed to be developed include Cu CMP and liner CMP. Finally, to manufacture Cu at high volume, it was necessary to ensure that all the unit processes were appropriately optimized and that the overall integration scheme yielded parts with a low level of defectivity. All this could be accomplished with a minimal disruption to the tooling requirements of a semiconductor fabrication plant.

References

[1] 1999 National Technology Roadmap for Semiconductors, Semiconductor Research Corporation and International Sematech.

[2] D. Edelstein et. al., *Tech. Digest IEEE IEDM*, p773 (1997)

[3] C. W. Kaanta et. al.,*Proc. 10th International VMIC*, 1993, pp 15-21.

[4] J. G. Ryan et. al., *IBM J. Res. Dev.*, 39, 4, p371 (1995)

[5] R. Rosenberg et. al., *Annu. Rev. Mater. Sci.*, 30, p229 (2000)

[6] D. Edelstein et. al., *IBM J. Res. Dev.*, 39, 4, p383 (1995)

[7] P. Solomon, *Proc. SPIE*, 947, p104 (1988)

[8] G. A. Sai-Halasz, *ICCD Dig.*, p230 (1992)

[9] D. Edelstein, *Proc. VMIC* 12, p301 (1995)

[10] C. K. Hu et. al., *Mater. Res. Soc. Symp.*, p507, (1998)

[11] J. Rickerby & J. H. G. Steinke, *Chem. Rev.*, 102, p1525 (2000)

[12] P. C. Andricacos et. al., *IBM J. Res. Dev.*, 42, 5, (1998)

[13] L. T. Romankiw, *US Patent 3,853,715*

[14] P. C. Andricacos & L. T. Romankiw, *Advances in Electrochemical Science and Engineering*, VCH Publishers, p227 (1994)

[15] M. Datta et. al., *Electrochem. Soc.* 142, p3779 (1995)

[16] M. M. Chow et. al., it US Patent 4,789,648

[17] C. K. Hu & J. M. E. Harper, *Mater. Phys.*, 52, 5, (1998)

[18] T. Ritzdorf et. al., *Proc. IEEE Int. Intercon. Tech. Conf.*, 166 (1998)

[19] C. Cabral Jr., et. al., *Proc. Adv. Metalliz. Conf.*, 81, (1998)

[20] C. Lingk et. al., *Proc. Adv. Metalliz. Conf.*, 89, (1998)

[21] H. Deligianni et al, *Proceedings of Symposium on Electrochemical Processing in ULSI Fabrication and Semiconductor Metal Deposition III, Electrochemical Society Proceedings*, 2000-8, 145, (2000).

[22] A.C. West et al, *Electrochem. and Solid State Lett.*, 4(7), C50-C53, (2001).

[23] T. P. Moffat et al, *Electrochem. and Solid State Lett.*, 4(4), C26-C29, (2001).

[24] T. P. Moffat et al, it J. Electrochem. Soc., 147 (12), 4524-4535, (2000).

[25] D. Josell et al, *Phys. Rev. Lett.*, 87(1):016102-1-016102-4, July (2001).

[26] P. Vereecken et al, *Symposium on Electrochemical Processing in ULSI Fabrication and Semiconductor Metal Deposition IV, Electrochemical Society Extended Abstracts*, Abstract No 517,(2002).

[27] C-K.Hu, S. Chang, M.B. Small, J.E. Lewis, *Proc. VMIC*, 181 (1986)

[28] L.A. Clevenger, et al., *J. Appl. Phys.*, 73, 300 (1992)

[29] B. Luther, et al., *Proc. VLSI Multi. Intl. Conf.*, 15 (1993)

[30] D. Edelstein, et al., *Tech. Digest of papers IEEE IEDM*, 773 (1997).

[31] K. Holloway, et al., *J. Appl. Phys.*, 71, 5433 (1992).

[32] D. Edelstein, et al., *Proc. Adv. Metalliz. Conf.*, (2001)

[33] C-K. Hu et al., *IEEE Inter. Interconnect Technology Conference (IITC)*, 267 (1999).

[34] D.S. Gardner and D. B. Fraser, *Proc. VLSI VMIC*, 287 (1995).

[35] S. Wong et al., *Mater. Res. Soc. Symp. Proc.*, 514, 75 (1998).

[36] J. M. E. Harper, *private communication*

[37] S. Rossnagel et al., *J. Vac. Sci. tech.*, B 14, 1819 (1996).

[38] C. Cabral Jr., et al., it European patent 00751566 A2 (1997) and it US Patent 6,291,885 (1991).

[39] L. Feinstein, et al., *Thin Solid Films*, 16, 129 (1973).

[40] E. G. Colgan and P.M. Fryer, *US Patents 5,221,449 (1993) and 5,281,485 (1994).*

[41] D. Edelstein, et al., *Proc. IITC*, (2001)

[42] R. Rosenberg, D.C. Edelstein, C-K. Hu, and K.P. Rodbell, *Annu. Rev. Mater. Sci.*, 2000, 30: 229-262.

[43] C-K. Hu, L. Gignac, E. Liniger, R. Rosenberg and A. Stamper, *Proc, of International Interconnect Tech. Conf.*, (2002) p.259.

[44] C-K. Hu, L. Gignac, E. Liniger, R. Rosenberg, *J. Electrochem. Soc.*, 149, G408 (2002).

[45] C-K. Hu, et al., *US Patent 6,380,075B1*, 2002.

[46] K. Ueno, et al., *AIP Conf. Proc., (USA)*, no.612 (2002) pp 49-60.

[47] J. Curry, et al., *22nd Annual IEEE Intl. Re. Phys. Symp. Proc.*, pp. 6-8, 1984.

[48] H. Okabayashi, *Mater. Sci. Eng.*, R11, pp.191-241, 1993

[49] T.D. Sullivan, *Ann. Rev. Mater. Sci.*, 26, pp. 333-364, 1996.

[50] T. Turner and K. Wendel, *23rd Ann. IEEE Inter. Rel. Phys. Symp. Proc.*, pp 142-147, 1985.

[51] E.T. Ogawa, et al., *40th Ann. IEEE Inter. Rel. Phys. Symp. Proc.*, pp. 312-312, 2002.

[52] J.M.E. Harper, et al., *J. Appl. Phys.*, 86, pp.2516-24, 1999.

[53] J. Koike, et al., *Sixth Int'l. Workshop on Stress Induced Phenomena in Metallization*, pp. 169-176, S.P. Baker, et al., editors, AIP 2002.

[54] Bai, et al., *Mater. Res. Soc. Symp. Proc.*, 403, pp501-506, 1996.

[55] R. Gonella et al., *Proc. 2000 IEEE Reliability Workshop*, pp.191-192, 2000.

[56] J. Noguchi, et al., *38th Ann. IEEE Intl. Rel. Phys. Symp.*, pp.339-343, 2000.

[57] D. C. Edelstein et. al., *Electrochem. Soc. Mtg.*, (1999)

[58] A. K. Stamper, W. A. Klaasen, and R. A. Wachnik, *Proc. Adv. Metallization Conf.*, Orlando, FL, Sept 28-30, pp9-15, 1999.

[59] S. M. Sze, *Semiconductor Physics and Devices*, Wiley-Interscience, New York, 2nd Edition. (1981).

[60] R. A. Wachnik et. al., *Proc. Adv. Metalliz. Conf.*, (2000).

Chapter 3

INTERCONNECT PARASITIC EXTRACTION OF RESISTANCE, CAPACITANCE, AND INDUCTANCE

Xiaoning Qi and Robert W. Dutton*
Sun Microsystems, Inc.
*Stanford University**

3.1 Introduction

As semiconductor technology continues to scale, wires, not devices, come to dominate the delay, power and area of microprocessors and ASIC designs. Increasing clock frequency combined with growing chip area results in the ratio of global wire delay to gate delay increasing at a super-linear rate. For sub-0.25 μm technology at gigahertz-scale clock frequencies, interconnects may exhibit transmission line behavior. This has spawned the need to accurately model the parasitics – resistance, capacitance and inductance – for on-chip wires.

In the analysis phase of design, lower metal layer's short on-chip wires can be modeled as lumped capacitive loads, and longer wires can be modeled as lossy RC transmission lines. Any wire whose resistance is small compared with the impedance of the circuit driving it can be considered short. Typically, wires under 1 mm are short, but resistance must be considered for all longer wires. Because of their relatively high resistivity, short length, and tight pitch, lower metal layers on-chip wires have inductance values that are sufficiently low to be safely ignored.

However, with clock frequencies reaching the multi-gigahertz regime, long global interconnects on upper metal layers may exhibit RLC transmission line effects. By using wider wires on upper metal layers for critical signal nets, such as clocks, and using copper interconnects, the wire resistance is reduced. As a result, the inductive impedance part, ωL, becomes comparable to the wire resistance, R. In addition, signals on wide buses may switch simultaneously in one direction or another, which can induce large inductive noise through mutual inductive coupling. For these interconnects, parasitic inductance can no longer be ignored, and needs to be carefully modeled.

As a rule of thumb, interconnects should be modeled as transmission lines if the signal rise time , t_r, is comparable to (or smaller than) the one-way signal propagation time delay through the signal path, t_d. Namely, if $t_r/t_d < 2.5$, lumped analysis is not appropriate and transmission lines model or distributed model should be used [20].

To extract resistance, R, for on-chip interconnects, skin and proximity effects should be considered [51]. These effects are frequency dependent, and will increase wire resistance at high frequencies. To model interconnect capacitance, both capacitance to the ground or substrate as well as capacitance to the neighboring wires need to be taken into account. Based on the definition of inductance, identification of current loops is necessary to calculate inductance. Because of multiple current return paths for on-chip interconnects, calculating wire inductance is more difficult than calculating capacitance. In extraction, multiple wires around the current-carrying wire need to be included for any possible current return paths. The substrate may also offer return paths for signals, therefore, its effects need to be included in the simulation. Skin and proximity effects also affect current return paths and decrease wire inductance at high frequencies. The frequency dependence of inductance is important, especially when there is a ground plane, substrate, or other conductive grids nearby the interconnects.

3.2 Electromagnetic Formulation

3.2.1 Maxwell's Equations

All classical electromagnetic phenomena are governed by a compact and elegant set of fundamental rules known as *Maxwell's equations* [8], [20]. Maxwell's equation are based on three experimentally established facts, namely Coulomb's law, Ampere's law (or the Biot-Savart law) and Faraday's law, as well as the principle of conservation of electric charge. The physical meaning of the equations is better perceived in the context of their integral forms, which are listed below. The physical quantities that appear are the electric field $\bar{\mathcal{E}}$, the magnetic flux density $\bar{\mathcal{B}}$, the electric flux density $\bar{\mathcal{D}}$, the magnetic field intensity $\bar{\mathcal{H}}$, electric current density $\bar{\mathcal{J}}$, and electric charge density $\tilde{\rho}$.

1. Faraday's law is based on the experimental fact that time-varying magnetic flux induces an electromotive force:

$$\oint_C \bar{\mathcal{E}} \cdot d\mathbf{l} = -\int_{S_C} \frac{\partial \bar{\mathcal{B}}}{\partial t} \cdot ds \tag{3.1}$$

   where the contour C encloses the surface S_C and the sense of the line integration over the contour C (i.e., the direction of $d\mathbf{l}$) must be consistent with the direction of the surface vector $d\mathbf{s}$ in accordance with the so-called right-hand rule.

2. Maxwell's second equation represents Gauss's law which is a mathematical expression of Coulomb's law. Coulomb's law states the experimental fact that electric charges attract or repel one another with a force inversely proportional to the square of the distance between them.

$$\oint_{S_V} \bar{\mathcal{D}} \cdot d\mathbf{s} = \int_V \tilde{\rho} dv \tag{3.2}$$

where the surface S_V encloses the volume V.

3. The third equation is a generalization of Ampere's law which states that the line integral of the magnetic field over any closed contour must equal the *total* current[1] enclosed by that contour:

$$\oint_C \bar{\mathcal{H}} \cdot d\mathbf{l} = \int_{S_C} \bar{\mathcal{J}} \cdot d\mathbf{s} + \int_{S_C} \frac{\partial \bar{\mathcal{D}}}{\partial t} \cdot d\mathbf{s} \tag{3.3}$$

This equation expresses the fact that time-varying electric fields produce magnetic fields. The first term on the right-hand side is the conduction-current whereas the second term is known as the displacement-current term. This equation is very important in understanding inductance related issues.

4. The last equation is based on the fact that there are no magnetic charges, hence magnetic field lines always close on themselves.

$$\oint_{S_V} \bar{\mathcal{B}} \cdot d\mathbf{s} = 0 \tag{3.4}$$

This equation is not completely independent for it can be derived from the Biot-Savart law [20].

The two constitutive relations $\bar{\mathcal{D}} = \epsilon\bar{\mathcal{E}}$ and $\bar{\mathcal{B}} = \mu\bar{\mathcal{H}}$ relate $\bar{\mathcal{E}}$ and $\bar{\mathcal{B}}$ to medium-independent quantities, $\bar{\mathcal{D}}$ and $\bar{\mathcal{H}}$.[2] The current density $\bar{\mathcal{J}}$ is given by $\bar{\mathcal{J}} = \bar{\mathcal{J}}_{source} + \bar{\mathcal{J}}_c$ where $\bar{\mathcal{J}}_{source}$ represents the source currents from which magnetic fields originate, and $\bar{\mathcal{J}}_c = \sigma\bar{\mathcal{E}}$ is the conduction current, which flows in a conducting media ($\sigma \neq 0$) whenever there is an electric field present.

The Maxwell equations can also be represented in differential form which is shown below for the case of time-harmonic (sinusoidal steady-state) conditions:

$$\nabla \times \mathbf{E} = -j\omega\mathbf{B} \tag{3.5}$$

$$\nabla \cdot \mathbf{D} = \rho \tag{3.6}$$

$$\nabla \times \mathbf{H} = \mathbf{J} + j\omega\mathbf{D} \tag{3.7}$$

$$\nabla \cdot \mathbf{B} = 0 \tag{3.8}$$

The $\mathbf{E}$, $\mathbf{D}$, $\mathbf{H}$, $\mathbf{B}$, ρ are complex phasors that do not vary with time.

3.2.2 Boundary Conditions

Electromagnetic boundary conditions can be derived using Maxwell's equations. The following boundary conditions are vital in understanding and solving capacitance and inductance problems.

1. $\bar{\mathcal{E}}_{1t} = \bar{\mathcal{E}}_{2t}$, where $\bar{\mathcal{E}}_{1t}$ and $\bar{\mathcal{E}}_{2t}$ are the tangential components of the electric field $\bar{\mathcal{E}}$ at two adjoining surfaces.

2. $\bar{\mathcal{H}}_{1t} = \bar{\mathcal{H}}_{2t}$, where $\bar{\mathcal{H}}_{1t}$ and $\bar{\mathcal{H}}_{2t}$ are the tangential components of the magnetic field $\bar{\mathcal{H}}$. However, if surface currents ($\bar{\mathcal{J}}_s$) exist, such as at the surface of a perfect conductor (i.e., $\sigma = \infty$): $\hat{\mathbf{n}} \times \bar{\mathcal{H}}_1 = \bar{\mathcal{J}}_s$. The field inside a perfect conductor is zero.

3. $\bar{\mathcal{D}}_{1n} - \bar{\mathcal{D}}_{2n} = \tilde{\rho}_s$, where $\bar{\mathcal{D}}_{1n}$ and $\bar{\mathcal{D}}_{2n}$ are the normal components of electric flux density $\bar{\mathcal{D}}$ across the interface. $\tilde{\rho}_s$ is the surface charge that exists on the interface.

4. $\bar{\mathcal{B}}_{1n} = \bar{\mathcal{B}}_{2n}$, where $\bar{\mathcal{B}}_{1n}$ and $\bar{\mathcal{B}}_{2n}$ are the normal components of magnetic field $\bar{\mathcal{B}}$ across the two regions.

The direction of the normal vector across the boundary is defined by the unit vector $\hat{\mathbf{n}}$ which is perpendicular to the interface and directed outward from medium two.

3.3 Resistance Extraction

3.3.1 Introduction

The *resistance* of an interconnect, by definition, is the ratio of potential difference of the two ends of a wire to the total current flowing into it. Resistance, which is originally defined under a steady electric current, can be evaluated using fundamental laws of electrostatics, specifically [20],

$$R \equiv \frac{\Phi_{12}}{I} = \frac{-\int_L \mathbf{E} \cdot d\mathbf{l}}{\int_A \sigma \mathbf{E} \cdot d\mathbf{s}} \tag{3.9}$$

where $\Phi_{12} > 0$ is the potential difference of the two ends. A is the cross-section of the wire, L is any path from low potential end to the high one. σ is the conductivity of the wire. The general way to extract resistance is to solve the typical electrostatic problem, which calculates electric field $\mathbf{E}$ based on *Poisson's equation* or *Laplace's equation* which solves the scalar potential

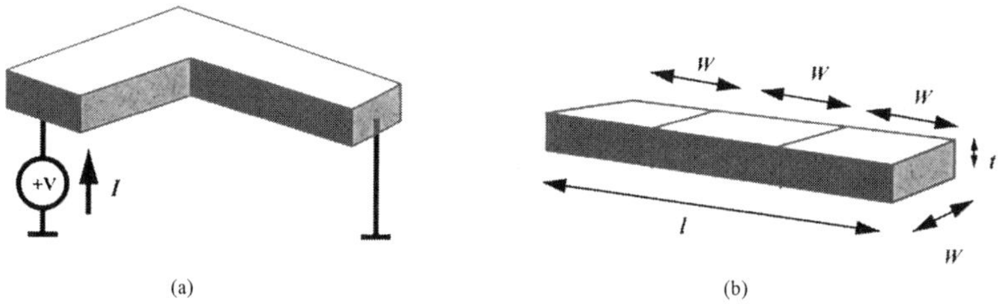

Figure 3.1. (a) To extract the resistance for a corner of an interconnect. (b) a straight, rectangular shape interconnect is divided into three squares to use sheet resistance .

Φ within conductors. If a wire cross-section is rectangular and the wire is straight, Equation (3.9) can be solved easily and reduces to a simple formula. Sheet resistance is also widely used in resistance extraction for VLSI on-chip interconnects.

3.3.2 Laplace's Equation

For a conductor with an arbitrary shape, for example, a corner of interconnect wires as shown in Figure 3.1(a), Poisson's equation is used to extract the conductor resistance. In a metal conductor with net charge being zero (i.e., a charge-free space), Poisson's equation reduces to Laplace's equation [20], which is

$$\nabla^2 \Phi = 0 \tag{3.10}$$

∇^2 represents the operation $\nabla \cdot \nabla(\cdot)$ which has different forms, depending on the coordinate system. By assuming specific boundary conditions, for example, the voltage of one end of the conductor is V and the other end is at potential zero, and $\frac{\partial \Phi}{\partial n} = 0$ for all the other surfaces, where n is the normal direction of the surfaces, Equation (3.10) within the conductor can be solved. Taking the negative gradient of Φ, namely, $\mathbf{E} = -\nabla \Phi$, the electric field is obtained, which in turn is used to calculate the current density $\mathbf{J} = \sigma \mathbf{E}$, where σ is the conductivity of the metal layer. The current density $\mathbf{J}$ can be further integrated across the proper cross-section area to calculate the total current, I, flowing into the conductor and in turn to determine the resistance according to $R = \frac{V}{I}$.

In generally, it is not easy to calculate Laplace's equation in a 3-D region. Numerical methods, such as Finite-Difference and Finite-Element Methods, etc., [49] are used to solve the equation and then to extract currents and resistances. Commercial field solvers include: Avanti! Raphael [4], Ansoft Maxwell [2] and Agilent HFSS (High-Frequency Structure Simulator) [1]. In some cases, a 3-D problem can be reduced to a 2-D problem by exploring the geometric symmetry properties of the conductors so that solving Laplace's equation becomes easier.

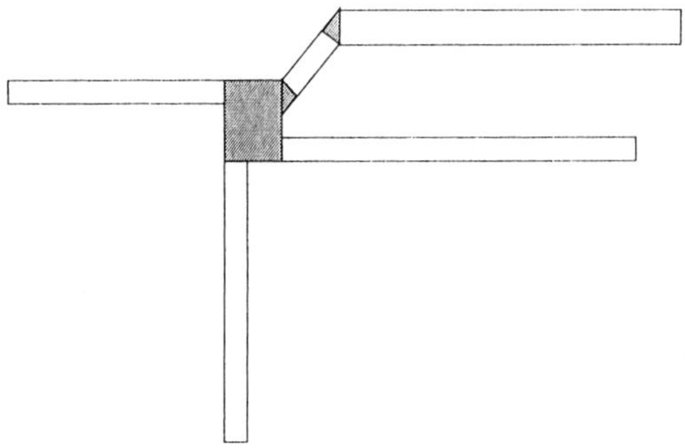

Figure 3.2. Irregular shapes can be cut into pieces to estimate the resistance based on sheet resistance.

3.3.3 Sheet Resistance

For more regular shapes of conductors, Equation (3.9) can be easily calculated. If it is a straight conductor with rectangular cross-section which is usually the case for on-chip interconnects, the following formula can be used to extract the resistance.

$$R = \rho \frac{l}{wt} \tag{3.11}$$

where l is the wire length, w is the wire width, t is the wire thickness, and ρ is resistivity of the metal wire. Because the thickness of a metal layer does not change for a given IC process, Equation (3.11) can be simplified to be $R = R_s \frac{l}{w}$ where $R_s = \rho/t$ is called sheet resistance. With the known wire width and length, the resistance of a wire can be extracted. Actually, by counting the number of squares (with w as its one edge) along the wire length, the resistance of the wire is $R = R_s \times n$ where n is the number of squares as shown in Figure 3.1(b). The sheet resistance is process- and layer-dependent. For current technology nodes, R_s is around 50 mΩ/square for metal layers one to four.

Square counting methods obtain exact resistance for straight conductors while for corners and irregular shaped conductors, it is necessary to cut them into squares or rectangular shapes, as shown in Figure 3.2. Equivalent numbers of squares for the non-rectangular shapes can be calculated in order to estimate the resistance. In some cases, small non-rectangular shapes can be ignored due to the fact that their resistance is small compared with the resistance of long, straight conductors. Currently, many commercial tools are able to cut interconnects into many regular pieces for such resistance calculations. For greater accuracy, Laplace's equation can be solved for these non-rectangular shapes such as corners, bends and vias. Field solvers can be used to extract these resistances, and typical cases can be analyzed. Via resistance is process dependent and usually may have a different conductivity.

3.3.4 Resistance Frequency Dependency

At high frequencies, current tends to distribute close to the surface of a conductor due to skin effects. Skin effect results from the eddy currents within the conductor which are induced by time varying magnetic fields and governed by the law of of induction – Faraday's law. The skin depth where the current flows near the surface of conductors is frequency dependent,

$$\delta = \frac{1}{\sqrt{\pi f \mu \sigma}} \tag{3.12}$$

where f is the frequency, μ and σ are the conductor's permeability and conductivity respectively. Because currents flow in a smaller cross-section of the conductors, the resistance at high frequencies increases compared to its dc values. Proximity effects which are due to the eddy currents within nearby conductors also affect the current distribution in the conductor, causing current to flow in such a manner that current loops with other conductors are minimized [50]. Proximity effects may also increase wire resistance at high frequencies. Field solvers can capture these resistance increases while the sheet resistance can not. Fortunately, in modern technologies, most on-chip wire cross-section dimensions are comparable to their skin depth, and no major errors are observed if a sheet resistance value is used for modeling. For wide wires on upper metal layers, high frequency resistance values may be extracted by cutting wires into many filaments along the wire length while assuming the current is uniformly distributed within one filament [55].

3.4 Capacitance Extraction

3.4.1 Introduction

For an isolated conductor which holds a charge Q at a certain potential, the charge-to-potential ratio is a constant. For a pair of conductors, which hold equal but opposite charges by applying a potential difference $\Phi_{12} = \Phi_2 - \Phi_1$ between them, the ratio of charge to the potential difference is also a constant. This constant is called *capacitance*. Actually, an isolated case can be considered as a special case with the second conductor being at "infinity" ($\Phi_2 = 0$). Capacitance is a measure of the ability of a conductor configuration to hold charge per unit applied voltage, or store electrical energy.

Consider a two conductor system, where capacitance is defined as

$$C \equiv \frac{Q}{\Phi_{12}} = \frac{\oint_S \mathbf{D} \cdot ds}{-\int_L \mathbf{E} \cdot dl}$$

and Φ_{12} represents the voltage difference between the two conductors, S is any surface enclosing the positively charged conductor and L is any path going from the negative to the positive conductor. The capacitance is a physical property of the two-conductor system. It depends on the geometry of the conductors and the permittivity of the medium between them.

Similar to resistance extraction, capacitance is solved as an electrostatic problem. Capacitance of two conductors can be calculated by (1) assuming charges $+Q$ and $-Q$ on conductors, and determining Φ_{12} in terms of Q, (2) assuming a Φ_{12} and determining Q in terms of Φ_{12}, or (3) calculating the energy stored in the electric field. In the first method, Gauss's law is used to calculate $\mathbf{E}$ from Q,[3] and by performing integration along any path between the two conductors, Φ_{12} can be calculated. In the second method, Poisson's equation or Laplace's equation may be used to calculate spatial potential Φ. By applying the boundary conditions, $\mathbf{E}$ and Q can be obtained [20] [8]. In the third method, capacitance is obtained using the fact that $C_{12} = 2W_{12}/\Phi_{12}^2$ where W_{12} is the electric energy stored in the field, and the field is solved by Poisson's equation [2]. These methods can be generalized to multiple conductor systems where a conductor is considered at a certain potential, one at a time, while all others are grounded. To numerically solve 3-D capacitance problems, Finite-Difference [59], Finite-Element [15], Boundary Element or Multipole-Accelerated Boundary Element Methods [33][31] are widely used. To achieve quick estimations, analytical formulae are also used in design and analysis [5].

In the regime below the tens-of-gigahertz range or at higher frequencies, capacitance has little frequency dependency because of the charge equilibrium, dictated by the *relaxation time*, which occurs on the order of 10^{-18} to 10^{-19} seconds for most metallic conductors [20].

3.4.2 Field Solvers

To extract the capacitance of multi-conductor systems, electric field solvers produce the most accurate results, which are often used as a "golden standard" for analytical formulae as well as in establishing libraries for pattern-matching extraction tools. Suppose there is a three conductor system, a 3×3 capacitance matrix is used to represent the self- and mutual capacitance between the conductors. Generally, the diagonal terms are self- capacitance values, including the capacitance of a conductor to the infinity where the potential is assumed to be zero, as shown in Figure 3.3. A matrix which represents the relationship between charges Q_i and potentials V_i is shown as following,

$$\begin{bmatrix} Q_1 \\ Q_2 \\ Q_3 \end{bmatrix} = \begin{bmatrix} C_{10}+C_{12}+C_{13} & -C_{12} & -C_{13} \\ -C_{12} & C_{20}+C_{12}+C_{23} & -C_{23} \\ -C_{13} & -C_{23} & C_{30}+C_{13}+C_{23} \end{bmatrix} \begin{bmatrix} V_1 \\ V_2 \\ V_3 \end{bmatrix} \tag{3.13}$$

As can be seen from the matrix, each capacitance term is calculated by applying a potential (i.e., 1 V) on one conductor and zero on all others. Charge on

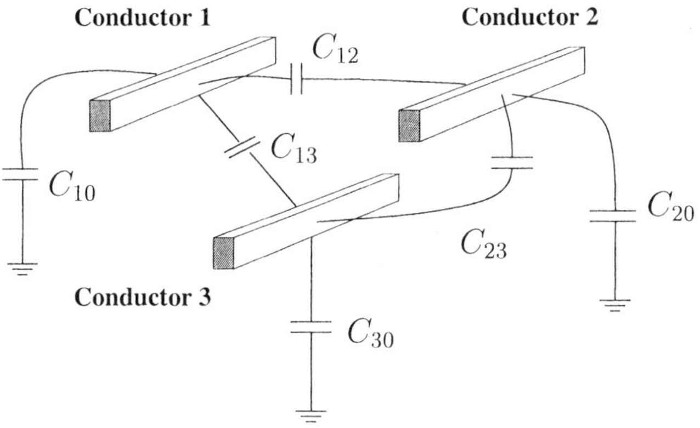

Figure 3.3. Three conductors and their capacitances.

each conductor can be solved from the electric field which, in turn, corresponds to the respective capacitance terms.

Volume-Based Methods. In volume-based analysis methods, Laplace's equation, $\nabla \cdot \epsilon \nabla \Phi = 0$ is solved outside the conductors, where ϵ is the dielectric constant of the media. One way to solve 3-D Laplace's equation numerically is to use Finite-Difference or Finite-Elements Methods which require volume meshing of the entire 3-D space which contains the conductors. The derivatives in the equation are approximated by finite-differences [49]. Boundary conditions, for example, potentials on conductor surfaces, are known. The volume-based method also needs to define boundary conditions on a bounding-box which contains the entire conductor structure. There are two kinds of boundary conditions for the box: 1) potential is zero everywhere on the boundary and 2) $dv/dn = 0$ where n is the normal direction of the surfaces of the box. Obviously, different boundary conditions yield different results. Generally, a bigger bounding box which includes the whole multi-conductor structure gives better results. For finite size of the box, the first kind of boundary condition may overestimate self-capacitance and underestimate coupling capacitance; while the second condition may underestimate self- capacitance and overestimate coupling capacitance. Volume-based methods are computationally very expensive due to the need to mesh the entire physical domain, but the method excels for complex variations of dielectrics in the media. Finite-difference and finite-elements usually end up with very large sparse matrices and require numerical techniques such as Incomplete Cholesky Conjugate-Gradient Method, which are used to improve the run time efficiency and memory storage [4]. Commercial tools which use Finite-Difference or Finite-Elements Methods include: Ansoft Maxwell [2], Agilent HFSS [1] and Avanti! Raphael [4].

Surface-Based Method. The surface-based method is also known as Integral Equation Fomulation Method. The Poisson equation can be written in terms of a Green's function, which is analogous to the circuit impulse response. In a free space, the Green's function is,

$$G(x, x') = \frac{1}{4\pi\epsilon \|x - x'\|} \tag{3.14}$$

where x, x' are the spatial positions in a 3-D reference space. The term $\|x - x'\|$ denotes the Euclidean distance between x and x'. The potential Φ can then be written as

$$\Phi = \int_{surfaces} G(x, x')\sigma(x')da' \tag{3.15}$$

where σ is the surface charge density. The quantity da' is the incremental surface area. If the conductor potentials are known, the charge density can be calculated by directly solving Equation (3.15). Once the surface charge density σ is obtained, the total charge on one conductor i is given by:

$$Q_i = \int_{i^{th} conductor\ surface} \sigma(x')da'. \tag{3.16}$$

Furthermore, the jth column of the capacitance matrix can be calculated by solving for the total charge on each of the conductors when the jth conductor is at unit potential and all other conductors are at zero potential. The resulting charge on conductor i, Q_i, is equal to C_{ij}.

To solve Equation (3.15) using either the panel methods [19] or the method of moments [17] explores the fact that the charge is restricted to the surface of the conductors. Rather than discretizing all of the free space, only the surface charge on the conductors is discretized. Commonly, the surfaces of m conductors in a free space are discretized into a total of n two-dimensional panels. For each panel k, an equation is written that relates the potential at the center of that k^{th} panel to the sum of the contributions to that potential from the charge distribution on all n panels. It is assumed that the surface charge density on each panel is constant. The formulation results in a dense linear system, which typically can be solved by some form of Gaussian elimination to compute the panel charges from a given set of panel potentials. Methods such as GMRES [46] can be used to improve the computational efficiency. An accurate approximation for the evaluation of potential, called the *multipole expansion*, is widely used [31]. The basic idea is that the details of the distribution of the charge, say in an inner circle of radius R, do not strongly affect the potential at the evaluation points *outside* an outer circle of radius r, provided $r \gg R$. Also *local expansions* can be applied to compute an approximation for the potential at the evaluation points in the *inner* circle by ignoring the details of the evaluation point distribution. By using multipole expansions, computing p potentials due to p panels is reduced to the order of p operations compared to doing the exhaustive computation that requires p^2 [31].

The surface-base method, like boundary elements, is usually faster than the volume-based method although it usually creates a dense matrx problem. The Boundary Element Method can also be applied to multiple dielectrics [32]. But dielectric interfaces need to be meshed into panels and their charge contributions need to be taken into account. Provided that good quality mesh is used, boundary elements can deal with arbitrary shape of conductors while the volume-based method usually has limitations in dealing with the shape of conductors (e.g., suitable for 3-D planar structures). Realistic 3-D geometry modeling requirements broaden the application domain for field solvers using the Boundary Elements Method [34]. FASTCAP [32] is a typical boundary element solver which exploits the multipole approximation.

Random Walk Method. A stochastic algorithm, called the Random Walk Method, was proposed by LeCoz and Iverson [26]. The basic idea is that the potential and electric field at the center of a domain, say a square in 2-D case or a cube in 3-D, can be represented by the respective boundary-integral solutions. For example, in the 2-D case, the integral can be estimated as:

$$V_{center} = \frac{1}{N} \sum_{i=1}^{N} V_{square}(x_i) \tag{3.17}$$

where x_i is the sampling points on the square boundary and N is the number of points. If the electric field is known on a surface bounding a conductor, also known as the Gaussian boundary, charges of the conductor enclosed by this surface boundary can be calculated using Gauss's law, which simply states that $Q = \oint_S \mathbf{D} \cdot ds$ where S is any closed surface enclosing Q. The capacitance can be obtained by definition $C = Q/V$. But if the electric field on the boundary is not known, another surface boundary needs to be formed, and so forth. The algorithm continues to construct maximal cubes until a conductor is encountered where the electric potential is known, and thus the electric field is known. When the electric field $\mathbf{E}$ is known on S_i, the charge Q on conductor i can be calculated and thus the self-capacitance can be determined. By sampling the potential on conductor j, the C_{ij} can be obtained. LeCoz and Iversion used a statistical method to estimate the electric field surrounding each conductor, whereby errors are cancelled during the Gaussian surface integration. Figure 3.4 shows a 2-D example where, S_i is a Gaussian boundary (actually a Gaussian-surface cross-section in 2-D case). Maximal squares are constructed. Each boundary S_i is decomposed into an electrode part which reaches conductors where potentials are known and a non-electrode part where new boundaries need to be constructed until they meet corresponding conductors.

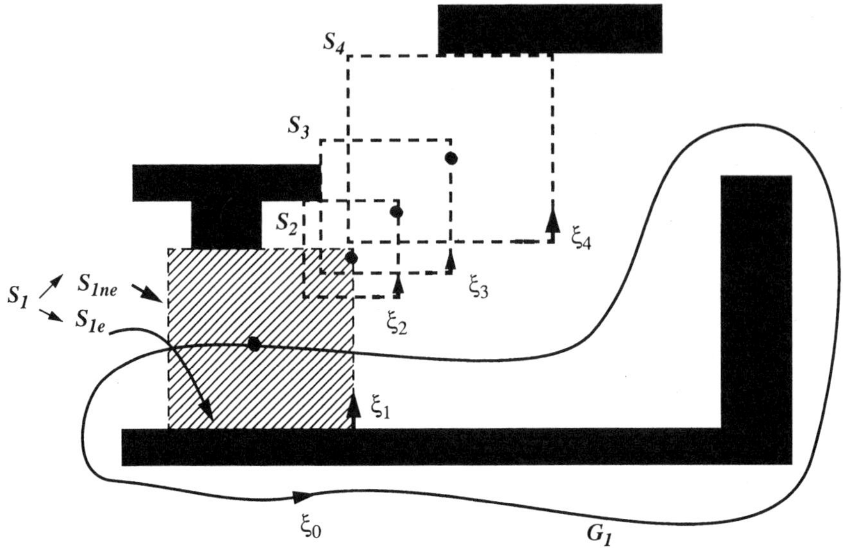

Figure 3.4. A 2-D example of random walk where subsequent maximal squares (initial one shaded) are constructed. The boundaries S_i are decomposed into an electrode part (e.g. S_{1e}) and a non-electrode part (e.g. S_{1ne}). G_1 is the Gaussian boundary for the enclosed conductor. ξ_i measures the length along any of the boundaries. These squares are centered on previous square boundary point (•).

The advantages of Random Walk Method include: its suitability to rectilinear geometries, statistical-error cancellation and no need to mesh geometries. It is also a good candidate for parallel computing . Commercial tools that use the Random Walk Method include the Random Logic Corp. QuickCap [37].

Comparing the three different field solver methods: the Random Walk Method is the best for calculating self-capacitance for a complicated net; Integral Formulation (Boundary Element) is most suitable for calculating small coupling capacitance while Volume Methods are good at dealing with multiple dielectrics.

3.4.3 Table-Based and Rule-Based Methods

In the process of VLSI chip design, especially for the giga-scale integration, it is not practical to use field solvers in the whole chip extraction because it is too computational expensive. Extraction tools for VLSI design usually use tables or rules generated based on field solver results, matching the geometry setup in a chip against the patterns in the tables to extract capacitance.

In a typical on-chip interconnect case, a wire has mutual capacitance with neighboring wires on the same layer as well as on layers above and below where fringing capacitance needs to be considered as shown in Figure 3.5,

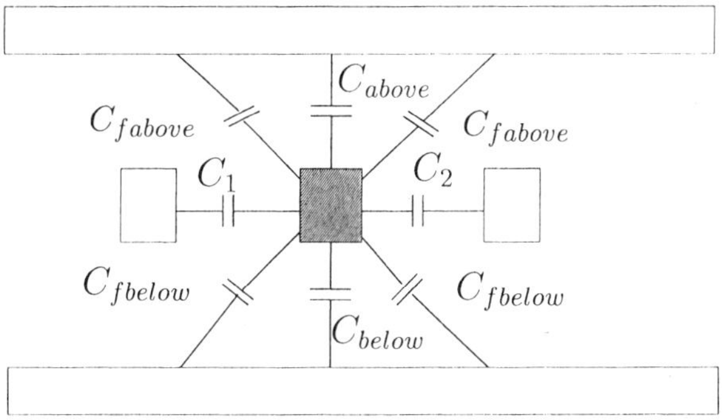

Figure 3.5. A typical interconnect capacitance models.

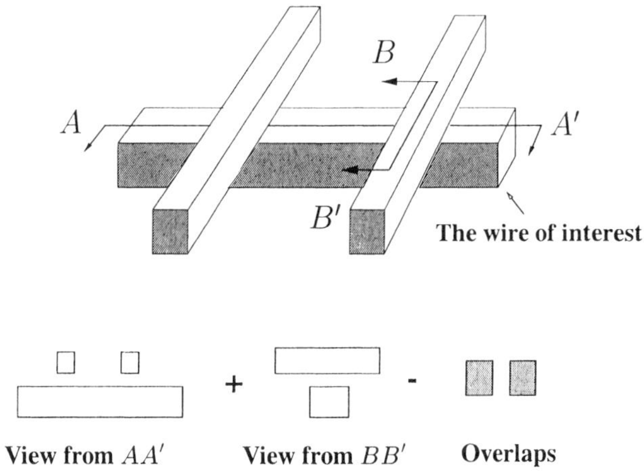

Figure 3.6. An interconnect 3-D structure is divided into two 2-D views which capacitance can be found in the table by multiplying proper interconnect length.

where C_1 and C_2 are lateral capacitance, C_{below} and C_{above} are overlap capacitances, and C_{fbelow} and C_{fabove} are fringing capacitances due to the electric field from edge of one conductor to the metal wires on other layers.

In a table-based method, field solvers are used to extract capacitance for some typical geometry patterns to establish a library for a certain process. In the extraction processes, on-chip interconnects are decomposed into primitives which match patterns in the library. A 3-D field solver is called for non-matched patterns that are found. For example in Figure 3.6, a three wire structure can be decomposed into two 2-D geometry views; capacitances of matched patterns are retrieved from the table which are then multiplied by the corresponding interconnect lengths. The capacitance which is counted twice needs to be subtracted from the result. A typical library for a process with three layers of metal contains tens-of-thousands of entries. Interpolation may also be used for the patterns which are not found in the library. Since a layout is cut into two orthogonal cross-sectional views as shown in Figure 3.6, this method is considered based on a quasi-3-D formulation. The advantage of this method is that good accuracy can be obtained with very fast speeds, and the library generation is done only once when the process parameters are defined. However, building a library involves huge effort; it is difficult to build a complete set of patterns.

Another quick and ad hoc method involves the estimation of capacitance, using some rules established by running 3-D field solvers. Rules are represented by many capacitance coefficients. The capacitance is obtained by multiplying the overlap area or length with corresponding coefficients. Boolean operations (AND, OR, etc.) can be used to identify individual configurations. This method ignores the lateral capacitance, and is often used in floorplanning, placement and routing tools. However, it is difficult to achieve a sufficient accuracy [57].

3.4.4 Analytical Formulae

In many cases, capacitance values can be estimated by analytical formulae. In the design phase, there is a compelling need for simple and fast approximation formulae to calculate capacitance values directly. These quick "back-of-envelope" calculations are especially useful for hand calculations.

The first basically capacitance calculation is the parallel plate formula given by

$$C = \epsilon \frac{S}{d} \tag{3.18}$$

where ϵ is the permitivity of the media. S is the overlap area of the parallel plates and d is the space separation of the two metal plates. As the IC technology advances, the thickness of interconnect wires can become larger than the width of the wire, which means that the fringing capacitance of interconnects can become comparable to the overlap plate capacitance, and needs to be accurately modeled as shown in Figure 3.5.

Because the capacitance to the "above" or "below" metal layers is usually modeled as capacitance to ground planes, since the activities at these two metal layers are not known, many investigations have focused on the formulae for wire-to-ground capacitance calculations. Chang [7] proposed an accurate formula in 1976 which used a conformal transformation to derive rather simple equations. Provided that $w \geq h$ holds, the accuracy is within one percent of the exact value, where w and h are the width and thickness of the interconnect wires. Yuan and Trick also presented a simple analytical formula which has a direct physical interpretation in 1982 [58]. They replaced the rectangular wire cross-section with an "oval" one. The final cross-section is composed of a rectangle and two half-cylinders shown in Figure 3.7. The resulting capacitance

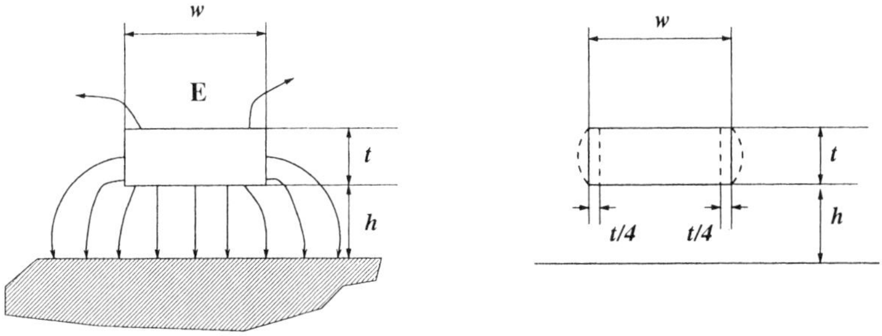

Figure 3.7. Wire electrical field and the geometry decomposition for capacitance calculation.

can be calculated as the sum of a parallel plate capacitor with width $w - t/2$ and a cylindrical one with radius, $r = t/2$, shown by Equation (3.19).

$$c = \epsilon \left[\frac{w - t/2}{h} + \frac{2\pi}{\ln\left(1 + \frac{2h}{t} + \sqrt{\frac{2h}{t}\left(\frac{2h}{t} + 2\right)}\right)} \right] \tag{3.19}$$

Provided that the conditions $w \geq t/2$ and $t \approx h$ hold, a maximum error of ten percent with respect to the Chang's formulation is obtained. In the case of $w < t/2$, an empirical formula is needed.

Sakurai and Tamaru [47] proposed a pure empirical formula based on evaluating numerical solutions. The formula was later improved by Meijs and Fokkema [29] by extending the empirical expression and simultaneously reducing the range of validity as shown in Equation (3.20). The first term of their formula describes the parallel plate capacitor and the other three terms represent side effects.

$$c = \epsilon \left[\left(\frac{w}{h}\right) + 0.77 + 1.06 \left(\frac{w}{h}\right)^{0.25} + 1.06 \left(\frac{t}{h}\right)^{0.5} \right] \tag{3.20}$$

The maximum deviation of the formula from Chang's equation is within two percent error when $w/h \geq 1$ and $0.1 \leq t/h \leq 4$, and reaches six percent when $w/h \geq 0.3$ and $t/h \leq 10$. A complete comparison of these formulae are presented in Barke's paper in [5].

3.5 Inductance Extraction

3.5.1 Introduction

Definition. Inductance is a single measure of the distribution of the magnetic field near and inside a current-carrying conductor. It is a property of the physical layout of the conductor, and is a measure of the ability of that conductor to *link magnetic flux*, or store magnetic energy.

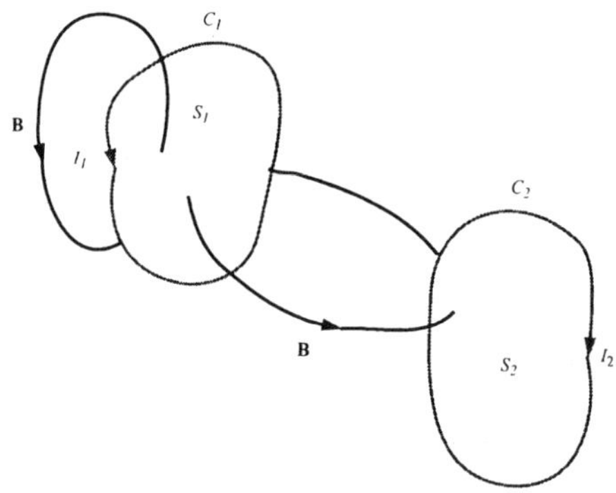

Figure 3.8. Magnetically coupled two circuit loops.

Consider two neighboring closed loops C_1 and C_2 as shown in Figure 3.8. If a current I_1 flows around the closed loop C_1, a magnetic field $\mathbf{B}_1$ is produced, and some of this magnetic field links with C_2. This magnetic flux , produced by the current I_1 flowing around C_1, is linked by the area S_2 enclosed by C_2 and can be designated as

$$\Psi_{12} = \int_{S_2} \mathbf{B}_1 \cdot d\mathbf{s}_2$$

If C_1 and C_2 consist of single turn loops respectively, the *mutual inductance* M_{12} is defined as [20]

$$M_{12} = \frac{\Psi_{12}}{I_1}$$

The magnetic flux produced by I_1 links the area S_1. The *self-inductance* is defined as

$$L_{11} = \frac{\Psi_{11}}{I_1}$$

where $\Psi_{11} = \int_{S_1} \mathbf{B}_1 \cdot d\mathbf{s}_1$. The *Neumann formula* [20] shows that

$$M_{12} = \frac{\mu_0 N_1 N_2}{4\pi} \oint_{C_1} \oint_{C_2} \frac{d\mathbf{l}_1 \cdot d\mathbf{l}_2}{R} \tag{3.21}$$

which indicates that $M_{12} = M_{21}$ since the dot product is commutative and the order in performing the line integrals can be interchanged. N_1 and N_2 are the number of turns in the loops C_1 and C_2, respectively.

Methods of Inductance Calculation. Although the inductances and mutual inductances of circuit elements, which are not associated with magnetic materials, are independent of the value of the current and dependent only on the geometry of the system[4], it is only for the simplest cases that these values can be calculated exactly. There are essentially four methods to calculate inductances.

1. The most direct method for calculating inductance is based on the definition of inductance and flux linkages. By the Biot-Savart law, the magnetic field dB, due to a differential current dI, at any point P of the field can be calculated.[5] Taking the integral of the entire current loop, the total B field at point P can be obtained. Take another integral over the surface enclosed by the current loop to calculate the magnetic flux, which is further divided by the current, the inductance for the current loop is determined. The magnetic flux that links a contour C may also be expressed in terms of the vector potential $\mathbf{A}$, where $\mathbf{B} = \nabla \times \mathbf{A}$ and

$$\Psi = \int_S \mathbf{B} \cdot d\mathbf{s} = \int_S \nabla \times \mathbf{A} \cdot d\mathbf{s} = \oint_c \mathbf{A} \cdot d\mathbf{l}$$

The last integral sometimes is more convenient to evaluate than $\int_s \mathbf{B} \cdot d\mathbf{s}$.[6]

2. Energy-based inductance calculation is another way to calculate the inductance of a circuit. The total energy W_m stored in a given steady-current configuration can be determined by integrating $\mathbf{B}$ over the entire volume V that surrounds it:

$$W_m = \frac{1}{2} \int_V \frac{B^2}{\mu_0} dv$$

The inductance can be determined according to the following expression:

$$L = \frac{2W_m}{I^2} \tag{3.22}$$

3. The Neumann formula, Equation (3.21), is the most general expression for finding the mutual inductance but it is not as simple to solve as the expression resulting from the use of the Biot-Savart law. For most cases it is not possible to perform the integrations in closed form. However, in such cases it is possible to obtain a numerical value or approximation by series expansions for specified cases.

4. Using some basic but fundamentally important inductance formulae is the fourth method to calculate inductance. Based on these formulae and basic circuit theory (Kirchhoff's current and voltage laws), formulae for generalized circuit structures can be obtained. Taking the integral of a formula for a basic structure can also lead to a new inductance formula. For example, a formula for the mutual inductance of cylindrical current sheets may be derived by the integration of the formula for the mutual inductance of coaxial circles, along the cylindrical length. However, it is necessary to select a suitable formula in which the terms involve the appropriate variable of integration (e.g., a length) [16].

As is stated in the inductance definition, a current loop needs to be identified in order to calculate the inductance. However, in practice this may be difficult for IC chips where current loops are not easily identified. The so-called partial inductance concept [43] assumes that the return current for a filament is at infinity, which is a mathematical concept but it has been proven that they can be used to calculate the physical loop inductance.

It is worthwhile to point out that the displacement current in a metallic conductor plays a limited role in calculating inductance. It is shown that $| J_c |_{max} / | J_d |_{max} \simeq 10^{17}/f$, where J_c and J_d are the conduction and displacement current densities, respectively, and f is the frequency [20]. The displacement current within a metallic conductor is completely negligible compared to the conduction current at frequencies below the optical range up to $\sim 10^{15}$ Hz.

Inductance Frequency Dependency. Frequency dependence of inductance results from the eddy currents in conductors, which are induced by the time-varying magnetic fields and governed by Faraday's law (Equation (3.1)). This varying magnetic field generates eddy currents inside of a conductor or other conductors nearby the current loop. The induced eddy currents flow in a direction that produces an opposing magnetic flux, which reduces the effective magnetic flux and therefore the inductance.

Skin and proximity effects result from eddy currents. In a conductor that is good but not perfect, an increasing magnetic field will penetrate the material to some extent. It will, in turn, induce voltage, and current will flow; the current will automatically distribute itself in such a way as to weaken the magnetic field and prevent the field from penetrating further into the conductor. If this magnetic field is generated by the conductor itself, then the phenomena is called the *"skin effect"*. If this magnetic field is generated by an adjacent time-varying current-carrying conductor, the phenomena is called the *"proximity effect"* – regardless of whether the first conductor carries current or not [50] [53]. Skin effects reduce wire inductance resulting from the reduction of the internal inductance of a conductor at high frequencies; proximity effects reduce wire inductance because currents in different conductors re-distribute themselves to form a smaller current loop at high frequencies[7] [51]. More generally, the skin-effect and proximity-effect eddy currents superimpose to form the total eddy current distribution [53].

If two conductors, parallel and close together, are carrying current in opposite directions, the current tends to concentrate at the nearer surfaces to minimize the current loop and resulting inductance [51]. In the case of a ground plane, the return current of a signal wire on the ground plane concentrates just beneath the signal wire at high frequencies to minimize the inductance loop due

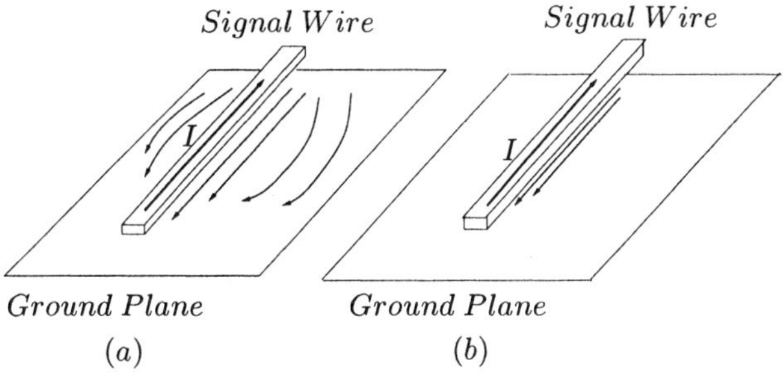

Figure 3.9. Ground plane return current (a) at low frequencies, and (b) at high frequencies.

to eddy currents while at low frequencies, the current spreads out to minimize the resistance of the return paths on the ground plane as shown in Figure 3.9.

Previous Work. The earliest work in calculating inductances can be traced back to Maxwell and Neumann in the late 1800's. By assuming that current returns from infinitely (i.e., far away), Neumann derived the mutual and self-inductance formulae for cylindrical wires [42]. Maxwell showed how to calculate mutual and self-inductance in several important cases by means of what he called the *geometrical mean distance* (GMD) – either of one conductor from another or of a conductor from itself [42]. Most of this early work was summarized by E. B. Rosa and F. W. Grover in their books [42]. A milestone contribution in on-chip inductance calculation method came with the work of A. E. Ruehli in the 1970's. Since it is very difficult to find return paths for on-chip interconnects due to their complexity, Ruehli extended Neumann's early work and proposed *Partial Inductance* or *Partial Element Equivalent Circuit* (PEEC) concepts [43] [44]. Conductor loops are divided into segments for which partial inductance is calculated. The partial inductances are then appropriately added to yield the desired loop inductance. Ruehli's work laid the foundation for many inductance extraction tools. Kamon and White used the partial inductance concepts and developed a program, FASTHENRY, to automate the calculation of inductance [21].

It was not until the mid-1990's that identification and modeling of on-chip transmission line effects were clearly identified as necessary for future scaling of interconnects [11][14] [40]. These papers pointed out the transmission line and inductive effects for on-chip interconnects using test chips and time domain analysis. Deutsch [13] [12], using transmission line theory, proposed conditional expressions to determine when transmission line and inductance effects are important for accurate delay and crosstalk prediction for on-chip interconnects. Kleveland [23] used frequency domain measurement of test chips to identify on-chip parasitic inductive effects for sub-0.25μm technology. In recent years, several microprocessor designs, for example, Intel's Itanium and

Compaq's Alpha chips, were reported to have required the use of on-chip inductance models [45] [56] [30] – both chips used co-planar waveguide structures, or ground planes for critical clock and signal wires to reduce inductance effects at the cost chip area.

Extraction methods based on applying a field solver to generate look-up tables and equivalent circuits for high frequencies have been reported [18] [24] [25]. He and Chang [18] show that without loss of accuracy, the extraction problem of n traces can be reduced to a number of one-trace and two-trace sub-problems. These sub-problems then can be solved via a table-based approach. In addition, a quick inductance screening process and use of rules to identify those inductive interconnects and victim wires were established by Lin [27]. In 1999, Restle [41] presented full-wave extraction and simulation methods. Effects such as overshoot, reflections, frequency dependent effective resistance and inductance were illustrated using the full-wave simulation tool. Another practical approach for extracting approximate inductances of on-chip interconnects was reported by Shepard [48]; signal and power/ground wires were modeled independently with inductive coupling localized by assuming currents only return via nearby power and ground wires.

In the following sections, inductance extraction using field solvers as well as analytical formulae benchmarked with measurements are presented. Ground plane effects on wire inductance are also discussed.

3.5.2 Field Solvers and 3-D Geometry Modeling

Field Solvers. Since the displacement current is negligible within metallic conductors, field solvers used to extract inductance usually apply the magneto-quasistatic assumption that the displacement current, $\epsilon\omega\mathbf{E}$ equals zero. Under this assumption, field solvers, such as Maxwell, solve the magnetic $\mathbf{B}$ field and $\mathbf{H}$ field by a Magnetostatic Field Simulator, and then use the energy method to calculate the self- and mutual inductance [2]. In each field simulation, one ampere of current is allowed to flow in a single conductor. No current is allowed to flow in any other conductors. The current returned through the ground conductors is determined by the field solution. The final extracted inductance matrix is a loop inductance matrix for a multi-conductor transmission line system, where the diagonal terms are self-loop inductances and the off-diagonal terms are mutual inductances between two conductor loops.

Another popular method to extract inductance is the Volume Integral Method which uses integral formulations of magnetoquasistatic analysis and calculates partial inductances, for example, FASTHENRY [21]. These integral formulations are derived by assuming sinusoidal steady-state and magnetoquisistatic conditions. Namely, the current density $\mathbf{J}$ can be represented by

$$\frac{\mathbf{J}(\mathbf{r})}{\sigma} + \frac{j\omega\mu}{4\pi}\int_{V'}\frac{\mathbf{J}(\mathbf{r}')}{|\mathbf{r}-\mathbf{r}'|}dv' = -\nabla\Phi(\mathbf{r}) \tag{3.23}$$

where σ is the conductivity, Φ is the scalar potential. By applying the current conservation equation, $\nabla \cdot \mathbf{J} = 0$, the current densities and the scalar potential can be solved using Equation (3.23). In order to properly capture the skin and proximity effects in the long and thin conductors, the cross-sections of the conductor are divided into a bundle of parallel filaments, where currents are assumed to be uniformly distributed within cross-sections of each filament. After discretization, Equation (3.23) can be written in matrix form

$$(R + j\omega L)I_b = \tilde{\Phi}_A - \tilde{\Phi}_B \tag{3.24}$$

where I_b is the vector of b filament currents.

$$R_{ii} = \frac{l_i}{\sigma a_i} \tag{3.25}$$

is the $b \times b$ diagonal matrix of filament dc resistance where l_i is the length of filament i and a_i is the cross-section. The inductance matrix

$$L_{ij} = \frac{\mu}{4\pi a_i a_j}\int_{V_i}\int_{V_j'}\frac{\mathbf{l_i}\cdot\mathbf{l_j}}{|\mathbf{r}-\mathbf{r}'|}dV'dV \tag{3.26}$$

is the $b \times b$ dense matrix of partial inductances, and $\tilde{\Phi}_A$ and $\tilde{\Phi}_B$ are the averages of the potential over the cross-sections of the filament faces [21]. The discretized equations can be reformulated using a mesh analysis approach which is solved iteratively using a preconditioned Generalized Minimal Residual (GMRES) algorithm that exploits mutlipole acceleration. Commercial tools, such as Raphael RI3, also use the Volume Integral Approach [4].

Layout-Based 3-D Geometry Modeling. Accurate and automatic 3-D geometry modeling is essential for chip level inductance modeling with field solvers. The inductance values extracted from the field solvers can then be used as a "golden standard" for comparison with simplified analytical inductance estimations. Layout-based 3-D geometry extraction can be developed to combine 2-D information from the layout design (e.g. GDS II files) with the layer thicknesses and other material parameters coming from the actual fabrication process to create a simplified 3-D geometry model. Figure 3.10 shows the program flow chart based on Synopsys Arcadia database. Electromagnetic simulators such as FASTHENRY [21] can then be used to extract the inductance for each 3-D structure. Automatic path searching is required to construct 3-D geometries of the selected nets since each net usually includes many

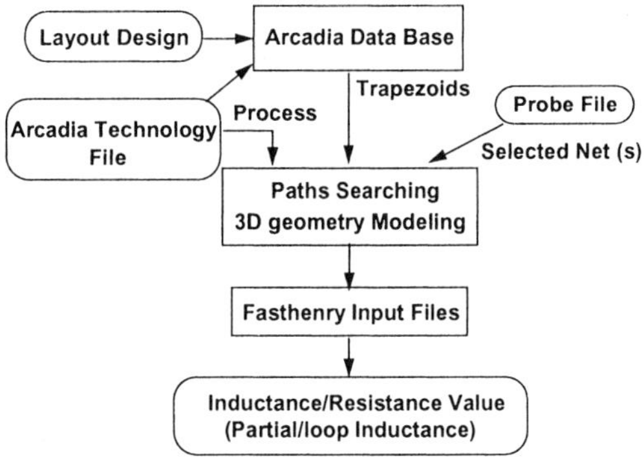

Figure 3.10. Program flow chart

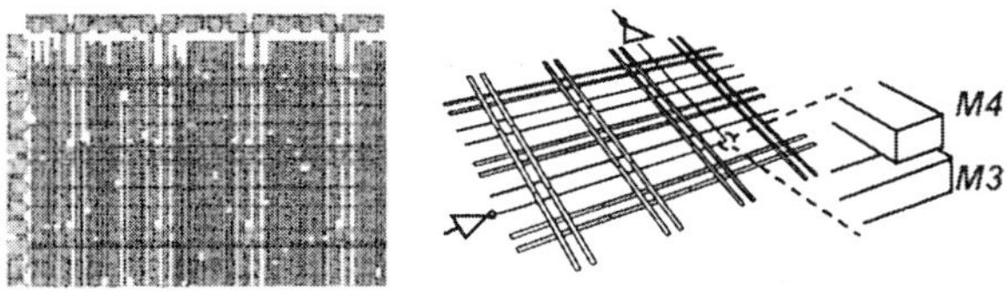

Figure 3.11. Extracted 3-D geometry with signal and power/ground wires from a test chip. A corner of metal 4 and metal 3 are magnified to demonstrate the 3-D effect. Driver and receiver are added for illustration.

branches. A data structure was designed to facilitate the searching process which is a depth-first search [36]. Figure 3.11 shows an example of the test chip layout and the extracted 3-D geometry which is subsequently used for electromagnetic simulations.

3.5.3 Analytical Formulae for Inductance Estimation

Although accurate inductance estimation based on field solver results is desirable, it is time consuming and memory intensive. It is therefore not practical to extract the inductance for a whole chip with this method. Instead, analytical formulae are desirable for providing design guidelines, as well as for screening and identifying nets with significant inductance for chip design and analysis. Based on electromagnetic field theory, analytical formulae for self and mutual inductance of wires and some typical interconnect structures including the coplanar waveguide can be derived. These formulae are based on the assumption that the return path of each wire is at an infinite distance (from the wire) as it is stated in PEEC method [43]; while the formulae for the coplanar

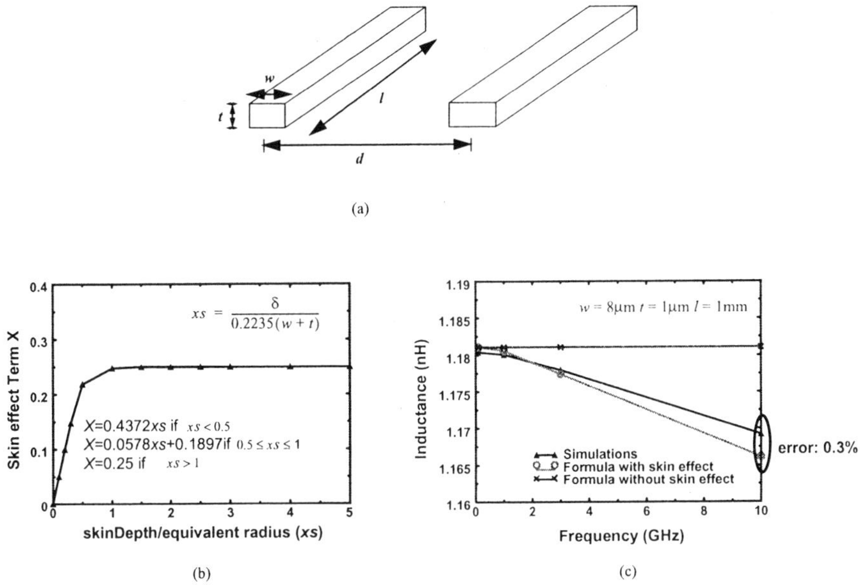

Figure 3.12. (a) Parameters in inductance formulae. (b) Fitted skin effect term, X. δ is the skin depth. (c) Comparison from the revised formula with the skin effect term.

waveguide considers a loop inductance in which current returns from the two adjacent ground wires.

Analytical Formulae for Self and Mutual Inductance. Self-inductance of a wire with a rectangular cross-section can be derived using electromagnetic field theory and the geometry mean distance approximation (G.M.D.)[42].

$$L_{self} = \frac{\mu_0 l}{2\pi}[\ln(\frac{2l}{w+t}) + \frac{1}{2} + 0.2235\frac{w+t}{l}] \tag{3.27}$$

Equation (3.27) is for the self-inductance when $l \gg (w+t)$ where l is the length of the wire. Parameters w and t are the width and height of the rectangular cross-section, respectively. Mutual inductance between two parallel wires with equal length can be shown to have the form

$$M = \frac{\mu_0 l}{2\pi}[\ln(\frac{2l}{d}) - 1 + \frac{d}{l}] \tag{3.28}$$

where d is the center-to-center distance, l the length, and $l > d$ as shown in Figure 3.12(a). The self-inductance is a nonlinear function of l, which means that it is not directly scalable with respect to length. In fact, it is super-linear when $l > (w+t)$.

To consider the skin effect, a frequency dependent term is added, as shown in Equation (3.29). This term can be represented by a Bessel Function [42]

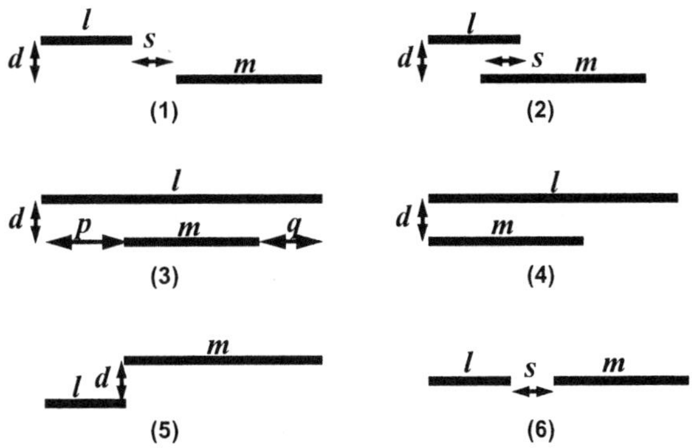

Figure 3.13. Six relative positions for mutual inductance. Wires in each case can be on the same layer or on different layers.

which is curve-fitted as shown in Figure 3.12(b) where δ is the skin depth at a particular frequency, X is a fitted parameter, and μ_r is the permeability of the conductor. All other parameters are the same as in Equation (3.27).

$$L_{self} = \frac{\mu_0 l}{2\pi}[\ln(\frac{2l}{w+t}) + \frac{1}{2} + 0.2235\frac{w+t}{l} - \mu_r(0.25 - X)] \qquad (3.29)$$

Figure 3.12(c) plots the comparison between the simulation and formulae, both with and without the skin effect. Equation (3.29) accurately captures the skin effect, though the frequency dependency is not large. The error falls within 1% for conductor widths between 1 μm to 50 μm.

Mutual Inductance of Two Parallel Wires with Unequal Length. To calculate wire inductance in a complex wiring environment or where self-inductance consists of several cascaded segments in sequence, the mutual inductance formulae of two parallel wires with unequal lengths must be derived. There are six different basic positions of two parallel wires which result in different mutual inductance formulae. Figure 3.13 illustrates these different cases; parameters l, m, p, q and s represent the wire lengths as well as wire overlap lengths. Parameter d is the center-to-center separation between two wires.

If the dimension of the wire is much smaller than the wavelength of the signal for the frequencies of interest, magnetic induction at every point of the field is in phase with the current. This condition can be met for wires on current integrated circuits at gigahertz frequencies (1 GHz in SiO_2 yields a wavelength of 15 cm which is much larger than the typical chip dimensions). The induced electromotive forces are in phase at all points. The magnetic flux linked with an interconnect may be considered as the summation of the fluxes, in phase under quasi-stationary conditions, contributed by the separate elements of the inducing circuit. As a result, the mutual inductance of a wire with the inducing wires is the algebraic sum of the mutual inductances of the separate elements of the inducing wires. For example, the mutual inductance of the two wires in CASE 4 can be calculated as,

$$M = \frac{1}{2}[(M_l + M_m) - M_{l-m}] \tag{3.30}$$

where M_l represents the mutual inductance of the two parallel wires with equal length of l and separation of d. (All mutual inductance between segments m, l and $l - m$ need to be considered to calculate M.) The other mutual terms follow in an analogous manner. Six additional formulae can be derived (shown below) where parameters l, m, d, p, q are as indicated in Figure 3.13.

CASE 1, 6:

$$M = \frac{\mu_0}{4\pi}\left[(l+s)\ln\left(\frac{l+m+s}{l+s}\right) + m\ln\left(\frac{l+m+s}{m+s}\right) + s\ln\left(\frac{s}{m+s}\right)\right]$$

$$s \neq 0$$

CASE 2:

$$M = \frac{\mu_0}{4\pi}\left[(l-s)\ln\left(\frac{l+m-s}{l-s}\right) + m\ln\left(\frac{l+m-s}{m-s}\right) + s\ln\left(\frac{4s\,(m-s)}{d^2}\right) - 2s\right]$$

$$l \neq s,\ m \neq s,\ d \neq 0$$

CASE 3:

$$M = \frac{\mu_0}{4\pi}\left[m\ln\left(\frac{4\,(m+p)\,(m+q)}{d^2}\right) + p\ln\left(\frac{m+p}{p}\right) + q\ln\left(\frac{m+q}{q}\right) - 2m\right]$$

$$d \neq 0,\ p \neq 0,\ q \neq 0$$

CASE 4:

$$M = \frac{\mu_0}{4\pi}\left[l\ln\left(\frac{l}{l-m}\right) + m\ln\left(\frac{4m\,(l-m)}{d^2}\right) - 2m + d\right]$$

$$l \neq m,\ d \neq 0$$

CASE 5:

$$M = \frac{\mu_0}{4\pi}\left[l\ln\left(\frac{l+m}{l}\right) + m\ln\left(\frac{l+m}{m}\right) - d\right]$$

Wires for which the lengths are greater than the wavelength can be partitioned into several segments so that mutual inductance can be calculated. Figure 3.14 shows the comparison of the model, for CASE 4, with results of field solver simulations. The formula gives an accurate estimation for reasonable separation distances. When the distance becomes comparable to the wire length, the formulae underestimate the mutual inductances. For on-chip interconnects, wires are usually thin and long compared to their separation in the simulation window. Therefore, use of the formulae does not result in appreciable errors. CASE 1 and CASE 6 share the same formula and their mutual inductances usually are negligible because the magnetic flux linking between the wires is very small. For CASE 5, if $d = 0$, it represents the situation in which two wires touch each other.

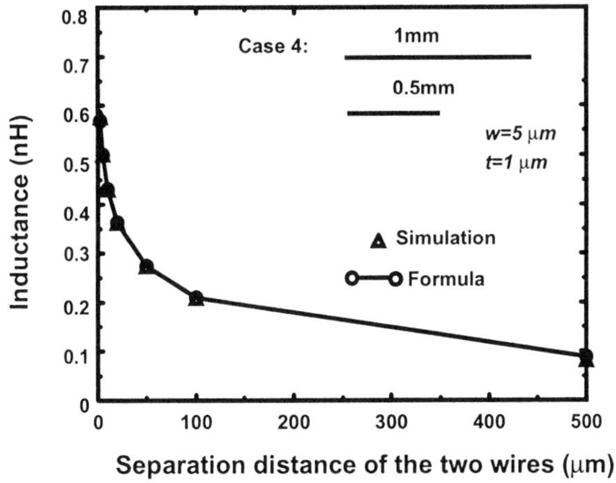

Figure 3.14. Formula and simulation comparison for mutual inductance of Case 4.

Calculation of Self-Inductance of an Entire Wire. If a wire consists of several segments, the self-inductance of the entire wire does not equal to the sum of the self-inductances of the segments; the existence of mutual inductances between segments of the wires must be considered. To compute the entire wire inductance, the self-inductance of all segments as well as mutual inductance between these segments should be included. It can be shown, using results from circuit theory, that the inductance of an entire wire constructed by several cascaded segments is as follows:

$$L_{self} = \sum_{i=1}^{N} l_i + \sum_{i=1}^{N} \sum_{j=i+1}^{N} 2k_{ij} M_{ij} \tag{3.31}$$

where N is the number of segments; l_i is the self inductance of segment i. The parameter M_{ij} is the mutual inductance between segment i and j. Term $k_{ij} = 0$ when segment i and j are orthogonal; $k_{ij} = 1$ when i and j have same current direction; $k_{ij} = -1$ when i and j have opposite current directions. With these calculated self- and mutual inductances, circuit simulation models such as used with SPICE can be constructed in support of circuit simulations for designs, especially in considering crosstalk analysis.

Analytical Formulae for Coplanar Waveguide Structure. In a coplanar waveguide, the signal wire is sandwiched between two ground wires (or more generally power/ground wires, since power wires are also ac ground to the signal wire) on the same layer. The electromagnetic field is confined between the signal and ground wires. Current flows through the signal wire and returns from the two ground wires. The structure is considered as a transmission line whose inductance per length can be calculated. To derive the analytical for-

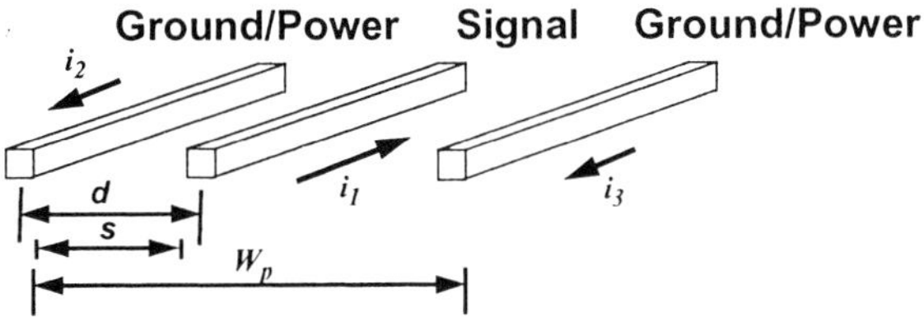

Figure 3.15. Coplanar waveguide structure: d is the center-to-center separation distance of the signal wire and the nearer ground/power wire. W_p is the ground/power wire pitch. S is the edge-to-edge spacing between the signal wire and the nearer ground/power wire. Wire connections of the two ground wires at the far-end and near-end are not shown for clarity.

mula for the inductance of this structure, the partial inductance for each wire is calculated, and combined using circuit theory. The loop inductance of the coplanar waveguide per unit length (Units: H/m) at high frequencies has the following form[8]

$$\hat{L}_{coplanar} = \frac{\mu_0}{2\pi}[\ln(\frac{\pi d}{w_{sig}+t}) + \frac{1}{2}\ln(\frac{\pi d}{w_{gnd}+t}) + \frac{1}{2}\ln(1-\frac{1}{\alpha}) + \frac{1}{2}\frac{\ln\frac{\pi w_p}{(\alpha-1)(w_{gnd}+t)}}{\ln\frac{\pi w_p}{w_{gnd}+t}}\ln(\alpha-1)] \tag{3.32}$$

where w_{sig} and w_{gnd} are the widths of the signal and ground wires, respectively. Parameter w_p is the ground wire pitch , and d is the center-to-center distance of the signal wire to the nearest ground wire (see Figure 3.15). Parameter α is the ratio of ground wire pitch, w_p, to d, and $\alpha \geq 2$. As long as the width of the signal wire is smaller than that of the ground wire, the ground wire width can be approximated by the signal wire width at high frequencies owing to proximity effects. If $2 \leq \alpha < 4$, the formula can be approximated by $\alpha = 2$ (the coplanar waveguide structure becomes symmetric); for α in the range $4 \leq \alpha \leq 100$, the last two terms in the Equation (3.32) can be approximated based on reasonable ground pitches and ground wire widths. The skin effect can also be included. Considering the rectangular cross-section typical for VLSI interconnects, the following simplified formulae can be used:

when $2 \leq \alpha < 4$

$$\hat{L}_{coplanar} = 0.3\ln(\frac{w+s}{w+t}\pi) - 0.1\ln 2 + 0.1I \tag{3.33}$$

when $4 \leq \alpha \leq 100$

$$\hat{L}_{coplanar} = 0.3\ln(\frac{w+s}{w+t}\pi) + 0.1 + 0.1I \tag{3.34}$$

The units for $\hat{L}_{coplanar}$ are nH/mm. Parameter s is the edge-to-edge spacing between the signal wire and the nearest ground wire, w is the width of signal

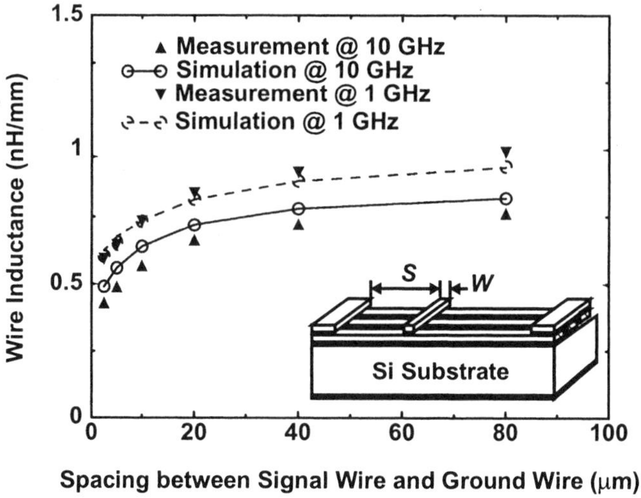

Figure 3.16. Measurement and simulation comparison: inductance vs. wire spacing. The signal wire width is 6 μm. The ground wire width is 16 μm.

wire and t is the thickness of the metal layer. Parameter I is the frequency dependent internal inductance given by $I = 0.75\tanh(\frac{2\delta\pi}{w+t})$, with $\delta = \frac{1}{\sqrt{\pi\sigma f\mu}}$, the skin depth of the material, where f is frequency, σ is the conductivity and μ is the permeability of the metal layer. (A term of $10^{-7}I$ can also be added to Equation (3.32) to model the skin effect.) Equation (3.34) is appropriate for a reasonable ground pitch (e.g., 250μm) and wire width (e.g., 1 to 80μm) while Equation (3.32) applies to all cases.

3.5.4 Experimental and Simulation Results

Wire Inductance. Test chips have been designed and fabricated to validate the above modeling results. A chip, referred to as the conventional test structure (chip), has coplanar waveguide structures on the fifth metal layer which is about 5 μm above the substrate. Measurement results and simulations for the coplanar conventional test structure are compared in Figure 3.16 [35]. Inductance increases monotonically with the *spacing* between the wire and the nearest ground wire of the coplanar structure. The pitch of the ground wires, w_p in Figure 3.15 is fixed in the test structure. Most current returns through the nearest ground wire for small spacings. For large spacings, return current is distributed between the two ground wires, which increases the current return loop and hence inductance. The inductance increase is less pronounced when spacing becomes large and reaches its maximum when the signal wire has equal distance to the two ground wires. In addition, wire inductance decreases monotonically with wire width since wider traces result in smaller loops with

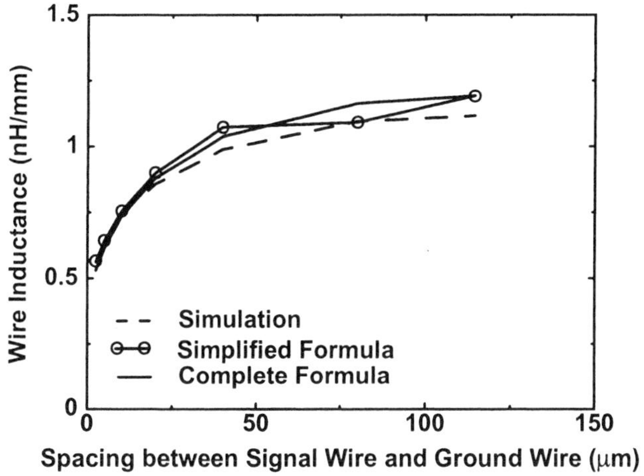

Figure 3.17. Formulae and simulation comparison for co-planar structure. The signal wire width is 6 μm. Frequency is 3.1 GHz.

an associated reduction in magnetic flux. At higher frequencies, inductance becomes smaller because of the skin and proximity effects. When frequencies increase from 1 GHz to 10 GHz, measured inductance is consistently smaller compared to simulated results, which indicates that smaller return paths exist for these test chips. The inductance is decreased due to the proximity effects between the wires as well as between the wire and the substrate at high frequencies, which is more evident from the measurements. Figure 3.17 compares the inductance results from the analytical formulae and simulations and demonstrates the accuracy of Equations (3.32)-(3.34) for the coplanar waveguide structure.

Substrate Effects on Wire Inductance. The substrate is usually grounded, and the electromagnetic field of the signal wires can be coupled into the substrate at high frequencies. Electric fields couple to the substrate, generating conduction and displacement currents that flow both laterally and vertically. Time-varying magnetic fields couple to the substrate, generating eddy currents that flow parallel to the devices above the substrate as shown in Figure 3.18. As a result, the substrate can offer a portion of the return paths for current. With the eddy currents, the substrate can reduce wire inductance, especially for substrates with high conductivity. This phenomenon is evident for the wires without intermediate metal layers between the wires and the substrate. In this case, there is no shielding effects of the electromagnetic field of the wires, which might, otherwise, prevent the electromagnetic waves from penetrating into the substrate. The substrate effects of reducing wire inductance at high frequencies are mainly due to the time-variant eddy current effects. To derive

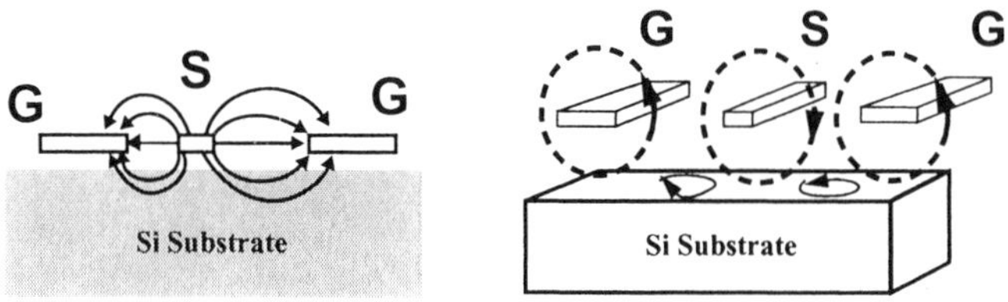

(a) (b)

Figure 3.18. (a) Electric field couples to the substrate. (b) Magnetic field couples to the substrate shown by dashed lines. Eddy currents are also shown within the substrate.

the exact formula for the substrate eddy current effects is difficult, however, approximations can be made to generate rough inductance estimation for some structures such as the coplanar waveguide. To model the substrate effect, it can be treated as a return path beneath the signal wire according to proximity effects at high frequencies. Since the ground wires are connected to the substrate with ohmic substrate taps along the whole length of each wire, the substrate currents beneath each ground wire are very small compared to the currents in the ground wires themselves, and thus can be ignored. Displacement current within the substrate is also ignored. The substrate skin depth is used to calculate the effective distance between signal and ground wires. Similar procedures as the derivation of Equation (3.32) can be used. The revised formula, including substrate effects, has the form (unit H/m):

$$\hat{L}_{total} = \hat{L}_{coplanar} - k\frac{\mu_0}{2\pi}\left(\ln\sqrt{(\left(\frac{s+\frac{w_{gnd}+w}{2}}{h+\frac{1}{2\sqrt{\pi f\sigma\mu_0}}}\right)^2} + 1 + A\right) \tag{3.35}$$

where $A = \frac{1}{2}ln\left(\frac{\pi(s+\frac{w_{gnd}+w}{2})}{2(w_{gnd}+t)}\right)$. The $\hat{L}_{coplanar}$ is the inductance without substrate effects, h is the distance of the metal layer (center) from the substrate, w_{gnd} is the width of the ground wires, σ is the conductivity of the substrate, f is the signal frequency, s, w and t are the same as that in Equations (3.33) and (3.34), and k is the percentage of current returned via the substrate, which is around 38% in this study. Parameter k is approximated by the dc current distribution under the coplanar structure. When the spacing between signal wire and ground wire is larger than 20 μm, with a ground wire pitch of 251 μm, analytical calculations without the substrate correction become inaccurate as seen in Figure 3.19. More current tends to return via the substrate (i.e., the proximity effect) which forms a smaller current loop, resulting in a reduction of wire inductance. With spacings larger than 40 μm, there is more than 18%

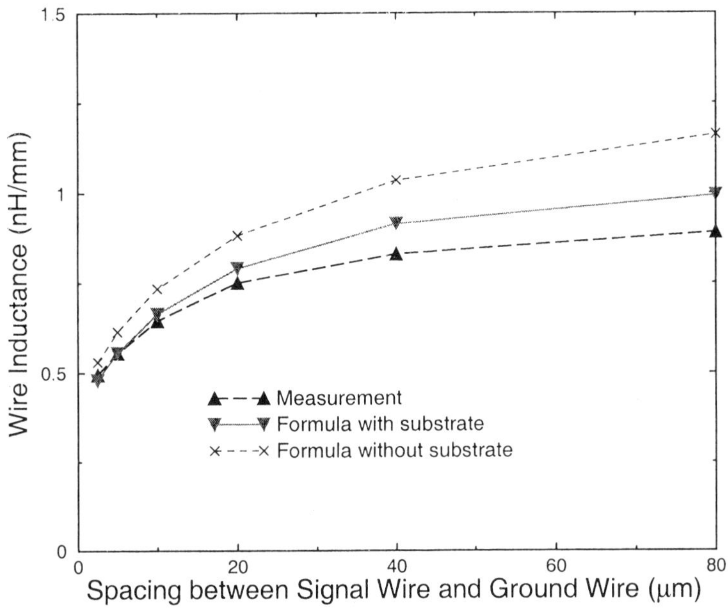

Figure 3.19. Wire inductance with substrate effects. The substrate resistivity, ρ_{sub}, is 0.015 Ω-cm.

reduction of wire inductance. This effect becomes more prominent for less resistive substrates.

Important design insight can be obtained from these analytical formulae. The spacing between the signal wire and the nearest ground wire is a critical factor that determines the loop inductance. Moreover, internal inductance is small compared to external inductance. Three critical factors that determine substrate effects are: the spacing between the signal wire and the nearest ground wire; the distance between signal wire and substrate and the substrate conductivity.

Inductance Modeling with Power/Ground Grids and Floating Grids. For VLSI digital chips, power and ground are usually distributed through grid structures in order to minimize IR drops as well as ground bounce. Due to proximity effects, the nearest power and ground wires provide the return paths for most of the signal current (and associated EM energy). Since grids perpendicular to the signal wires do not contribute to the signal wire inductance (their mutual inductance is zero), only grids parallel to the signal wires are needed to calculate the signal wire inductance. Simulations and measurements show that, with the substrate excluded, multiple parallel (with signal wires) ground wires in ground grids reduce signal wire inductance because of the multiple

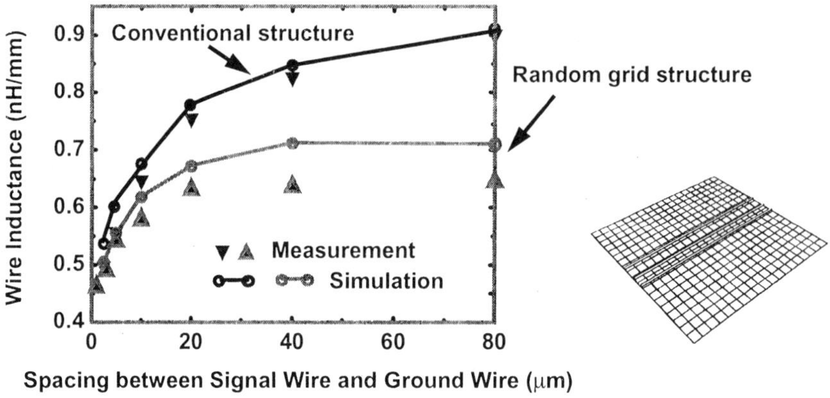

Figure 3.20. Measurement and simulation: wire inductance at 3 GHz is reduced with floating grids on M1 and M2. The signal wire width is 5 μm. A random grids structure is shown on the right.

parallel current return paths. However, if the substrate is included, the nearest ground wires and substrate dictate the current return paths. In this case, multiple parallel ground wires, configured in grids, cannot effectively reduce the signal wire inductance.

In a digital chip environment, if there are dense grids or a ground plane on some metal layers, they may offer current return paths for the signal current. A ground plane can shield electromagnetic fields and also offer current return paths for the signal current. In these cases, the signal wire inductance can be reduced significantly. Figure 3.20 compares the results of a conventional test chip with the results from a test chip which has coplanar waveguide structures on the fifth metal layer, and dense grids on the first and second metal layers across the chip. The chip is referred to as the random grid structure. For the conventional structure, inductance increases as the spacing to the nearest ground increases while for the random grid structure, the inductance saturates for spacings around 20 μm and becomes independent of the spacing at larger separations. The magnetic flux generated by eddy currents in the random grids oppose the magnetic field change, thus reducing the magnetic flux of the signal wires, and effectively reducing the inductance of the signal loop. The dense grids effectively change the electromagnetic field configuration in a manner similar to a continuous ground plane. As a result, wire inductance is primarily determined by the spacing from the wire to the grids. Each cell of the grids is about 15 μm x 15 μm. When the spacing between the signal wire and the nearest ground wire exceeds about 15 μm, the signal loop couples to the grids, and the effective magnetic flux is reduced. Figure 3.21 shows the frequency dependence of the inductance. As the frequency goes up, the inductance decreases due to the time varying nature of the eddy currents. The same phenomena can

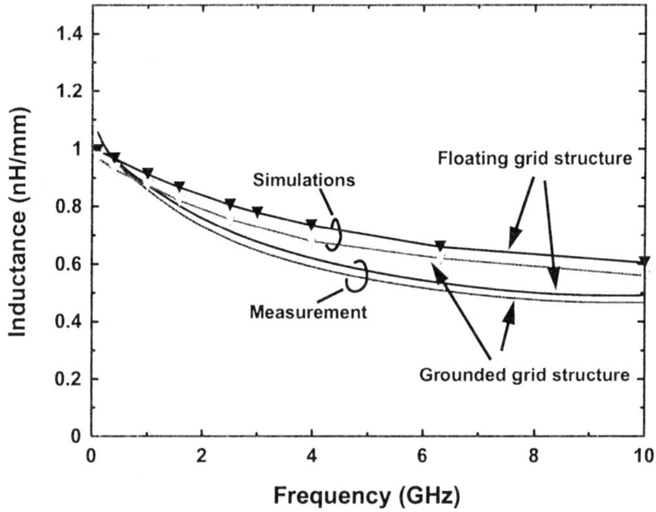

Figure 3.21. Frequency dependency of wire inductance with floating or grounded grids on M1 and M2 layers.

be observed in the simulations and measurements if the grid structure is replaced by a continuous ground plane. Whether the grid structure is floating or grounded does not make much difference in the value for wire inductance due to the eddy current effects.

3.5.5 Ground Plane Effects on Wire Inductance

For VLSI chips, a ground plane sometimes is specifically designed and inserted between metal layers to shield electromagnetic coupling in order to control the wire inductance of critical signal, clock and power nets. A densely populated grid structure, also exhibits ground plane effects. A ground plane alters electromagnetic fields due to the boundary conditions at the surface of the plane which must be satisfied. One major ground plane effect is that wire inductance is significantly reduced because of more confined electromagnetic fields, effectively shielding the signals. There are several previous studies of ground plane effects on wire inductance for PCBs. For example, microstrip lines have been studied by many groups. However, due to the nature of the problem, an exact analytical formula is not possible for the general case of a microstrip of width w with finite thickness t [28]. Some results for PCBs can not be directly used for on-chip interconnects because of geometry assumptions used in the derivation.

Analytical Formulae. Inductance estimations can be obtained from formulations of the characteristic impedance for transmission lines and their relationship with distributed inductances. For a transmission line consisting of perfect conductors[9], the inductance of the transmission line with characteristic impedance Z_0 (in ohms) is [22]

$$L = KZ_0 = \frac{1}{3}\sqrt{\epsilon_r} \times 10^{-4} \mu\text{H/cm} \tag{3.36}$$

where ϵ_r is the relative dielectric constant of a uniform dielectric enclosed by a transmission line. The characteristic impedance of a pair of parallel wires over a ground plane which has the same current direction is [22]

$$Z_{oe} = \frac{69}{\sqrt{\epsilon_r}} \log_{10} \left[\frac{4H}{d} \sqrt{1 + \left(\frac{2H}{D} \right)^2} \right] \tag{3.37}$$

and two wires with current in the opposite direction is

$$Z_{oo} = \frac{276}{\sqrt{\epsilon_r}} \log_{10} \left[\frac{2D}{d} \frac{1}{\sqrt{1 + \left(\frac{D}{2H} \right)^2}} \right] \tag{3.38}$$

where H is the height of the wires over the ground plane (from the center of the wires), d is their diameter, and D is the center-to-center wire separation. The two equations hold when $d \ll D, H$.[10] Considering that the inductance of the transmission line is comprised of the inductance of the individual wires and of the mutual inductance M between wires, the transmission line inductance may be expressed as

$$L_{oo} = KZ_{oo} = 2L - 2M \tag{3.39}$$

$$L_{oe} = KZ_{oe} = \frac{L}{2} + \frac{M}{2} \tag{3.40}$$

L and M then can be solved from Equations (3.36)-(3.40),

$$L = 4.59 \log_{10} \frac{4H}{d} \text{nH/cm} \tag{3.41}$$

$$M = 2.30 \log_{10} \left[1 + \left(\frac{2H}{D} \right)^2 \right] \text{nH/cm} \tag{3.42}$$

Considering the rectangular cross-section for on-chip wires, one obtains

$$L = 0.2 \ln \frac{2\left(h + \frac{t}{2}\right)\pi}{w + t} \text{nH/mm} \tag{3.43}$$

$$M = 0.2 \ln \left[1 + \left(\frac{2\left(h + \frac{t}{2}\right)}{D} \right)^2 \right] \text{ nH/mm} \tag{3.44}$$

where h is the wire height (from the bottom of the wire) to the ground plane, w and t are the width and thickness of the metal wire. To model the case when $h = 0$, an extra term Δ needs to be added in Equation (3.43). As $h = 0$, the magnetic flux enclosed by the loop consisting of the wire and ground plane is zero when L in Equation (3.43) should be equal zero. Δ can be determined by imposing the condition that

$$\ln \frac{2(h + \frac{t}{2} + \Delta)\pi}{w + t} |_{h=0} = 0$$

and then solving for $\Delta = \frac{w+t}{2\pi} - \frac{t}{2}$. Substitution of Δ back into Equation (3.43) gives

$$L = 0.2 \ln \left(\frac{2h\pi}{w + t} + 1 \right) \text{ nH/mm} \tag{3.45}$$

In addition, formulae can be derived by using the Image Current Method [8]. By assuming a system of perfect conductors, it can be shown that the loop inductance of the wire over a ground plane is

$$L = \frac{\mu_0}{2\pi} \left[\ln \left(\frac{2h}{w + t} + 0.22 \right) + \frac{3}{2} \right] \text{ H/m} \tag{3.46}$$

where w and t are the width and thickness of the metal wire. The self-inductance formula for a wire over a ground plane can also be derived based on a geometrical factor concept where self-inductance, capacitance, resistance and characteristic impedance are all represented by a geometrical factor Γ [54] [6]. For the case of a rectangular cross-section wire over a ground plane , the geometrical factor $\Gamma = \frac{w}{h}$, where w is the width of a wire and h is the wire height from the bottom of the wire to the ground plane. It can be shown that

$$L = \frac{\mu}{K_L} \left(\frac{h}{w} \right) \text{ H/m} \tag{3.47}$$

where K_L is an inductive fringing factor which is curve-fitted as

$$K_L = -0.0264 \left(\frac{h}{w} \right)^2 + 1.5618 \frac{h}{w} + 1.3745$$

Simulation and Experiment. Results from the analytical calculations are validated by comparing them with simulations using the commercial software package Maxwell [2]. Figure 3.22 shows the self-inductance calculated from

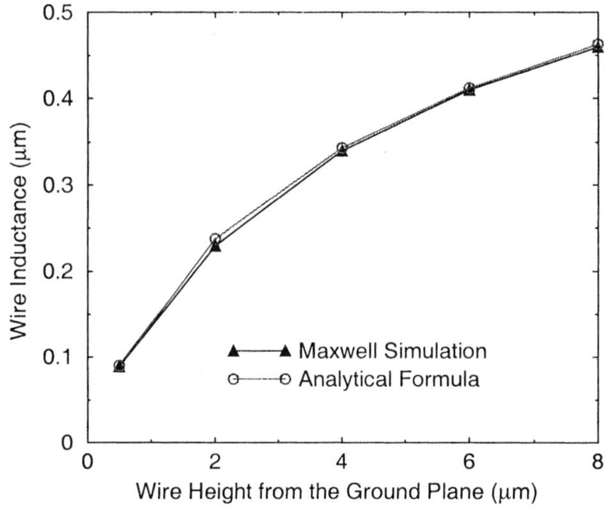

Figure 3.22. A wire over a ground plane : Equation (3.45) results are compared with Maxwell simulation results. Wire height is the distance from the bottom of the wire to the top surface of the ground plane. The signal wire width is 5 μm and the thickness is 0.5 μm for both the signal wire and ground plane.

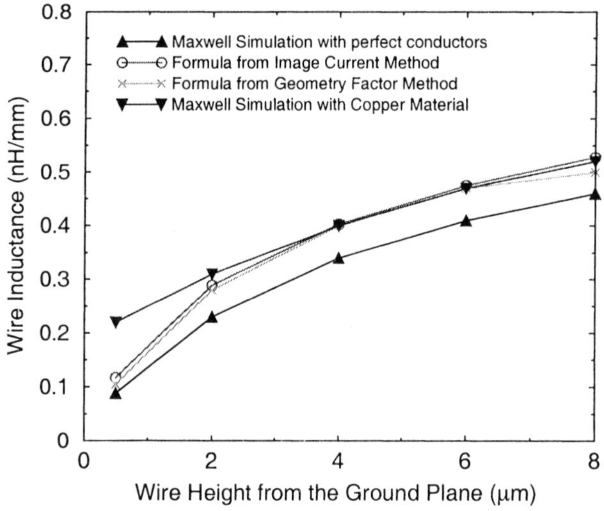

Figure 3.23. A wire over a ground plane : Results from Equations (3.46) and (3.47) are compared with Maxwell simulation results. The signal wire width is 5 μm and the thickness is 0.5 μm for both the signal wire and ground plane. Simulations with copper material at 3 GHz are also plotted.

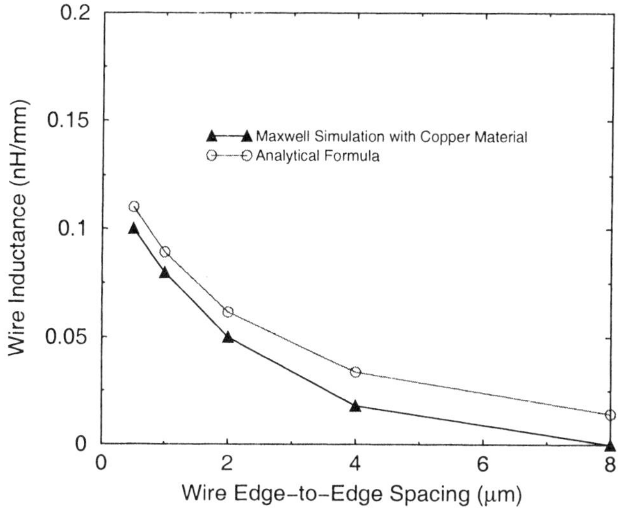

Figure 3.24. Two wires over a ground plane : Mutual inductance results from Equation (3.44) are compared with Maxwell simulations with copper material at 3 GHz. The width of the two wires is 5 μm.

Equation (3.45) and the Maxwell simulation results. An excellent agreement has been achieved with this formula. Figure 3.23 shows self-inductance calculations from Equations (3.46) and (3.47). Wire inductance increases as the height of the wire to the ground plane increases because it results in larger current loop. Simulation results at 3 GHz with a non-perfect conducting material (i.e., copper material) are also plotted in Figure 3.23, showing that the perfect conductor material results give a very good approximation to the non-perfect metal conductor case at high frequencies. Figure 3.24 plots the mutual inductance calculated from Equation (3.44) and the Maxwell simulation results. The mutual inductance decreases quickly as wire separation increases – much faster than the mutual inductance of two wires without a ground plane. This is due to the fact that the mutual inductance of two wires over a ground plane is inversely proportional to the square of the wire separation D in Equation (3.44).

In terms of a coplanar waveguide structure over a ground plane, the loop inductance is no longer a function of the spacing between the signal wire and the nearest ground wire as presented in the Figure 3.25 from test chip measurements. Because it forms a smaller loop, the inductance is mainly dictated by the height of the signal wire to the ground plane, not the spacing to the ground wires in the co-planar structure. In this case, controlling wire spacing can no longer control the loop inductance.

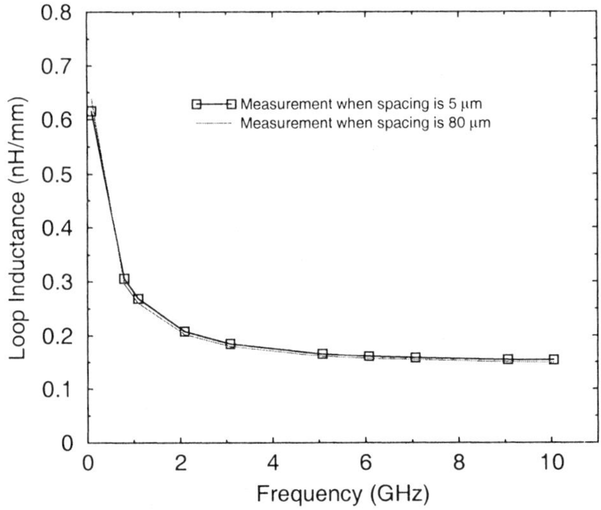

Figure 3.25. Loop inductance of a coplanar waveguide structure does not change with the spacing of the signal wire to the nearest ground wire. The signal wire width equals 5 μm. The width of the two ground wires is 16 μm.

3.6 Summary

Interconnects have become a major bottleneck for IC performance in the era of gigascale integration. There are strong needs to accurately and efficiently extract interconnect parasitic resistance, capacitance and inductance of on-chip interconnects in order to predict signal timing performance, signal integrity, and power consumption. Parasitics can be extracted by electromagnetic field solvers[11] which provide highly accurate RLC values. For chip design and analysis, analytical formulae and look-up tables are usually more efficient, but accuracy is a serious concern. With circuit clock and data frequencies operating in the gigahertz range, and faster signal rise and fall times in modern IC technologies, on-chip inductive effects become increasing problematic for interconnect delay and signal integrity, and need to be modeled accurately. Physical inductance modeling using field solvers as well as analytical formulae is presented, which includes effects of 3-D geometry, the overall complexity of multi-conductor environment and substrate effects.

Acknowledgments

Authors would like to thank Dr. Bendik Kleveland, Dr. Gaofeng Wang, Dr. Zhiping Yu, Prof. S. Simon Wong, Dr. Tak Young and Takeshi Furusawa for their helpful discussions on on-chip inductance modeling.

Notes

1. The total current, $\bar{\mathcal{J}}_{total} = \bar{\mathcal{J}} + \frac{\partial \bar{\mathcal{D}}}{\partial t}$. The first term on the right hand side is the sum of the conduction current and the source current while the second term represents the displacement current.

2. See Sections 4.10 and 6.8 in [20] for the detailed discussion of electromagnetic fields in material media.

3. The surface charge density, $\rho_s = \epsilon \mathbf{E}$.

4. Inductance frequency dependence will be discussed shortly

5. The Biot-Savart law states: the $\mathbf{B}$ field vector at any point P identified by the position vector $\mathbf{r}$, due to a differential current element $Id\mathbf{l}'$ located at position $\mathbf{r}'$, is

$$d\mathbf{B_P} = \frac{\mu_0 I d\mathbf{l}' \times \hat{\mathbf{R}}}{4\pi R^2}$$

where $\hat{\mathbf{R}}$ is the unit vector pointing from the location of the current element to the field point P, and $R = |\mathbf{r} - \mathbf{r}'|$ is the distance between them. See [20].

6. The last integral transform is based on Stokes's theorem.

7. Figure 3.9 can serve as a simple example.

8. The frequency is high enough so that internal inductance can be ignored. The complete derivation is not shown due to the chapter length. See [34].

9. So it is a lossless transmission line.

10. This assumption is remedied by adding a Δ term for small H case. See Equation (3.45) discussion.

11. Currently most field solvers use quasi-static electromagnetic field models which are sufficient for on-chip interconnects as the wavelength of signal frequency is larger than the interconnect lengths. For the models considering the field retardation, interested readers can read papers [10][9].

References

[1] Agilent Technologies, Palo Alto, CA *www. agilent.com.*

[2] Ansoft Corporation, Pittsburgh, PA, *Maxwell 2D Parameter Extractor – User's Reference*, Feb. 1993.

[3] Ansoft Corporation, Pittsburgh, PA, *Maxwell Quick 3D Parameter Extractor – User's Reference*, Dec. 1994.

[4] Avant! Corporation, Fremont, CA, *Raphael - Reference Manual.*

[5] Erich Barke, "Line-to-Ground capacitance calculation for VLSI: A comparison", *IEEE Transactions on CAD of Integrated Circuits and Systems*, pp.295-298, Vol. 7, No. 2, Feb, 1988.

[6] I. Catt, *Digtal Hardware Design*, Macmillan, London, 1979.

[7] W. H. Chang, "Analytic IC-metal-line capacitance formulas", *IEEE Trans. on Microwave Theory Tech.*, Vol. 24, pp. 608-611, 1976; also Vol. 25, p. 712, 1977.

[8] David K. Cheng, *Field and Wave Electromagnetics*, Addison-Wesley Publishing Company, Menlo Park, California, 1983.

[9] E. Chiprout, H. Heeb, M.S. Nakhla and A. E. Ruehli, "Simulating 3-D retarded interconnect models using complex frequency hopping

(CFH)", *Proceedings of IEEE Int. Conf. on Comp. Aided Des.*, pp. 66-72, Nov. 1993.

[10] Jane Cullum, Albert Ruehli and Tong Zhang, "A method for reduced-order modeling and simulation of large interconnect circuits and its application to PEEC models with retardation", *IEEE Trans. on Circuits and Systems II*, Vol. 47, pp.261-273, Apr. 2000.

[11] A. Deutsch, H. Harrer, C. W. Surovic, G. Hellner, D. C. Edelstein, R. D. Goldbaltt, G. A. Biery, N. A. Greco, D. M. Foster, E. Crabbe, L. T. Su and P. W. Coteus, "Functional high-speed characterization and modeling of a six-layer copper wiring structure and performance comparison with aluminum on-chip interconnections," *IEEE Technical Digest of International Electron Devices Meeting*, pp. 295-298, Dec., 1998.

[12] A. Deutsch, G. V. Kopcasy, V. A. Ranieri, J. K. Cataldo, E. A. Galligan, W. S. Graham, R. P. McGouey, S. L. Nunes, J. R. Paraszczak, J. J. Ritsko, R. J. Serino, D. Y. Shih, and J. S. Wilczynski,"High-speed signal propagation on lossy transmission lines," *IBM J. Research Development*, Vol. 34, No. 4, July, 1990.

[13] A. Deutsch, G. V. Kopcasy, P. J. Restle, H. H. Smith, G. Katopis, W. D. Becker, P. W. Coteus, C. W. Surovic, B. J. Rubin, R. P. Dunne, Jr., T. Gallo, K. A. jenkins, L. M. Terman, R. H. Dennard, G. A. Sai-Halasz, B. L. Krauter, and D. R. Knebel, "When are transmission-line effects important for on-chip interconnects?", *IEEE Trans. Microwave Theory Tech.*, vol. 45, No. 10, pp. 1836- 1846, Oct. 1997.

[14] A. Deutsch, Howard H. Smith, Christopher W. Surovic, Gerard V. Kopcsay, David A. Webber, Paual W. Coteus, George A. Katopis, W. Dale Becker, Allan H. Dansky, George A. Sai-Halasz and Phillip J. Restle, " Frequency-dependent crosstalk simulation for on-chip interconnects," *IEEE Transaction on Advanced Packaging*, pp. 292-308, Vol. 22, No. 3, Aug., 1999.

[15] A.J. van Genderen and N.P. van der Meijs, "Using articulation nodes to improve the efficiency of finite-element based resistance extraction," *IEEE Proceedings of Design Automation Conference*, pp.758-763, 1996.

[16] Frederick W. Grover, *Inductance calculations - Working formulas and table*, Dover Publications, Inc., New York, NY, 1973.

[17] R. F. Harrington, *Field Computation by Moment Methods*, Macmillan, New York, 1968.

[18] L. He, N. Chang, S. Lin, and O. S. Nakagawa,"An efficient inductance modeling for on-chip interconnects", *Proceedings of IEEE 1999 Custom Integrated Circuits Conference*, pp. 457-460, 1999.

[19] J. L. Hess and A. M. O. Smith, "Calculation of potential flow about arbitrary bodies", *Progress in Aerospace Sciences*, Vol. 8, pp.1-138, 1966.

[20] U. S. Inan and A. S. Inan, *Engineering Electromagnetics*, Addison-Wesley, Menlo Park, CA, pp. 606-607, 1999.

[21] M. Kamon, M. J. Tsuk and J. K. White, " FASTHENRY: a multipole-accelerated 3D inductance extraction program", *IEEE Trans. Microwave Theory and Techniques*, pp.1750, 1994.

[22] A. G. Kandoian et al, *Reference Data for Radio Engineers*, Howard W. Sams & Co., Inc., New York, the Fifth Edition, 1968.

[23] B. Kleveland, X. Qi, L. Madden, R. Dutton and S. Wong, "Line inductance extraction and modeling in a real chip with power grid", *Technical Digest of IEEE International Electron Devices Meeting (IEDM'99)*, pp. 901, Dec. 1999.

[24] B. Krauter and S. Mehrotra, "Layout based frequency dependent inductance and resistance extraction for on-chip interconnect timing analysis", *Proceedings of Design Automation Conference* pp. 303-308, 1998.

[25] B. Krauter, S. Mehrotra and V. Chandramouli, "Including inductive effects in interconnect timing analysis," *Proceedings of Design Automation Conference*, pp. 445-452, June, 1999.

[26] Y. L. LeCoz and R. B. Iverson, "A stochastic algorithm for high speed capacitance extraction in integrated circuits", *Solid-State Electronics*, Vol. 35, No. 7, pp. 1005-1012, 1992.

[27] S. Lin, N. Chang, and Sam Nakagawa,"Quick on-chip self- and mutual-inductance screen", *Proceedings of IEEE 2000 International Symposium on Quality Electronic Design*, pp. 513-520, 2000.

[28] Richard E. Matick, *Transmission Lines for Digital and Communication Networks*, IEEE Press, Piscataway, NJ, 1969.

[29] N.V.D. Meijs and J. T. Fokkema, "VLSI circuit reconstruction from mask topology", *Integration*, Vol. 2, No. 2, pp. 85-119, 1984.

[30] Shannon V. Morton, "On-chip inductance issues in multiconductor systems," *Proceedings of Design Automation Conference*, June, 1999.

[31] K. Nabors and J. White, "FastCap: A multipole accelerated 3-D capacitance extraction program", *IEEE Transactions on Computer Aided Design of Integrated Circuits and Systems*, pp. 1447-1459, Vol. 10, No. 11, Nov., 1991.

[32] K. Nabors and J. White, "Multipole-Accelerated Capacitance Extraction Algorithm for 3-D Structures with Multiple Dielectrics", *IEEE Transactions on Circuits and Systems – I; Fundamental Theory And Applications*, pp. 946-954, Vol. 39, No. 11, Nov., 1992.

[33] Q. Ning, P. M. Dewilde and F. L. Neerhoff, " Capacitance coefficients for VLSI multilevel metalization lines", *IEEE Transactions on Electron Devices*, pp.644-649, Vol. 3, No. 34, 1987.

[34] Xiaoning Qi, *High Frequency Characterization and Modeling of On-Chip Interconnects and RF IC Wire Bonds*, Ph.D. Thesis, Stanford University, June, 2001.

[35] Xiaoning Qi, Bendik Kleveland, Zhiping Yu, S. Simon Wong, Robert W. Dutton and Tak Young, "On-Chip Inductance Modeling of VLSI Interconnects", *2000 IEEE Solid-State Circuits Conference: Digest of Technical Papers*, pp. 172-173, Feb. 2000.

[36] Xiaoning Qi, Gaofeng Wang, Zhiping Yu, Robert W. Dutton, Tak Young and Norman Chang, "On-chip inductance modeling and RLC extraction of VLSI interconnects for circuit simulation", *IEEE Custom Integrated Circuits conference*, pp. 487-490, May, 2000.

[37] http://www.randomlogiccorp.com.

[38] P. J. Restle, "High speed interconnects: a designers perspective", *ICCAD'98 Tutorial: Interconnect in high speed designs: problems, methodologies and tools*, Nov. 1998.

[39] P. J. Restle and Alina Deutsch, "Designing the best clock distribution network", *Digest of technical papers of 1998 Symposium on VLSI Circuits*, pp. 2-5, 1998.

[40] P. J. Restle, K. A. Jenkins, A. Deutsch and P. W. Cook "Measurement and modeling of on-chip transmission line effects in a 400 MHz microprocessor," *IEEE Journal of Solid-State Circuits* Vol. 33, No. 4, April, 1998.

[41] P. J. Restle, Albert Ruehli and Steven G. Walker, "Dealing with inductance in high-speed chip design," *1999 Proceedings of Design Automation Conference*, June, 1999.

[42] E. B. Rosa and F. W. Grover, "Formulas and tables for the calculation of mutual and self- inductance," Government Printing Office, 1916.

[43] A. E. Ruehli," Inductance calculations in a complex integrated circuit environment", *IBM J. Research Development*, pp. 470-481, Sept. 1972.

[44] A. E. Ruehli,"Equivalent circuit models for three-dimensional multiconductor systems", *IEEE Trans. Microwave Theory and Tech.* Vol. 22, No. 3, March, 1974.

[45] S. Rusu, S. Tam, " Clock generation and distribution for the first IA-64 microprocessor", *Digest of 2000 IEEE International Conference on Solid-State Circuits Conference*, pp. 176-177, 2000.

[46] Y. Saad and M. H. Schultz, " GMRES: a generalized minimal residual algorithm for solving nonsymmetric linear systems," *SIAM J. Scientific and Statistical Computing*, Vol. 7, pp. 856-869, July, 1986.

[47] T. Sakurai and K. Tamaru, "Simple formulas for two- and three- dimensional capacitance", *IEEE Trans. Electron Devices*, vol. 30, pp. 183-185, 1983.

[48] Kenneth L. Shepard and Zhong Tian, "Return-limited inductance: a practical approach to on-chip inductance extraction," *IEEE Transactions on CAD of Integrated Circuits and Systems*, pp. 425-346, Vol. 19, No. 4, April 2000.

[49] P. P. Silverster and R. L. Ferrari, *Finite Elements for Electrical Engineers*, Cambridge University Press, 1983.

[50] Hugh H. Skilling, *Fundamentals of Electric Waves*, John Wiley & Sons, New York, NY, 1948.

[51] Hugh H. Skilling, *Electric Transmission Lines - Distributed Constants, Theory and Applications*, McGraw-Hill Book Company, Inc., New York, NY, 1951.

[52] http://www.sonnetusa.com.

[53] J. A. Tegopoulos and E. E. Kriezis, "Eddy currents in linear conducting media", Lesevier Science Publishing Company Inc., New York, NY, 1985.

[54] Charles S. Walker, *Capacitance, Inductance and Crosstalk Analysis*, Artech House, Boston, 1990.

[55] W. T. Weeks and L. L. Wu and M. F. McAllister and A. Singh, "Resistive and inductive skin effect in rectangular conductors", *IBM Journal of Research and Development*, Vol. 23, No. 6, Nov., 1979.

[56] T. Xanthopoulos, D. Balley, A. Gangwar, M. Gowan, A. Jain and B. Prewitt, "The design and analysis of the clock distribution network for a 1.2 GHz Alpha microprocessor," *2001 Digest of International Solid-State Circuit Conference*, pp.402-403, Feb., 2001.

[57] Tak Young, "Practical RC Extraction Techniques", *Tutorial-A at IEEE Design Automation Conference*, June, 1998.

[58] C. P. Yuan and T. N. Trick, "A simple formula for the estimation of the capacitance of two-dimensional interconnects in VLSI circuits", *IEEE Electron Device Lett.*, Vol, 3, pp. 391-393, 1982.

[59] A. H. Zemanian, R. R. Tewarson, C.P. Ju and J.F. Jen, "Three-dimension capacitance computations for VLSI/ULSI Interconnects", *IEEE Transactions on CAD of Integrated Circuits and Systems*, pp.1319-1326, Vol. 12, No. 8, 1989.

Chapter 4

DISTRIBUTED RC AND RLC TRANSIENT MODELS

J. Davis, A. Naeemi, and J. Meindl
School of Electrical and Computer Engineering
Georgia Institute of Technology

4.1 Introduction

Compact expressions for response time and crosstalk of coupled, distributed resistance capacitance (RC) networks driven by a step input source voltage with arbitrary source resistance are widely referenced in the current literature [1], [2], [3]. This chapter will show a derivation of the solutions for the PDEs that describe distributed RC and RLC networks. These models are derived from very regular IC interconnect structures, and can be used on a wide range of wire lengths and wire dimensions for high-speed semi-global and global interconnects.

For distributed RC models, general solutions of two and three coupled lines that includes the effects of a finite rise-time source voltage are presented. These solutions best describes the transient response of semi-global GSI wires where wire resistive effects are significant but inductive effects are not. A closed-form expression for the peak crosstalk voltage reveals a strong dependence of crosstalk on interconnect length, scaling, driver impedance, and material properties for high-speed interconnects with non-negligible driver switching time.

In addition, models for high-speed global interconnects must incorporate distributed self and mutual inductance to accurately estimate interconnect time delay and crosstalk in a multilevel network for multi-GHz GSI system [4]. This chapter also shows the derivation of compact solutions for worst-case time delay and crosstalk of two and three coupled wires that are modeled as a distributed *RLC* networks. Simplified expressions for the time delay and crosstalk are also presented. Finally, the impact of non-ideal return paths on the electrical characteristics of GSI interconnects are also discussed.

4.2 Distributed RC Models

Global and semi-global wires in current complex digital designs have electrical characteristics that are severely influenced by distributed resistance. As digital processors' clock frequencies move into the gigahertz regime, inductance will play an important role in long interconnects, but there will always be a range of wire lengths in a future GSI product that will be best modeled with a distributed RC network. This section will review a derivation of a complete model for single and coupled distributed RC networks.

4.2.1 Single Line Solution

To understand the transient characteristics of high-speed wires, we first examine a single line model that has only capacitive coupling to surrounding ground planes. For multi-conductor lines, this would describe the transient solution *if* all lines were switching simultaneously or if a line was shielded by a coplanar ground lines to reduce time delay variations and crosstalk. Therefore, this single line solution gives the best-case solution for a GSI wire with significant distributed resistance.

Transient Expressions. Using Kirchoff's voltage law (KVL) on a differential segment of a distributed RC line (see Figure 4.1) gives the following differential equation relating the current to the spatial derivative of the voltage as

$$I(x,t)r = -\frac{\partial}{\partial x}V(x,t) \tag{4.1}$$

where r is the resistance per unit length of the line.

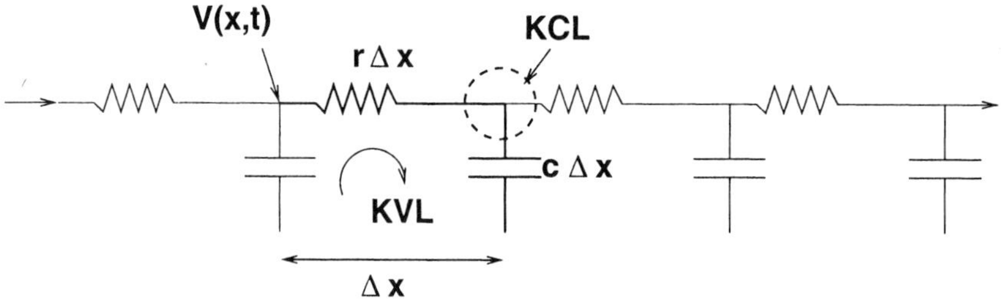

Figure 4.1. Distributed RC segment

Using Kirchoff's current law (KCL) on a differential segment of a distributed RC line (see Figure 4.1) gives the following differential equation relating the spatial derivative of the current to the time derivative of the voltage as

$$\frac{\partial}{\partial x}I(x,t) = -c\frac{\partial}{\partial t}V\left(x,t\right) \tag{4.2}$$

where c is the capacitance-per-unit length of the line.

Taking the spatial derivative of equation (4.1), and combining (4.1) with (4.2) gives the standard diffusion equation,

$$\frac{\partial^2}{\partial^2 x}V(x,t) = rc\frac{\partial}{\partial t}V\left(x,t\right). \tag{4.3}$$

To find a solution that is appropriate for GSI interconnects, an approximation to CMOS boundary conditions are used as seen in [3]. To approximate the boundary condition of a CMOS driver, the transistor output impedance is modeled as a linear resistor, R_s. The effect of the CMOS receiver on interconnect transients is captured by the inclusion of a load capacitance, C_L, that terminates the line.

In the Laplace domain, the solution to the voltage at the end of the line (at $x = L$) is a straight forward derivation. The full expression for a single line with non-zero source impedance, R_s, load capacitance, C_L, and $V_{in}(t)$ excitation voltage is given by

$$V(L,s) = \frac{V_{in}(s)}{(\sqrt{rcs}(\frac{C_L}{c} + \frac{R_s}{r})\sinh\sqrt{sRC} + \cosh(\sqrt{sRC})(R_sC_Ls + 1))} \tag{4.4}$$

where r is the distributed resistance per unit length, c is the distributed capacitance per unit length, L is the interconnect length, $R = rL$, and $C = cL$. Assuming that $V_{in}(s)$ is a finite ramp source voltage whose Laplace transform is given by $\frac{V_{dd}}{s^2 T_{rise}}\left(1 - e^{-sT_{rise}}\right)$, the Laplace domain voltage at the end of the line can be written as

$$\frac{V(L,s)}{V_{dd}} = \frac{1}{T(s)s^2}\left(1 - e^{-sT_{rise}}\right) \tag{4.5}$$

where $T(s)$ is equal to the denominator in (4.4). To determine the time domain expression of (4.5), use a partial fraction expansion of $\frac{1}{T(s)s^2}$, which is

$$\frac{1}{s^2 T\left(s\right)} = \frac{A_0}{s^2} + \frac{A_1}{s} + \frac{B_0}{s - b_0} + \frac{B_1}{s - b_1} + ... + \frac{B_k}{s - b_k}. \tag{4.6}$$

To find the residues (i.e. A_0, A_1, B_0,B_k, etc...) in (4.6) multiply both sides of (4.6) by $s^2 T\left(s\right)$ giving

$$1 = A_0 T\left(s\right) + A_1 s T\left(s\right) + s^2 T\left(s\right)\sum_{k=1}^{n}\frac{B_k}{s - b_k}. \tag{4.7}$$

Setting $s = 0$ in (4.7) gives A_o as

$$A_0 = \frac{1}{T\left(s=0\right)} = 1. \tag{4.8}$$

To solve for A_1 in (4.7), substitute a power series expansion for $T(s)$ (= $\sum_{i=1}^{i=\infty} t_i s^i$) into (4.7), which gives

$$1 = A_0 \left(\sum_{k=0}^{\inf} t_k s^k \right) + A_1 \left(\sum_{k=0}^{\inf} t_k s^{k+1} \right) + s^2 T(s) \sum_{k=1}^{n} \frac{B_k}{s - b_k}. \tag{4.9}$$

In (4.9) there are only two terms linear in s. Therefore, using the method of coefficients [5], the summation of the terms that are linear in s must equal zero, that is

$$A_0 t_1 s + A_1 t_o s = 0. \tag{4.10}$$

Letting $A_0 = 1$ and solving for A_1 gives

$$A_1 = -\frac{t_1}{t_o} \tag{4.11}$$

where t_0 is defined as the first term in the power series expansion of $T(s)$ and is equal to $T(0) = 1$. The parameter t_1 is the coefficient to the linear term in the power series expansion of $T(s)$, which is equal to $\frac{\partial T}{\partial s}\left(s=0\right)$. The derivative of $T(s)$ is given by

$$\begin{aligned} &\frac{\sinh\sqrt{sRC}}{2\sqrt{sRC}}(RC + RC_L + R_s C + RCR_s C_L s) \\ &+0.5\cosh\sqrt{sRC}(2R_s C_L + RC_L + R_s C). \end{aligned} \tag{4.12}$$

Evaluating (4.12) at $s = 0$ with L'Hospital's rule gives

$$t_1 = \frac{\partial T(s=0)}{\partial s} = R\left(C_L + \frac{1}{2}C\right) + R_s\left(C + C_L\right). \tag{4.13}$$

Making the substitution of (4.13) into (4.11) gives

$$A_1 = -\left(R\left(C_L + \frac{C}{2}\right) + R_s\left(C + C_L\right)\right). \tag{4.14}$$

To solve for B_k in (4.7), let the $s \to b_k$, where b_k is the kth root of $T(s)$, then (4.7) simplifies with L'Hospital's rule to

$$1 = \lim_{s \to b_k} s^2 T\left(s\right) \frac{B_k}{s - b_k} = b_k^2 \frac{\frac{\partial T}{\partial s}\left(s=b_k\right) B_k}{1}. \tag{4.15}$$

Therefore, B_k is given by

$$B_k = \frac{1}{b_k^2 \frac{\partial T(s=b_k)}{\partial s}}. \tag{4.16}$$

Finally, the kth root of T(s) (i.e. b_k) is obtained by setting $T(s) = 0$, or

$$\left(\sqrt{rcs}\left(\frac{C_L}{c} + \frac{R_s}{r}\right)\sinh\sqrt{sRC} + \cosh\left(\sqrt{sRC}\right)(R_sC_Ls + 1)\right) = 0. \tag{4.17}$$

This is a transcendental equation for the roots of T(s). The final implicit equation for b_k is given by

$$\tanh\sqrt{b_kRC} = -\frac{(R_sC_Lb_k + 1)}{\sqrt{RCb_k}\left(\frac{C_L}{C} + \frac{R_s}{R}\right)}. \tag{4.18}$$

This expression is the same expression that Sakurai encounters in his work [3]. Sakurai observes that all roots of $T(s)$ are negative; therefore, he defines an expression for the negative root to be

$$b_k = -\frac{\sigma_k}{RC} \tag{4.19}$$

where σ_k is always positive. Substituting (4.19) into (4.18) gives

$$\tanh j\sqrt{\sigma_k} = j\frac{(RC - \sigma_kR_sC_L)}{\sqrt{\sigma_k}(RC_L + R_sC)} \tag{4.20}$$

where $j = \sqrt{-1}$. Using $\tanh(jx) = j\tan(x)$, (4.20) becomes

$$\tan\sqrt{\sigma_k} = \frac{(RC - \sigma_kR_sC_L)}{\sqrt{\sigma_k}(RC_L + R_sC)}. \tag{4.21}$$

According to [3], the largest solution to this equation is

$$\sigma_0 = \frac{1.04RC}{R_sC_L + R_sC + RC_L + (2/\pi)^2RC}. \tag{4.22}$$

Using (4.16) and (4.22) an expression for B_0 is given by

$$B_o = -\frac{K_o}{\sigma_o}RC = \frac{RC}{\sigma_0}1.01\frac{R_sC + RC_L + RC}{R_sC + RC_L + \frac{\pi}{4}RC} \tag{4.23}$$

where K_0 is the residue for a step response excitation voltage and is approximated in [3].

The partial fraction expansion of the Laplace domain is now fully determined, and the time domain expansion of (4.6) can be approximated by

$$\frac{1}{s^2T(s)} \rightarrow\approx t - \left(R\left(C_L + \frac{1}{2}C\right) + R_s(C + C_L)\right) + B_0e^{-\frac{\sigma_0 t}{RC}}. \tag{4.24}$$

This time domain expansion, call it $P(t)$ for a moment, is related to the voltage at the end of a single distributed RC line by

$$V(t) = (P(t)u(t) - P(t - T_{rise})u(t - T_{rise}))\,. \tag{4.25}$$

Substituting (4.24) into (4.25) gives the final approximation for the voltage transients at the end of a single line as

$$\frac{V(t)}{V_{dd}} = \left(\begin{array}{c} \left(t - \left(R\left(C_L + \frac{C}{2}\right) + R_s\left(C + C_L\right)\right) + B_0 e^{-\frac{\sigma_O}{RC}t}\right) \frac{u(t)}{T_{rise}} \\ - \left(\begin{array}{c} t - T_{rise} - \left(R\left(C_L + \frac{C}{2}\right) + R_s\left(C + C_L\right)\right) \\ + B_0 e^{-\frac{\sigma_O}{RC}(t - T_{rise})} \end{array} \right) \frac{u(t-T_{rise})}{T_{rise}} \end{array} \right). \tag{4.26}$$

In the limit as T_{rise} approaches zero, (4.26) becomes $\frac{0}{0}$. Using L'Hos-pital rule, (4.26) is shown to converge to the step response solution as derived in [3], and is given by

$$\frac{V\,(t, L)}{V_{dd}} = \left(1 - 1.01 \frac{R_T + C_T + 1}{R_T + C_T + \frac{\pi}{4}} e^{-\frac{1.04\frac{t}{RC}}{\left(R_T C_T + R_T + C_T + \left(\frac{2}{\pi}\right)^2\right)}} \right). \tag{4.27}$$

The more general expression in (4.26) is compared to HSPICE simulation in Figure 4.2 for a single line with a 200Ω driver resistance and a 1 ns rise-time excitation voltage. The expression in (4.26) can also be written as a piecewise defined function for additional physical insight. The following expression corrects for a slight discontinuity between the two regions so that this function is continuous.

Region I: $t \leq T_{rise}$

$$\frac{V(t)}{V_{dd}} = \left(t - B_0(1 - e^{\frac{-\sigma_O t}{RC}})\right) \frac{1}{T_{rise}} \tag{4.28}$$

Region II: $t > T_{rise}$

$$\frac{V(t)}{V_{dd}} = \left(1 - B_o e^{\frac{-\sigma_O t}{RC}} \frac{(-1 + e^{\frac{\sigma_O T_{rise}}{RC}})}{T_{rise}}\right) \tag{4.29}$$

Time Delay Expressions. In this section, a general expression for the time, t_ν at which the end-of-line voltage is equal to νV_{dd} is presented. If $t_\nu \leq T_{rise}$ then (4.28) is used to find an implicit expression for t_ν given by

$$t_\nu + B_0 e^{\frac{-\sigma_0 t_\nu}{RC}} = T_{rise}\nu + B_0. \tag{4.30}$$

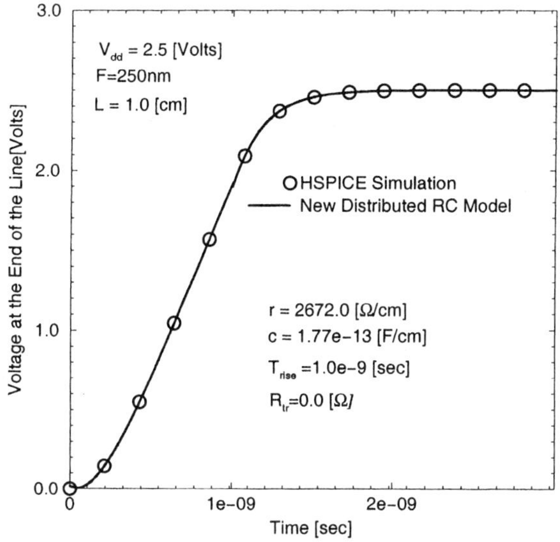

Figure 4.2. Closed-form expression for single RC line is compared to HSPICE simulation of 500 lumped elements with a 1ns rise time

For very long input rise times, t_ν simplifies to

$$t_\nu \approx T_{rise}\nu. \tag{4.31}$$

However, assuming that $t_\nu > T_{rise}$, then the time delay expression can be derived from (4.29) and is given by

$$t_\nu = \frac{RC}{\sigma_0} \ln[\frac{Bo(e^{\frac{-\sigma_0 T_{rise}}{RC}} - 1)}{(1-\nu)T_{rise}}]. \tag{4.32}$$

In the limit as T_{rise} approaches zero, (4.32) becomes

$$t_\nu = \left(R_s C_L + R_s C + R C_L + \left(\frac{2}{\pi}\right)^2 RC\right) \ln\left(\frac{1}{1-\nu}\right) \tag{4.33}$$

This is almost exactly the result of Sakurai; however, he introduces a slight correction to his model by adding $0.1RC$

$$t_\nu = \left(R_s C_L + R_s C + R C_L + \left(\frac{2}{\pi}\right)^2 RC\right) \ln\left(\frac{1}{1-\nu}\right) + 0.1RC \tag{4.34}$$

Adding this same factor to the above expression gives

$$t_\nu = \left(R_sC_L + R_sC + RC_L + \left(\frac{2}{\pi}\right)RC\right)\ln[\frac{Bo(e^{\frac{\sigma_0 T_{rise}}{RC}} - 1)}{(1-\nu)T_{rise}}] + 0.1RC. \tag{4.35}$$

Letting $\nu = 0.5$ in (4.35) gives a general expression for the 50% time delay of the distributed RC line as

$$t_{0.5} = \left(R_sC_L + R_sC + RC_L + \left(\frac{2}{\pi}\right)RC\right)\ln[\frac{Bo(e^{\frac{\sigma_0 T_{rise}}{RC}} - 1)}{0.5T_{rise}}] + 0.1RC. \tag{4.36}$$

For $T_{rise} = 0$ the 50% time delay ($\nu = 0.5$) then becomes

$$t_{0.5} = 0.69315\,(R_sC_L + R_sC + RC_L) + 0.38092RC. \tag{4.37}$$

4.2.2 Coupled Line Solutions

Transient Expression. In general, signal propagation along highly resistive and parallel interconnects can be modeled as a distributed RC network. The partial differential equations that describe the voltage at any time, t, and any position,x, along a distributed RC network is given by

$$\frac{\partial^2 \tilde{V}}{\partial x^2} = r\,[C]\,\frac{\partial \tilde{V}}{\partial t} \tag{4.38}$$

where $\tilde{V} = \begin{bmatrix} V_1 \\ V_2 \\ \vdots \\ \vdots \\ V_n \end{bmatrix}$, $[C]$ is the capacitance matrix of the n-conductor system, and r is the distributed resistance per unit length of each wire. The eigenvectors and the eigenvalues of the capacitance matrix can be used to find the transformations to decouple the system of PDEs in (4.38). For example, multiplying (4.38) by the transpose of a constant column vector, $\tilde{e}$, gives

$$\frac{\partial^2\left(\tilde{e}^T\tilde{V}\right)}{\partial x^2} = r\left(\tilde{e}^T\,[C]\right)\frac{\partial \tilde{V}}{\partial t}. \tag{4.39}$$

Using the matrix identity $AB = (B^TA^T)^T$ (4.39) can be written as

$$\frac{\partial^2\left(\tilde{e}^T\tilde{V}\right)}{\partial x^2} = r\left([C]^T\,\tilde{e}\right)^T\frac{\partial \tilde{V}}{\partial t}. \tag{4.40}$$

If we assume that that $\tilde{e}$ is one of the n eigenvectors of the nxn capacitance matrix, then this is also the eigenvector of the transpose of the capacitance matrix because of the symmetry of this matrix (e.g. $c_{ij} = c_{ji}$). Therefore, (4.40) simplifies to

$$\frac{\partial^2 \left(\tilde{e}^T \tilde{V}\right)}{\partial x^2} = r \left(\lambda \tilde{e}\right)^T \frac{\partial \tilde{V}}{\partial t}. \tag{4.41}$$

With slight algebraic manipulation this expression becomes

$$\frac{\partial^2 \left(\tilde{e}^T \tilde{V}\right)}{\partial x^2} = r\lambda \frac{\partial \left(\tilde{e}^T \tilde{V}\right)}{\partial t}. \tag{4.42}$$

The product of $\tilde{e}^T \tilde{V}$ gives the transformation that decouples the couple PDEs in (4.38). Each resulting decoupled partial differential equation has the same form as the single line solution with an altered capacitance value, which is given by the corresponding eigenvalue of the capacitance matrix.

Two Conductor Example. For the two conductor example, we will focus for a moment on the crosstalk between two interconnects. Therefore, the initial voltage of both lines in this example are zero. At time $t = 0$, the active line begins to charge to a high voltage, while the quiescent line tries to maintain a low voltage. Capacitive coupling will cause unwanted charge to build up on the quiescent line, and the peak voltage at the end of the quiescent line will be referred to as the crosstalk voltage. The partial differential equations that describe two coupled distributed RC networks are given in matrix notation as

$$\frac{\partial^2}{\partial x^2} \begin{bmatrix} V_1(x,t) \\ V_2(x,t) \end{bmatrix} = r \begin{bmatrix} c_g + c_m & -c_m \\ -c_m & c_g + c_m \end{bmatrix} \frac{\partial}{\partial t} \begin{bmatrix} V_1(x,t) \\ V_2(x,t) \end{bmatrix} \tag{4.43}$$

where c_g is the ground capacitance of each wire and c_m is the mutual capacitance between each wire.

Two unnormalized eigenvectors of this capacitance matrix are

$$e_1 = \begin{bmatrix} 1 \\ 1 \end{bmatrix} e_2 = \begin{bmatrix} 1 \\ -1 \end{bmatrix} \tag{4.44}$$

The corresponding eigenvalues are given by $\lambda_1 = c_g$ and $\lambda_2 = c_g + 2c_m$. This leads to two transformation that decouple the PDEs in (4.3). The two transformations are

$$V_+ = \tilde{e}_1^T \begin{bmatrix} V_1 \\ V_2 \end{bmatrix} = V_1 + V_2 \tag{4.45}$$

$$V_- = \tilde{e}_2^T \begin{bmatrix} V_1 \\ V_2 \end{bmatrix} = V_1 - V_2 \tag{4.46}$$

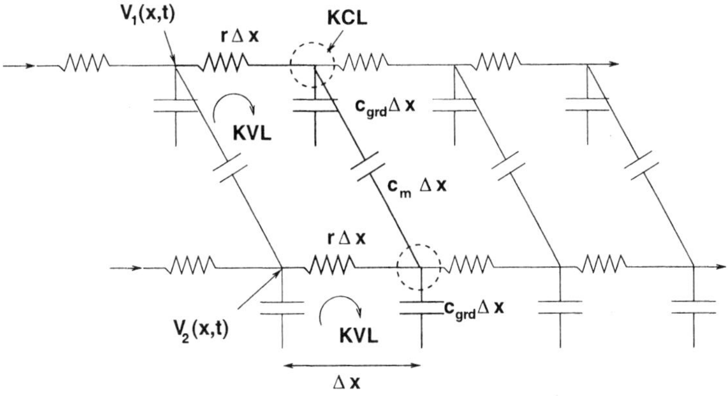

Figure 4.3. Coupled Differential Segment using RC model

where V_+ and V_- are referred to as the plus and minus mode, respectively. The partial differential equations for each mode are described by

$$\frac{\partial^2}{\partial x^2} V_+ (x,t) = rc_g \frac{\partial}{\partial t} V_+ (x,t) . \tag{4.47}$$

and

$$\frac{\partial^2}{\partial x^2} V_- (x,t) = r (c_g + 2c_m) \frac{\partial}{\partial t} V_- (x,t) \tag{4.48}$$

The boundary conditions for the plus and minus modes are similar to the single line boundary conditions. The active line is driven by a ramped input response and the quiet line is connected through an transistor resistance to ground; therefore, the the boundary conditions at $x = 0$ for the plus and minus modes are given by

$$V_+ (x = 0, t) = V_1 (x = 0, t) + V_2 (x = 0, t) \tag{4.49}$$

$$V - (x = 0, t) = V_1 (x = 0, t) - V_2 (x = 0, t) \tag{4.50}$$

The boundary conditions of the plus and minus modes at $x = L$ are given by

$$I_+(x = L, t) = I_1 (x = L, t) + I_2(x = L, t) \tag{4.51}$$

$$I_-(x = L, t) = I_1 (x = L, t) - I_2(x = L, t) \tag{4.52}$$

Because the partial differential equations and the boundary conditions have the same form of a single line solution with adjusted values of the capacitance, the solution found in the previous section is used to determine the transient voltages at the ends of the active and quiescent lines using the following two relationships

$$V_1 = \frac{V(t, C = C_g, B_0 = B_0^+) + V(t, C = C_g + 2C_m, B_0 = B_0^-)}{2} \tag{4.53}$$

$$V_2 = \frac{V(t, C = C_g, B_0 = B_0^+) - V(t, C = C_g + 2C_m, B_0 = B_0^-)}{2} \tag{4.54}$$

where the function V is defined in (4.26), C is the total lumped ground capacitance, C_m is the total lumped mutual capacitance, $C^- = C + 2Cm$, and B_0^+ is

$$B_o^+ = 1.01 \frac{1}{\sigma_o^+} \frac{R_s C + RC_L + RC}{\frac{R_s}{R} + \frac{C_L}{C} + \frac{\pi}{4}}; \tag{4.55}$$

B_0^- is

$$B_o^- = 1.01 \frac{1}{\sigma_o^-} \frac{R_s C^- + RC_L + RC^-}{\frac{R_s}{R} + \frac{C_L}{C^-} + \frac{\pi}{4}}; \tag{4.56}$$

σ_o^+ is

$$\sigma_o^+ = \frac{(1.04)\, RC}{R_s C_L + R_s C + RC_L + RC(2/\pi)^2}; \tag{4.57}$$

and σ_o^- is:

$$\sigma_o^- = \frac{(1.04)\, RC^-}{R_s C_L + R_s C^- + RC_L + (2/\pi)^2 RC^-}; \tag{4.58}$$

Simplifying (4.54), the transient noise voltage at the end of a quiet line is piecewise defined and written in the following form:

Region I: $t \leq T_{rise}$

$$\frac{V_2}{V_{dd}} = \frac{C_m\,(R + 2R_s) + B_0^+ e^{-\frac{\sigma_o^+}{RC}t} - B_0^- e^{-\frac{\sigma_o^-}{RC^-}t}}{2T_{rise}} \tag{4.59}$$

Region II: $t > T_{rise}$

$$\frac{V_2}{V_{dd}} = \frac{-B_0^- e^{-\frac{\sigma_o^-}{RC^-}t}(-e^{\frac{\sigma_o^-}{RC^-}T_{rise}} + 1) + B_0^+ e^{-\frac{\sigma_o^+}{RC}t}(1 - e^{\frac{\sigma_o^+}{RC}T_{rise}})}{2T_{rise}} \tag{4.60}$$

This compact expression is compared to HSPICE simulation of 500 lumped elements in Figure 4.4.

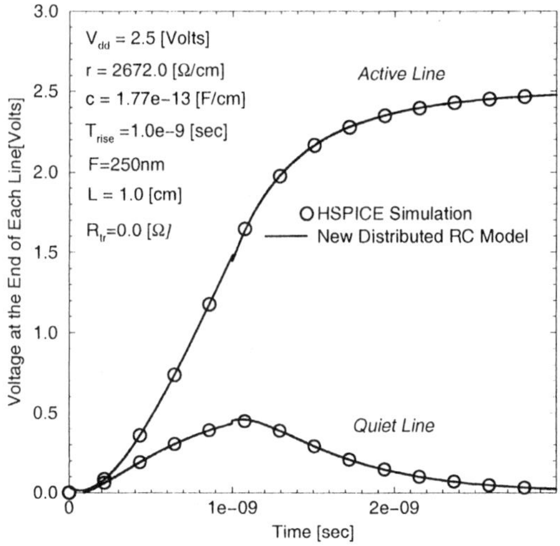

Figure 4.4. New closed-form expressions for coupled RC lines are compared to HSPICE simulation of 500 lumped elements

Three Conductor Example. With three parallel conductors there are two interesting switching conditions that can occur. First, the worst-case crosstalk occurs when the outer two lines switch high while the inner line remains quiescent. The worst- case time delay on the inner line occurs when it switches from low to high and the outer lines switch from high to low.

The differential equations that describe the transient voltage along the three conductors are written in matrix form [6] as

$$\frac{\partial^2}{\partial x^2}\begin{bmatrix} V_1(x,t) \\ V_2(x,t) \\ V_3(x,t) \end{bmatrix} = r \begin{bmatrix} \sum\limits_{i=0,2,3} c_{1i} & -c_{12} & -c_{13} \\ -c_{21} & \sum\limits_{i=0,1,3} c_{2i} & -c_{23} \\ -c_{31} & -c_{32} & \sum\limits_{i=0,1,2} c_{2i} \end{bmatrix} \frac{\partial}{\partial t}\begin{bmatrix} V_1(x,t) \\ V_2(x,t) \\ V_3(x,t) \end{bmatrix} \tag{4.61}$$

where r is the distributed resistance of each line, and c_{10}, c_{20}, and c_{30} are the ground capacitances of each conductor, and c_{xy}, where $y \neq 0$ or $y \neq x$ is the capacitance between the xth and y^{th} conductor.

To simplify the above equation, it is first assumed that $c_g = c_{10} = c_{20} = c_{30}$ and $c_m = c_{12} = c_{21} = c_{23} = c_{32}$. These conditions are satisfied because of the geometrical symmetry of the three conductors in this model. From the previous section, we know that we can decoupled these PDEs by using a transformation that comes directly from the eigenvalue and eigenvectors of the capacitance matrix. Three eigenvectors of this 3x3 capacitance matrix are

$$e_1 = \begin{bmatrix} 1 \\ 1 \\ 1 \end{bmatrix} e_2 = \begin{bmatrix} -1 \\ 0 \\ 1 \end{bmatrix} e_3 = \begin{bmatrix} 1 \\ -2 \\ 1 \end{bmatrix} \quad (4.62)$$

and the corresponding eigenvalues are $\lambda_1 = c_g$, $\lambda_2 = c_g + c_m$, and $\lambda_3 = c_g + 3c_m$.

In addition, for the worst-case conditions we assume that $V_1(x,t) = V_3(x,t)$. This condition is satisfied by having identical boundary and initial conditions on the outer two conductors. To emphasize the distinction between the outer conductors and the inner conductor, we define $V_o(x,t)$ and $V_i(x,t)$ as

$$V_o(x,t) = V_1(x,t) = V_3(x,t) \quad (4.63)$$

$$V_i(x,t) = V_2(x,t) \quad (4.64)$$

With this additional simplification, the two transformations that decouple the PDEs are

$$V_{sum}(x,t) = V_i(x,t) + 2V_o(x,t) \quad (4.65)$$

$$V_{diff}(x,t) = V_o(x,t) - V_i(x,t). \quad (4.66)$$

Because the boundary conditions on these two lines are also symmetric, then known single line solutions are used for $V_{sum}(x,t)$ and $V_{diff}(x,t)$.

The solutions for the inner and outer conductor are found by

$$V_o = \frac{V_{sum} + V_{diff}}{3} \quad (4.67)$$

and

$$V_i = \frac{V_{sum} - 2V_{diff}}{3}. \quad (4.68)$$

To illustrate the symmetry in the boundary conditions, consider the boundary conditions for the inner and outer conductors with zero source impedance. For worst-case crosstalk, the outer two conductors switch high and the inner conductor stays low.

$$V_o(0,t) = V_s(t) \quad (4.69)$$

$$V_i(0,t) = 0 \quad (4.70)$$

Using (4.65) and (4.66) gives

$$V_{sum}(0,t) = 2V_{in}(t) \quad (4.71)$$

and

$$V_{diff}(0,t) = V_{in}(t) \quad (4.72)$$

At $x = L$ all currents of the physical wires are assumed to be zero, which means that the sum and difference modes' currents at $x = L$ are zero as well.

As illustrated, the boundary conditions for the plus and minus mode are the same as a single line with an input voltage of $V_{in}(t)$ for the minus mode and $2V_{in}(t)$ for the plus mode. The worst-case crosstalk transient expressions for the three conductor two ground plane model come directly from the single line solutions with the aforementioned boundary conditions. The transient expression for the inner quiet line with the outer two conductors switching high is given by

$$V_Q(L,t) = \frac{2}{3}\left(V_{fin}\left[L,t,c=2c_g\right] - V_{fin}\left[L,t,c=2c_g-3c_m\right]\right). \quad (4.73)$$

The worst-case time delay conditions occur when the inner line is switching from low to high and the outer two lines are switching from high to low. Using these new boundary and initial conditions gives the following transient expression

$$V_A(L,t) = \frac{4}{3}V_{fin}\left[L,t,c=2c_g+3c_m\right] - \frac{1}{3}V_{fin}\left[L,t,c=2c_g\right]. \quad (4.74)$$

These expressions are compared to HSPICE simulation of 500 lumped elements in Figure 4.5.

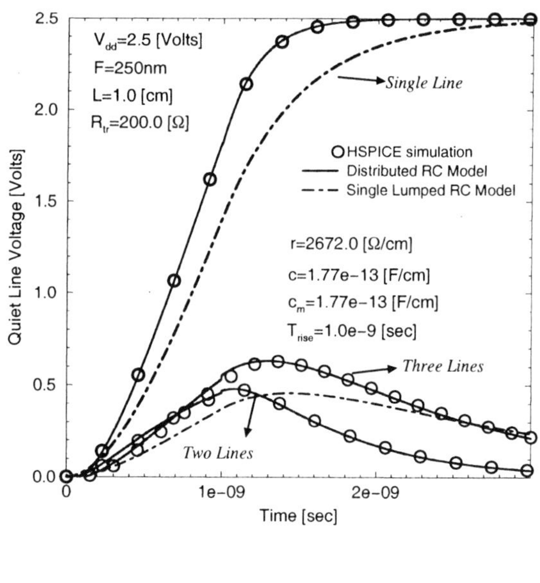

Figure 4.5. New closed-form expressions for coupled RC lines are compared to HSPICE simulation of 500 lumped elements and lumped RC models

Peak Crosstalk Voltage. Assuming that the peak noise voltage occurs *at a time greater than* T_{rise}, then the peak noise voltage is determined from taking the derivative of (4.60). Setting the derivative equal to zero gives

$$\frac{d}{dt}V_2 = \frac{V_{dd}}{2T_{rise}}\left[\begin{array}{c} \frac{\sigma_O^-}{RC^-}e^{-\frac{\sigma_O^-}{RC^-}t}B_0^-\left(-e^{\frac{\sigma_O^-}{RC^-}T_{rise}}+1\right) \\ -\frac{\sigma_O^+}{RC}e^{-\frac{\sigma_O^+}{RC}t}B_0^+\left(-e^{\frac{\sigma_O^+}{RC}T_{rise}}+1\right)\end{array}\right] = 0. \qquad (4.75)$$

Solving for t in (4.75) gives the time at which the crosstalk voltage is at a maximum value and is given by

$$t_{peak} = \frac{1}{\left(\frac{\sigma_O^+}{RC}-\frac{\sigma_O^-}{RC^-}\right)}\ln\left[\frac{C^-}{C}\frac{\sigma_o^+}{\sigma_o^-}\frac{B_0^+\left(-e^{\frac{\sigma_O^+}{RC}T_{rise}}+1\right)}{B_0^-\left(-e^{\frac{\sigma_O^-}{RC^-}T_{rise}}+1\right)}\right]. \qquad (4.76)$$

Substituting (4.76) into (4.60) gives the general peak crosstalk voltage expression, which is

$$\frac{V_{peak}}{V_{dd}} = \frac{\chi B_o^+(1-e^{\frac{\sigma_O^+T_{rise}}{L^2rc^+}})(1-\frac{\sigma_o^+c^-}{\sigma_o^-c^+})}{T_{rise}}\left(\frac{\sigma_o^+c^-B_o^+(1-e^{\frac{T_{rise}\sigma_O^+}{L^2rc^+}})}{\sigma_o^-c^+B_o^-(1-e^{\frac{T_{rise}\sigma_O^-}{L^2rc^-}})}\right)^{\frac{-\frac{\sigma_O^+}{c^+}}{(\frac{\sigma_O^+}{c^+}-\frac{\sigma_O^-}{c^-})}} \qquad (4.77)$$

where $B_o^{\pm} = B_o\,(c_s = c^{\pm})$, $\sigma_o^{\pm} = \sigma_o\,(c_s = c^{\pm})$, $\chi = \frac{1}{2}$,$c^+ = c_g$, and $c^- = c_g + 2c_m$ for the peak crosstalk voltage between two parallel lines; and $\chi = \frac{2}{3}$, $c^+ = 2c_g$, and $c^- = 2c_g + 3c_m$ for the worst-case peak crosstalk voltage between three parallel lines.

Example of $T_{rise} \approx 0$. If T_{rise} is equal to zero, then (4.77) converges to an expression that has been derived by [3] and is given by

$$\frac{V_p}{E_1} = 1.01\frac{R_sC+RC_L+RC}{R_sC+RC_L+\frac{\pi RC}{4}}\left(\frac{C}{(C+2C_c)}\right)^{\frac{1}{2}\frac{C}{C_c}}\left[\frac{C_c}{C+2C_c}\right] \qquad (4.78)$$

It is also shown in [3] that by making $R_s = 0$ and $C_L = 0$ the (4.78) can simplify to

$$\frac{V_p}{E_1} = 1.01\frac{R_sC+RC_L+RC}{R_sC+RC_L+\frac{\pi RC}{4}}\left(\frac{C}{(C+2C_c)}\right)^{\frac{1}{2}\frac{C}{C_c}}\left[\frac{C_c}{C+2C_c}\right] \approx \frac{1}{2}\frac{C_c}{C+C_c}. \qquad (4.79)$$

Moreover, additional approximations to (4.77) bring physical insight into the noise problem for semi-global GSI wires. Assuming that the rise time is much greater than the intrinsic wire charging time (i.e. $t_{peak} > T_{rise} > rL\left(c + 2c_m\right)L$), then (4.77) becomes

$$V_{peak} = \beta \frac{V_{dd}}{T_{rise}} 1.01 \left(\frac{R_s + rL\left(\frac{2}{\pi}\right)^2}{1.04rL} \right) \frac{rL + R_s}{\frac{R_s}{rL} + \frac{\pi}{4}} \left(C_m\right), \tag{4.80}$$

where $\beta = 1$ for the two parallel lines and $\beta = 2$ for three parallel lines. If the driver resistance is negligible, then (4.80) becomes

$$V_{peak} = 0.501\beta \frac{V_{dd}}{T_{rise}} \left(rc_m\right) L^2. \tag{4.81}$$

On the other hand, if the driver resistance is significantly larger than the interconnect resistance then (4.80) becomes

$$V_{peak} = 0.971\beta \frac{V_{dd}}{T_{rise}} R_s c_m L. \tag{4.82}$$

The result in (4.81) is compared to the exact expression (4.77) and HSPICE simulation in Figure 4.6.

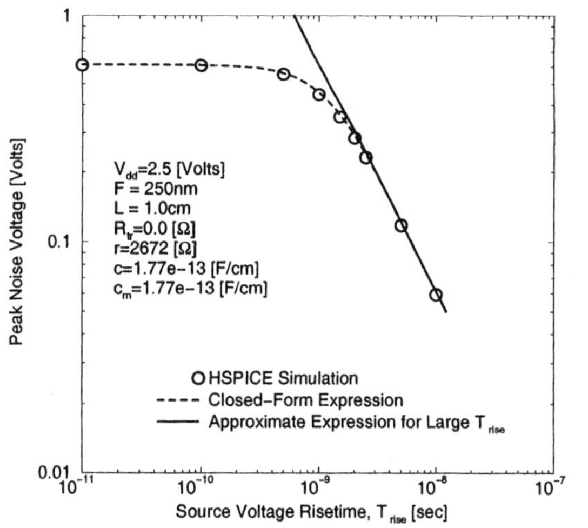

Figure 4.6. New peak crosstalk voltage in (4.77) and approximate expression in (4.81) compared to HSPICE simulation of 500 lumped elements

4.3 Distributed RLC Models

High-speed global interconnect models must incorporate self and mutual inductance to accurately estimate interconnect time delay and crosstalk for global wires in multi-GHz GSI systems. Compact solutions for the transient response of the PDEs that describe the worst-case time delay and crosstalk of two and three coupled distributed *RLC* interconnect models are rigorously derived in this section with the assumption that the voltage rise time is much smaller than the time of flight of the signal. In addition, simple closed-form expressions for time delay, crosstalk, and overshoot are presented.

4.3.1 Single Line Solution

Significant work has been done by [3] to develop simplified expressions for the step response of distributed RC networks. The significant contribution of the solutions in [3] is to provide physical insight and enable rapid calculations. In this section, a compact single line solution for distributed RLC networks that model high-speed global wires are presented.Specifically, the solution of the transient voltage along a semi-infinite line is first derived. This solution is then used to determine the transient response of a single distributed RLC line of finite length.

Semi-Infinite Line. The transient response of a single semi-infinite distributed RLC interconnect with arbitrary source impedance is first determined. The partial differential equation (PDE) that describes the voltage along a single distributed *RLC* line is given by

$$\frac{\partial^2}{\partial x^2}V(x,t) = lc\frac{\partial^2}{\partial t^2}V(x,t) + rc\frac{\partial}{\partial t}V(x,t) \tag{4.83}$$

where r is the distributed resistance per unit length, l is the distributed inductance per unit length, and c is the distributed capacitance per unit length.

Using a single-sided Laplace transform of $V(x,t)$ and assuming $V(x,t=0)=0$, the partial differential equation in (4.83) becomes an ordinary differential equation and is given by

$$\frac{\partial^2}{\partial x^2}V(x,s) = lcs^2V(x,s) + rcsV(x,s). \tag{4.84}$$

The general solution to (4.84) in the Laplace domain is

$$V(x,s) = Ae^{-x\sqrt{lc}\sqrt{s\left(s+\frac{r}{l}\right)}} + Be^{x\sqrt{lc}\sqrt{s\left(s+\frac{r}{l}\right)}}. \tag{4.85}$$

For a semi-infinite line, B must be zero so that the solution of (4.85) is well-behaved and finite at infinity. The coefficient A is determined from the boundary condition at $x=0$ where $V_{inf}(x=0,s)$ is equal to the input voltage, $V_{in}(s)$, minus the voltage across the source impedance given by

$$V(x=0,s) = V_{in}(s) - R_sI(x=0,s) \tag{4.86}$$

where $V_{in}(s) = \frac{Vdd}{s}$ for a step input response, R_s is the source impedance, and $I(x=0,s)$ is the current at the beginning of the line. The expression for the current is given from a telegrapher's equation as

$$-\frac{\partial}{\partial x}V\left(x,t\right) = I\left(x,t\right)r + l\frac{\partial}{\partial t}I\left(x,t\right). \tag{4.87}$$

Using (4.87), the current in the Laplace domain is given by

$$I\left(x,t\right) = \frac{-\frac{\partial}{\partial x}V\left(x,t\right)}{\left(r+ls\right)} = A\frac{\sqrt{lc}\sqrt{s\left(s+\frac{r}{l}\right)}}{\left(r+ls\right)}e^{-x\sqrt{lc}\sqrt{s\left(s+\frac{r}{l}\right)}}. \tag{4.88}$$

Substituting (4.85) and (4.88) into (4.86) gives

$$A = \frac{V_{dd}}{s} - R_s A\frac{\sqrt{lc}\sqrt{s\left(s+\frac{r}{l}\right)}}{\left(r+ls\right)}. \tag{4.89}$$

Solving for A in (4.89) and making a substitution into (4.85) gives the voltage in the Laplace domain as

$$V\left(x,s\right) = \frac{V_{dd}}{s}\frac{Z\left(s\right)}{Z\left(s\right)+R_s}e^{-x\sqrt{lc}\sqrt{s\left(s+\frac{r}{l}\right)}} \tag{4.90}$$

where Z(s) is the lossy characteristic impedance and is given by

$$Z\left(s\right) = \sqrt{\frac{r+sl}{sc}} = Z_o\sqrt{\frac{s+\frac{r}{l}}{s}} \tag{4.91}$$

and $Z_o = \sqrt{\frac{l}{c}}$ is the lossless characteristic impedance. Substituting (4.91) into (4.90) gives the voltage along the semi-infinite line in the Laplace domain

$$V\left(x,s\right) = \frac{V_{dd}}{s}\frac{Z_o\sqrt{\frac{s+\frac{r}{l}}{s}}}{Z_o\sqrt{\frac{s+\frac{r}{l}}{s}}+R_s}e^{-x\sqrt{lc}\sqrt{s\left(s+\frac{r}{l}\right)}} \tag{4.92}$$

To determine a time domain solution, the following transformation is used [5]

$$\frac{\sigma^{\nu}e^{-k\sqrt{s^2-\sigma^2}}}{\sqrt{s^2-\sigma^2}\left(\sqrt{s^2-\sigma^2}+s\right)^{\nu}} \rightarrow \left(\frac{t-k}{t+k}\right)^{\frac{\nu}{2}} I_{\nu}\left(\sigma\sqrt{t^2-k^2}\right)u_o\left(t-k\right) \tag{4.93}$$

where $u_0(t)$ is a unit step function and $I_\nu[-]$ is a νth order modified Bessel function. To get (4.92) in a form that the transformation in (4.93) can be used, (4.92) is rearranged to be

$$V\left(x,s\right) = \frac{V_{dd}}{s}\frac{e^{-x\sqrt{lc}\sqrt{s\left(s+\frac{r}{l}\right)}}}{\sqrt{s\left(s+\frac{r}{l}\right)}}\sqrt{s\left(s+\frac{r}{l}\right)}\frac{Z_o\sqrt{\frac{s+\frac{r}{l}}{s}}}{Z_o\sqrt{\frac{s+\frac{r}{l}}{s}}+R_s} \tag{4.94}$$

Two transformations used by Oliver Heaviside in [7] are used to obtain the time domain solution. The first transformation is given by

$$s \to s - \sigma. \tag{4.95}$$

Performing this transformation on s in the function $V(x,s)$ produces a new function $V'(x, s-\sigma)$ and is related in the time domain to $V(x,t)$ by

$$V(x,t) = e^{-\sigma t} V'(x,t). \tag{4.96}$$

Using the transformation in (4.95) on (4.94) and simplifying gives

$$V'(x,s) = V_{dd} \frac{e^{-x\sqrt{lc}\sqrt{s^2-\sigma^2}}}{\sqrt{s^2-\sigma^2}} \left[\frac{s+\sigma}{s-\sigma} \frac{Z_o}{Z_o\sqrt{\frac{s+\sigma}{s-\sigma}} + R_s} \right]. \tag{4.97}$$

The second transformation that is helpful for finding the time domain solution is given by

$$s = \frac{\sigma}{2}\left(a + a^{-1}\right) \tag{4.98}$$

Making a temporary substitution of (4.98) into the bracketed terms in (4.97) gives

$$V'(x,s) = V_{dd} \frac{e^{-x\sqrt{lc}\sqrt{s^2-\sigma^2}}}{\sqrt{s^2-\sigma^2}} \left[\frac{\frac{\sigma}{2}\left(a+a^{-1}\right)+\sigma}{\frac{\sigma}{2}\left(a+a^{-1}\right)-\sigma} \frac{Z_o}{Z_o\sqrt{\frac{\frac{\sigma}{2}(a+a^{-1})+\sigma}{\frac{\sigma}{2}(a+a^{-1})-\sigma}} + R_s} \right]. \tag{4.99}$$

Simplifying (4.99) gives

$$V'(x,s) = V_{dd} \frac{e^{-x\sqrt{lc}\sqrt{s^2-\sigma^2}}}{\sqrt{s^2-\sigma^2}} \frac{(a+1)^2}{(a-1)^2} \frac{Z_o}{Z_o\frac{(a+1)}{(a-1)} + R_s}. \tag{4.100}$$

The final expression in terms of s and a is given by

$$V'(x,s) = V_{dd} \frac{Z_o}{(Z_o+R_s)} \frac{e^{-x\sqrt{lc}\sqrt{s^2-\sigma^2}}}{\sqrt{s^2-\sigma^2}} \left\{ \frac{(a+1)^2}{a^2 - 2\frac{R_s}{(Z_o+R_s)}a + \frac{R_s-Z_o}{Z_o+R_s}} \right\}. \tag{4.101}$$

Factoring the term in the curly brackets in (4.101) leads to

$$V'(x,s) = V_{dd} \frac{Z_o}{(Z_o+R_s)} \frac{e^{-x\sqrt{lc}\sqrt{s^2-\sigma^2}}}{\sqrt{s^2-\sigma^2}} \left\{ \frac{(a+1)^2}{(a-a_+)(a-a_-)} \right\}. \tag{4.102}$$

Using a partial fraction expansion of the term in the curly bracket gives

$$V'(x,s) = \begin{array}{l} V_{dd}\frac{Z_o}{(Z_o+R_s)}\frac{e^{-x\sqrt{lc}\sqrt{s^2-\sigma^2}}}{\sqrt{s^2-\sigma^2}}\times \\ \left\{1+\frac{(a_{+}+1)^2}{(a_{+}-a_{-})(a-a_{+})}-\frac{(a_{-}+1)^2}{(a_{+}-a_{-})(a-a_{-})}\right\} \end{array} \quad (4.103)$$

where

$$a_{\pm} = \frac{R_s \pm Z_o}{Z_o + R_s}. \quad (4.104)$$

Making the substitution of (4.104) into (4.103) and letting $\Gamma = \frac{R_s - Z_o}{R_s + Z_o}$ simplifies to

$$V'(x,s) = \begin{array}{l} V_{dd}\frac{Z_o}{(Z_o+R_s)}\frac{e^{-x\sqrt{lc}\sqrt{s^2-\sigma^2}}}{\sqrt{s^2-\sigma^2}}\times \\ \left\{1+\frac{1}{\left(\frac{2Z_o}{Z_o+R_s}\right)}\left(\frac{4}{(a-1)}-\frac{\left(\frac{2R_s}{Z_o+R_s}\right)^2}{(a-\Gamma)}\right)\right\}. \end{array} \quad (4.105)$$

To determine the power series representation of (4.105) in a, the following series definition is used

$$\frac{1}{1-x} = \sum_{k=0}^{k=\infty} (x)^k. \quad (4.106)$$

Therefore (4.105) can simplify to

$$\begin{aligned} \frac{V'(x,s)}{V_{dd}} &= \frac{Z_o}{(Z_o+R_s)}\frac{e^{-x\sqrt{lc}\sqrt{s^2-\sigma^2}}}{\sqrt{s^2-\sigma^2}}\times \\ &\left\{1+\frac{1}{1-\Gamma}\sum_{k=1}^{k=\infty}\left(\frac{1}{a}\right)^k\left(4-(1+\Gamma)^2\Gamma^{k-1}\right)\right\}. \end{aligned} \quad (4.107)$$

Solving for a in (4.98) gives

$$a = \frac{1}{\sigma}\left(s+\sqrt{s^2-\sigma^2}\right). \quad (4.108)$$

Making this substitution into (4.107) gives the final usable form of the solution in the Laplace domain and is given by

$$\begin{aligned} \frac{V'(x,s)}{V_{dd}} &= \frac{Z_o}{(Z_o+R_s)}\frac{e^{-x\sqrt{lc}\sqrt{s^2-\sigma^2}}}{\sqrt{s^2-\sigma^2}}\times \\ &\left\{1+\frac{1}{1-\Gamma}\sum_{k=1}^{k=\infty}\frac{\sigma^k\left(4-(1+\Gamma)^2\Gamma^{k-1}\right)}{\left(s+\sqrt{s^2-\sigma^2}\right)^k}\right\} \end{aligned} \quad (4.109)$$

Using the transformation presented in (4.93) on (4.109) gives the time domain expression $V'(x,t)$ as

$$\frac{V'(x,t)}{V_{dd}} = \frac{Z_o u_o(t - x\sqrt{lc})}{(Z_o + R_s)} \times \tag{4.110}$$
$$\left\{ \begin{array}{l} I_0\left(\sigma\sqrt{t^2 - \left(x\sqrt{lc}\right)^2}\right) \\ +\frac{1}{1-\Gamma}\sum\limits_{k=1}^{k=\infty}\left(\frac{t-x\sqrt{lc}}{t+x\sqrt{lc}}\right)^{\frac{k}{2}} I_k\left(\sigma\sqrt{t^2 - \left(x\sqrt{lc}\right)^2}\right)\left(4 - (1+\Gamma)^2\,\Gamma^{k-1}\right) \end{array} \right\}$$

Because of the transformation $s \to s - \sigma$, (4.96) is utilized to determine the final expression for the voltage at a position x down the semi-infinite line, which is given by

$$\frac{V_{inf}(x,t)}{V_{dd}} = \frac{Z_o u_o(t - x\sqrt{lc})e^{-\frac{r}{2l}t}}{(Z_o + R_s)} u_o\left(t - x\sqrt{lc}\right) \times \tag{4.111}$$
$$\left\{ \begin{array}{l} I_0\left(\sigma\sqrt{t^2 - \left(x\sqrt{lc}\right)^2}\right) \\ +\frac{1}{1-\Gamma}\sum\limits_{k=1}^{k=\infty}\left(\frac{t-x\sqrt{lc}}{t+x\sqrt{lc}}\right)^{\frac{k}{2}} I_k\left(\sigma\sqrt{t^2 - \left(x\sqrt{lc}\right)^2}\right)\left(4 - (1+\Gamma)^2\,\Gamma^{k-1}\right) \end{array} \right\}$$

Letting the resistance in (4.111) approach zero ($r = 0$) gives

$$\frac{V_{\text{inf}}(x,t)}{V_{dd}} = \frac{Z_o u_o(t - x\sqrt{lc})}{(Z_o + R_s)} \times \tag{4.112}$$
$$\left\{ \begin{array}{l} I_0\,(0) \\ +\frac{1}{1-\Gamma}\sum\limits_{k=1}^{k=\infty}\left(\frac{t-x\sqrt{lc}}{t+x\sqrt{lc}}\right)^{\frac{k}{2}} I_k\,(0)\left(4 - (1+\Gamma)^2\,\Gamma^{k-1}\right) \end{array} \right\} u_o\left(t - x\sqrt{lc}\right)$$

Because $I_0(0) = 1$ and $I_k(0) = 0$ for $k > 0$, then (4.112) becomes a traveling wave solution for a lossless line and is given by

$$V_{\text{inf}}(x,t) = V_{dd}\frac{Z_o}{(Z_o + R_s)} u_o\left(t - x\sqrt{lc}\right). \tag{4.113}$$

Letting $t = x\sqrt{lc}$ in (4.113) gives the expression for the voltage wave front traveling down a lossy infinite line,

$$V_{\text{inf}}(x,t) = V_{dd}\frac{Z_o}{(Z_o + R_s)} e^{-\frac{rx}{2Z_o}} \tag{4.114}$$

The expression in (4.114) is a traditional low-loss transmission line solution; however, the compact expressions presented in this section are used to obtain a more accurate approximation of the transient response close to the wave front on a distributed RLC line.

Near Wavefront Approximation Example. To derive a near-wave-front expression, the expansion of a zero order Bessel function is used and given by

$$I_o(x) = 1 + \frac{x^2}{2^2 1!} + \frac{x^4}{2^4 2!} + ... \tag{4.115}$$

In addition, if the argument of the modified Bessel function is *much less* than the order of the modified Bessel function, then the modified Bessel function can be approximated by [5]

$$I_k(x) \approx \frac{1}{!k}\left(\frac{x}{2}\right)^n \tag{4.116}$$

where $k >> x$. Substituting (4.115) and (4.116) into (4.111) and simplifying gives

$$\frac{V_{inf}(x,t')}{V_{dd}} = \frac{Z_o}{Z_o + R_s} e^{-\frac{R}{2Z_o}t'} u_o\left(t'-1\right) \times \tag{4.117}$$

$$\left\{ \begin{array}{l} 1 + \frac{\left(\frac{R}{2Z_o}\sqrt{t'^2-1}\right)^2}{2^2 1!} + ... \\ + \frac{1}{(1-\Gamma)} \sum\limits_{k=1}^{k=\infty} \frac{1}{k!}\left(\frac{R}{4Z_o}\right)^k (t'-1)^k \left(4 - (1+\Gamma)^2 \Gamma^{k-1}\right) \end{array} \right\}$$

where $t' = \frac{t}{\sqrt{lcx}}$.

The summation over k in (4.117) is the difference of two exponential functions. To derive its exact form, a new function, $f(t')$, is defined that is the value of the summation from $k = 0$ to infinity, which is

$$\begin{aligned} f(t') &= \sum_{k=0}^{k=\infty}\left(\frac{R}{4Z_o}\right)^k (t'-1)^k \left(4 - (1+\Gamma)^2 \Gamma^{k-1}\right) \\ &= 4e^{\frac{R}{4Z_o}(t'-1)} - \frac{(1+\Gamma)^2}{\Gamma} e^{\frac{R}{4Z_o}\Gamma(t'-1)} \end{aligned} \tag{4.118}$$

Therefore the value of the summation from $k = 1$ to infinity is determined by the subtraction from (4.118) of the $k = 0$ term, which gives

$$\sum_{k=1}^{k=\infty}\left(\frac{R}{4Z_o}\right)^k (t'-1)^k \left(4 - (1+\Gamma)^2 \Gamma^{k-1}\right) = f(t') - f(1) \tag{4.119}$$

where $f(t')$ is defined in (4.118) and $f(1) = 4 - \frac{(1+\Gamma)^2}{\Gamma}$.

Assuming that the zero order Bessel function is approximately 1, then a simplified expression for the transient voltage near the wave front is given by

$$\frac{V_{inf}(x,t')}{V_{dd}} = \frac{Z_o}{R_s + Z_o} e^{-\frac{R}{2Z_o}t'} u_o(t'-1) + e^{-\frac{R}{2Z_o}t'}\frac{1}{2}(f(t') - f(1))u_o(t'-1) \tag{4.120}$$

The first term of (4.120) is a fast rising attenuated traveling wave solution. The second term is a slow rising waveform that is more indicative of traditional distributed RC solutions. This near-wave-front approximation is compared to the exact compact solution in Figure 4.7 and is labeled "no Bessel approx."

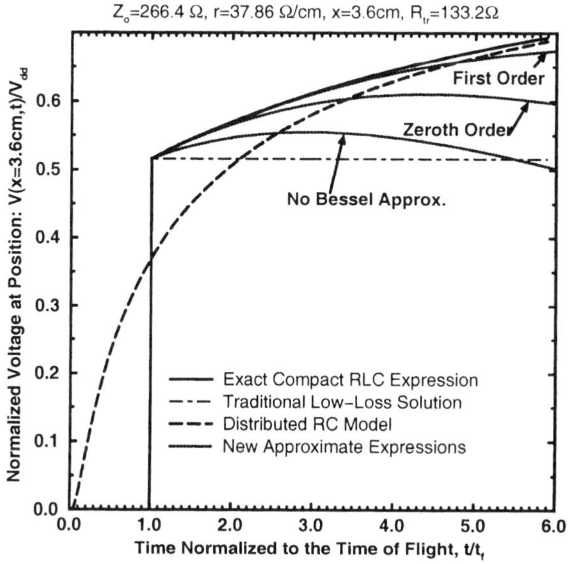

Figure 4.7. Semi-Infinite line approximations

To capture transient behavior further from the edge of the wave front, zero order and first order modified Bessel function approximations are used. Using (4.111) and (4.118) the zeroth and first order modified Bessel function approximations, respectively, are

$$\frac{V_{inf}(x,t')}{V_{dd}} = \left[\begin{array}{l} \frac{Z_o}{R_s+Z_o} e^{-\frac{R}{2Z_o}t'} I_o\left(\frac{R}{2Z_o}\sqrt{t'^2-1}\right) \\ +\frac{1}{2} e^{-\frac{R}{2Z_o}t'} \left(f(t') - f(1)\right) \end{array}\right] u_o(t'-1) \qquad (4.121)$$

and

$$\begin{aligned} \frac{V_{inf}(x,t')}{V_{dd}} = & \left[\frac{Z_o}{R_s+Z_o} e^{-\frac{R}{2Z_o}t'} I_o\left(\frac{R}{2Z_o}\sqrt{t'^2-1}\right)\right. \qquad (4.122) \\ & +\frac{1}{2} e^{-\frac{R}{2Z_o}t'} \left(\frac{t'-1}{t'+1}\right)\left(4-(1+\Gamma^2)\right) I_1\left(\frac{R}{2Z_o}\sqrt{t'^2-1}\right) \\ & \left.+\frac{1}{2} e^{\frac{-R}{2Z_o}t'} \left(-(t'-1)f'(1) + f(t') - f(1)\right)\right] u_o(t'-1). \end{aligned}$$

These additional approximations are compared to the exact compact model, a distributed RC model, and a traditional low-loss model in Figure 4.7. Physically, the addition of each modified Bessel function in (4.111) provides greater accuracy further from the wave front. In addition, Figure 4.7 illustrates that the distributed RC model significantly overestimates the 50% time delay for this example.

Finite Line. The most appropriate boundary conditions for a global GSI interconnect is given by a finite line with an arbitrary source impedance and an open circuit load termination. Using a reflection diagram as seen in Figure 4.8 the expression for the voltage at the end of line is given by

$$V_{fin}\left(L,s\right) = 2V_{\text{inf}}\left(L,s\right) + 2\sum_{n=1}^{q} V_{\text{inf}}\left((2n+1)L,s\right)\left(\frac{R_s - Z\left(s\right)}{R_s + Z\left(s\right)}\right)^n \tag{4.123}$$

where $Z(s)$ is defined in (4.91), n is the reflection number, and q is the maximum reflection number as seen in Figure 4.8.

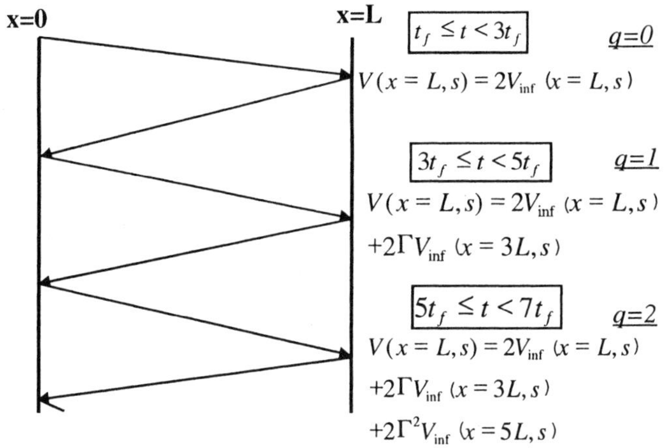

Figure 4.8. Reflection diagram for distributed RLC line

Substituting (4.91) and using the transformation in (4.95) gives

$$V'_{fin}\left(L,s\right) = 2V'_{\text{inf}}\left(L,s\right) + 2\sum_{n=1}^{q} V'_{\text{inf}}\left((2n+1)L,s\right)\left(\frac{R_s - Z_o\sqrt{\frac{s+\sigma}{s-\sigma}}}{R_s + Z_o\sqrt{\frac{s+\sigma}{s-\sigma}}}\right)^n \tag{4.124}$$

Using (4.98) in (4.124) leads to

$$V'_{fin}(L,s) = 2V'_{\text{inf}}(L,s) + 2\sum_{n=1}^{q} V'_{\text{inf}}((2n+1)L,s)\left(\frac{R_s - Z_o\frac{a+1}{a-1}}{R_s + Z_o\frac{a+1}{a-1}}\right)^n. \tag{4.125}$$

Simplifying further and using the definition of Γ gives

$$V'_{fin}(L,s) = 2V'_{\text{inf}}(L,s) + 2\sum_{n=1}^{q} V'_{\text{inf}}((2n+1)L,s)\,\Gamma^n\left(\frac{1-\frac{1}{a\Gamma}}{1-\frac{\Gamma}{a}}\right)^n. \tag{4.126}$$

To determine the series solution of (4.126), the series definition of the following expression must be determined.

$$\left(\frac{1-x}{1-y}\right)^n \tag{4.127}$$

Assuming that n is an integer, then the term in the numerator is evaluated using a finite binomial expansion [5] and is given by

$$\begin{aligned}(1-x)^n &= 1 - nx + \frac{n(n-1)}{2!}x^2 - \frac{n(n-1)(n-2)}{3!}x^3 + \ldots \\ &= \sum_{i=0}^{n}\left(\frac{n!}{i!(n-i)!}x^i\right).\end{aligned} \tag{4.128}$$

Likewise an infinite series expansion of $(1-y)^{-n}$ is given by [5]

$$\begin{aligned}(1-y)^{-n} &= 1 + ny + \frac{n(n+1)}{2!}y^2 + \frac{n(n+1)(n+2)}{3!}y^3\ldots \\ &= \sum_{j=0}^{\infty}\frac{(n-1+j)!}{j!\,(n-1)!}y^j\end{aligned} \tag{4.129}$$

The final series evaluation of (4.127) is determined by combining (4.128) and (4.129) which gives

$$\left(\frac{1-x}{1-y}\right)^n = \sum_{i=0}^{n}\sum_{j=0}^{\infty}\frac{n\,(n-1+j)!}{i!j!\,(n-i)!}(-1)^i x^i y^j \tag{4.130}$$

where the compact summation notation is valid for $n > 0$.

Using (4.130) in (4.126) gives

$$\begin{aligned}V'_{fin}(x=L,s) &= 2V'_{\text{inf}}(L,s) + \\ &2\sum_{n=1}^{q}\sum_{i=0}^{n}\sum_{j=0}^{\infty}\frac{n\,(n-1+j)!}{i!j!\,(n-i)!}(-1)^i V'_{\text{inf}}((2n+1)L,s)\,\Gamma^{n-i+j}\left(\frac{1}{a}\right)^{i+j}\end{aligned} \tag{4.131}$$

Adding $V'_{inf}(x,s)$ from (4.107) gives the final usable form of the Laplace domain solution, which is

$$V'_{fin}(x=L,s) = 2V'_{\text{inf}}(L,s) + 2V_{dd}\frac{Z_o}{(Z_o+R_s)} \times$$
$$\sum_{n=1}^{q}\sum_{i=0}^{n}\sum_{j=0}^{\infty}\frac{n(n-1+j)!}{i!j!(n-i)!}(-1)^i\,\Gamma^{n-i+j}\frac{e^{-(2n+1)L\sqrt{lc}\sqrt{s^2-\sigma^2}}}{\sqrt{s^2-\sigma^2}}\times$$
$$\left\{\left(\frac{1}{a}\right)^{i+j}+\frac{1}{1-\Gamma}\sum_{k=1}^{k=\infty}\left(\frac{1}{a}\right)^{k+i+j}\left(4-(1+\Gamma)^2\,\Gamma^{k-1}\right)\right\} \tag{4.132}$$

Using the transformation in (4.93) and (4.132), the final time domain expression is given by

$$V_{fin}(L,t) = 2V_{\text{inf}}(L,t) +$$
$$2V_{dd}\frac{Z_o}{(Z_o+R_s)}e^{-\frac{r}{2l}t}\sum_{n=1}^{q}\sum_{i=0}^{n}\sum_{j=0}^{\infty}\frac{n(n-1+j)!}{i!j!(n-i)!}(-1)^i\,\Gamma^{n-i+j}\times$$
$$\left\{\begin{array}{l}\left(\frac{t-(2n+1)L\sqrt{lc}}{t+(2n+1)L\sqrt{lc}}\right)^{\frac{i+j}{2}} I_{i+j}\left(\sigma\sqrt{t^2-\left((2n+1)L\sqrt{lc}\right)^2}\right)\\ +\frac{1}{1-\Gamma}\sum\limits_{k=1}^{k=\infty}\left(\frac{t-(2n+1)L\sqrt{lc}}{t+(2n+1)L\sqrt{lc}}\right)^{\frac{i+j+k}{2}}\times\\ I_{i+j+k}\left(\sigma\sqrt{t^2-\left((2n+1)L\sqrt{lc}\right)^2}\right)\left(4-(1+\Gamma)^2\,\Gamma^{k-1}\right)\end{array}\right\}$$
$$\times u_o\left(t-(2n+1)L\sqrt{lc}\right) \tag{4.133}$$

where q, which is defined as the maximum reflection number can be written as a function of time according to

$$q = \left\langle 0.5\left(\frac{t}{x\sqrt{lc}}+1.0\right)\right\rangle - 1.0 \tag{4.134}$$

where the notation $\langle x\rangle$ is defined as the decimal truncation of x (i.e. $<2.7>=2$).

This new expression is compare to HSPICE simulation of 1, 10, 50 and 500 lumped RLC elements in Figure 4.9-4.12 with $Z_o = 266.5\Omega$, $R_s = 133.2\Omega$, r=37.87 Ω/cm, and $L = 3.6$ cm. This series of figures illustrates that as the number of lumped elements is increased in the HSPICE simulation, HSPICE results converge to the compact distributed RLC solutions. Unlike HSPICE, the compact expressions can be used with various searching algorithms to calculate directly without intermediate steps interconnect characteristics such as time delay and overshoot.

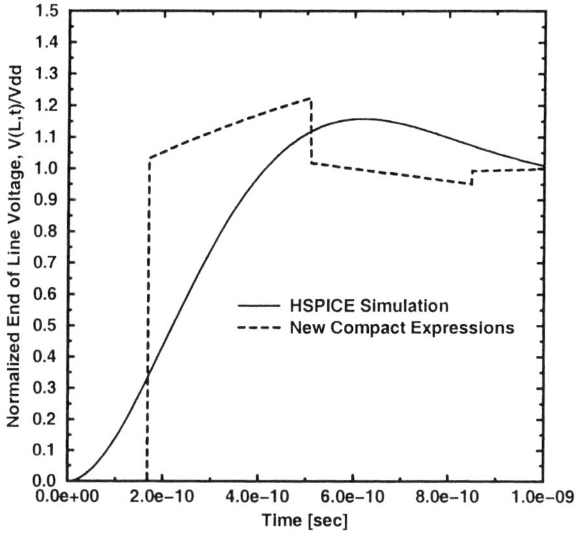

Figure 4.9. Compact expression compared to HSPICE simulation of 1 lumped *RLC* element

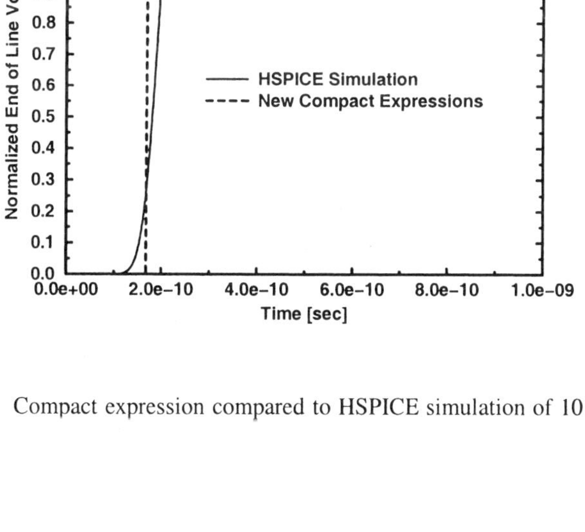

Figure 4.10. Compact expression compared to HSPICE simulation of 10 lumped *RLC* element

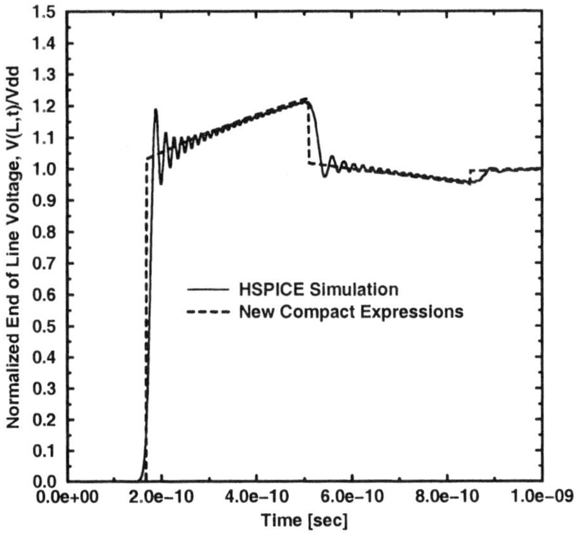

Figure 4.11. Compact expression compared to HSPICE simulation of 50 lumped RLC element

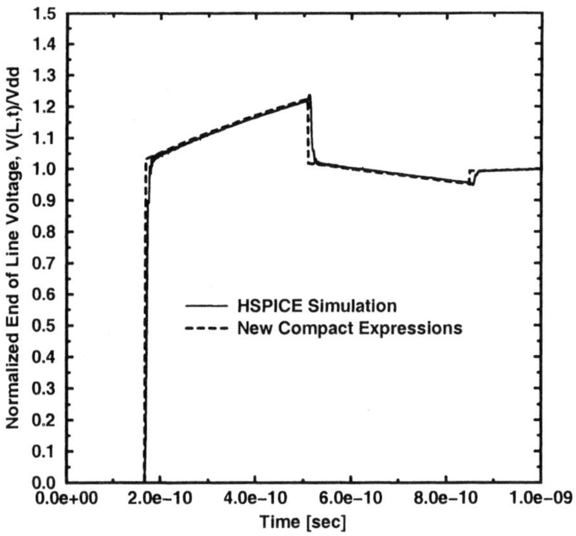

Figure 4.12. Compact expression compared to HSPICE simulation of 500 lumped RLC element

Example $R_{tr} = Z_o$ Simplification. In addition, special cases are explored of (4.133) to provide further insight into the distributed RLC interconnect operation. For example, consider (4.133) when the source resistance is equal to the lossless characteristic impedance of the line, Z_o. The reflection coefficient, Γ, in this case becomes zero. The only summation terms that survive in (4.133) are when $\Gamma^0 = 1$ is satisfied. This condition occurs when

$$n - i + j = 0 \tag{4.135}$$

Because $n \geq i$ and $j \geq 0$ then (4.135) is satisfied only when $i = n$ and $j = 0$. Making this substitution into (4.133) gives the following simplification when the source resistance is equal to the lossless characteristic impedance.

$$V_{fin}(x = L, t) = 2V_{\inf}(x = L, t) + V_{dd}e^{-\frac{r}{2l}t} \sum_{n=1}^{q} (-1)^n$$
$$\begin{pmatrix} \left(\frac{t-(2n+1)L\sqrt{lc}}{t+(2n+1)L\sqrt{lc}}\right)^{\frac{n}{2}} I_n\left(\sigma\sqrt{t^2 - \left((2n+1)L\sqrt{lc}\right)^2}\right) \\ +3\left(\frac{t-(2n+1)L\sqrt{lc}}{t+(2n+1)L\sqrt{lc}}\right)^{\frac{n+1}{2}} I_{n+1}\left(\sigma\sqrt{t^2 - \left((2n+1)L\sqrt{lc}\right)^2}\right) \\ +4\sum_{k=2}^{k=\infty}\left(\frac{t-(2n+1)L\sqrt{lc}}{t+(2n+1)L\sqrt{lc}}\right)^{\frac{n+k}{2}} I_{n+k}\left(\sigma\sqrt{t^2 - \left((2n+1)L\sqrt{lc}\right)^2}\right) \end{pmatrix}$$
$$\times u_o\left(t - (2n+1)L\sqrt{lc}\right) \tag{4.136}$$

where $0^0 = 1$. As mentioned previously, n has the interpretation of being the reflection number where $2V_{inf}(x = L, t)$ is the first reflection ($n = 0$), and the summation terms associated with n=1,2,3,... are the later reflections that begin at $3t_f$,$5t_f$, $7t_f$... (where $t_f = L\sqrt{lc}$), respectively. (4.136) differs from traditional *lossless* transmission line theory in which the source reflection coefficient absorbs all power at the source end of the transmission line after the first reflection. A distributed RLC interconnect, however, prevents this type of perfect matching because the voltage and current ratio are out of phase and their ratio changes with time. In lossless transmission line theory, the ratio of the voltage to the current is always a constant and equal to the lossless characteristic impedance, which allows perfect impedance matching.

Even though perfect matching is not possible, it can be assumed that the first reflection provides significant information about the transient characteristics of this matched interconnect circuit.

Single Reflection Approximation. Using expression (4.133) and (4.122) and normalizing the time variable, t, to the time of flight (i.e. $t = t'\sqrt{lc}L$), the single reflection approximation is given by

$$\frac{V_{fin}(x=L,t')}{V_{dd}} = 2e^{-\frac{R}{2Z_o}t'}\left[\frac{Z_o}{R_s+Z_o}I_o\left(\frac{R}{2Z_o}\sqrt{t'^2-1}\right)+\right.$$
$$\frac{1}{2}\left(\frac{t'-1}{t'+1}\right)\left(4-(1+\Gamma^2)\right)I_1\left(\frac{R}{2Z_o}\sqrt{t'^2-1}\right)$$
$$\left.+\frac{1}{2}(-(t'-1)f'(1)+f(t')-f(1))\right]\times \tag{4.137}$$
$$\left(u_o\left(t-L\sqrt{lc}\right)-u_o\left(t-3L\sqrt{lc}\right)\right)+V_{dd}u_o\left(t-3L\sqrt{lc}\right)$$

This expression gives the detailed transient response of an interconnect in a range up to three time-of-flights of the signal on the line. This is verified in Figure 4.13 in which (4.137) is compared to the exact compact expression for a finite line. The first reflection approximation provides useful information on time delay and peak overshoot as clearly demonstrated in Figure 4.13. The new approximations are also compared to a distributed RC model that significantly underestimates the 50% time delay by around 50% and significantly overestimates the 90% response time of the interconnect by around 50%. In addition, the distributed RC model does not predict the 20% overshoot on this interconnect.

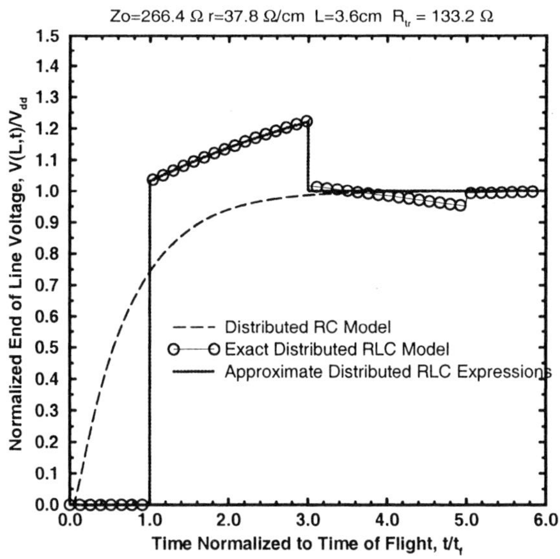

Figure 4.13. Finite line first reflection approximation compared to complete compact model ($Z_o = 266.5\Omega, r = 37.87\Omega/cm, L = 3.6cm, R_s = 133.2\Omega$)

Simplified Voltage Overshoot Expressions. The complete compact models provide great flexibility to calculate a variety of parameters on a distributed RLC line including 50% time delay and maximum overshoot; however, approximate expressions for time delay and maximum overshoot are also desirable for the multilevel network designer.

First, a closed-form expression for the peak overshoot is proposed. Figure 4.13 illustrates that the peak overshoot on a single line occurs at $t = 3t_f$, where t_f is the time of flight of an electromagnetic signal traveling down the line. Generalizing this assumption, the first reflection approximation is used to determine a closed-form expression for peak overshoot. Substituting $t = 3L\sqrt{lc}$ into the first three Bessel functions in (4.111) gives

$$\frac{V\left(L, t=3t_f\right)}{V_{dd}} = 2\frac{Z_o}{\left(Z_o+R_s\right)}e^{-\frac{3rL}{2Z_o}}\left\{I_0\left(\frac{rL}{2Z_o}\sqrt{8}\right)+\right. \tag{4.138}$$

$$\left.(\Gamma+3)\left(\frac{1}{2}\right)^{\frac{1}{2}} I_1\left(\frac{rL}{2Z_o}\sqrt{8}\right)+\frac{(\Gamma(\Gamma+3)+4)}{2}I_2\left(\frac{rL}{2Z_o}\sqrt{8}\right)\right\}$$

This expression should be valid as long as its result is greater than one. If the result is less than one, then it can be assumed that there is no overshoot; therefore, an expression for the peak overshoot is given by

$$\frac{V_{overshoot}}{V_{dd}} = \max(1, \frac{V\left(L, t=3t_f\right)}{V_{dd}}) \tag{4.139}$$

where $V\left(L, t=3t_f\right)/V_{dd}$ is defined in (4.138) and max(-) is a function that returns the maximum value of its arguments. The results of this simplified expression are compared to the compact RLC expression for various values of Z_o, R_s, and R in Figure 4.14. Figure 4.14 shows that this approximate expression very accurately describes the peak overshoot on a distributed RLC interconnect.

Simplified Time Delay Expression. Time delay is defined as the time at which the voltage at the end of an interconnect reaches 50% of its steady-state value. Sakurai has rigorously derived an expression for the time delay of a distributed RC line and it has the following form [3]:

$$\tau = 0.693R_s cL + 0.377rcL^2. \tag{4.140}$$

Using a low loss model, Sakurai's model can be improved. Setting (4.114) equal to $0.5V_{dd}$, and solving for the resistance to lossless characteristic impedance ratio gives the following condition for time of flight interconnect operation as

$$\frac{R}{Z_o} \leq 2\ln\left[\frac{4Z_o}{R_s+Z_o}\right]. \tag{4.141}$$

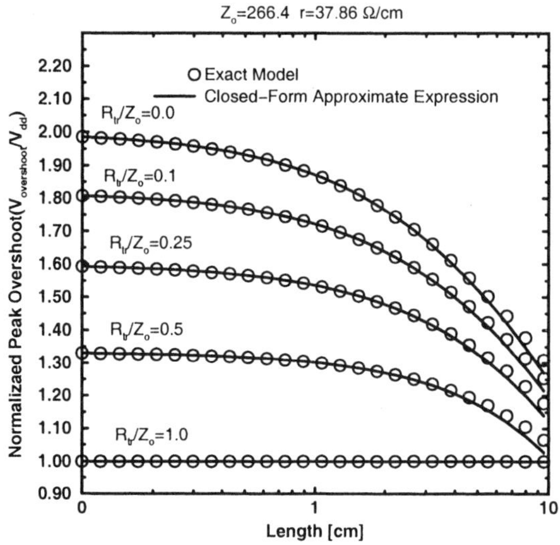

Figure 4.14. Verification of simplified overshoot expression

If the source resistance is greater than $3Z_o$ in (4.141), then the voltage launched at $x = 0$ on the transmission line is less than $0.25V_{dd}$. Therefore, the voltage at the end of the line with an open circuit termination is less than $0.5V_{dd}$ at the time of flight regardless of the interconnect resistance. This means that (4.141) is valid *only* for $R_s < 3.0Z_o$. In addition, if the source impedance is matched to the characteristic impedance, the interconnect resistance must be less than

$$R \leq 2ln\left[2Z_o\right] = 1.39Z_o \tag{4.142}$$

to have a time delay equal to the time of flight.

An improvement to Sakurai's model is, therefore, given by

Region I: $\frac{R}{Z_o} \leq 2\ln\left[\frac{4Z_o}{R_s+Z_o}\right]$ AND $R_s < 3Z_o$

$$\frac{\tau}{t_f} = 1.0 \tag{4.143}$$

Region II: $\frac{R}{Z_o} \geq 2\ln\left[\frac{4Z_o}{R_s+Z_o}\right]$ OR $R_s > 3Z_o$

$$\frac{\tau}{t_f} = 0.693\frac{R_s}{Z_o} + 0.377\frac{r\ell}{Z_o} \tag{4.144}$$

This closed-form expression for time delay is compared to the exact compact RLC expression in Figure 4.15. The simplified expression provides less than 5% error when R_s/Z_o is less than 0.2 or R/Z_o is greater than 2.3. To obtain more accurate results outside this region, the full compact distributed RLC expressions are needed.

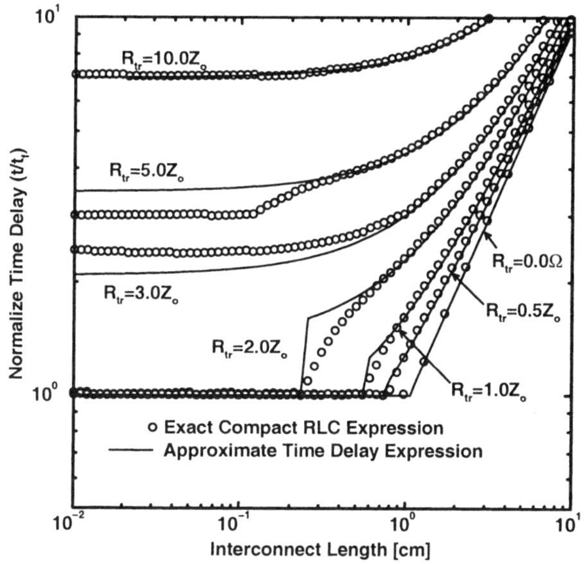

Figure 4.15. Comparison of closed-form time delay expression to exact results using the compact RLC model for various interconnect lengths and driver impedances ($Z_o = 266.5\Omega, r = 37.87\Omega/cm$)

4.3.2 Coupled Line Solutions

In general, signal propagation down a multi-conductor distributed RLC network can be written in matrix form as

$$\frac{\partial^2 V}{\partial x^2} = r\,[C]\,\frac{\partial V}{\partial t} + [L]\,[C]\,\frac{\partial^2 V}{\partial t^2} \tag{4.145}$$

where $[C]$ and $[L]$ are the capacitance and inductance matrix of the multi-conductor system. To decouple these PDE the capacitance and inductance matrix must share the same eigenvectors. In general, this only occurs for the one and two coupled line case. However, if ideal return paths are provided, then a quasi-TEM approximation can be made and this expression can simplify to

$$\frac{\partial^2 V}{\partial x^2} - r\,[C]\,\frac{\partial V}{\partial t} + \frac{1}{\nu^2}\frac{\partial V}{\partial^2 t} \tag{4.146}$$

The implication of making the assumption of $[L][C] = \frac{1}{\nu^2}[I]$ for on-chip global interconnects are discussed in the last section in this chapter. However, if this approximation can be made then the eigenvector of the capacitance matrix can be used to decoupled this PDEs – just as in the case of the distributed RC lines.

Two Conductor Example. The partial differential equations that describe two coupled distributed RLC lines are given by

$$\frac{\partial^2 \tilde{V}}{\partial x^2} = r \begin{bmatrix} c_g + c_m & -c_m \\ -c_m & c_g + c_m \end{bmatrix} \frac{\partial \tilde{V}}{\partial t} + \begin{bmatrix} l_s & l_m \\ l_m & l_s \end{bmatrix} \begin{bmatrix} c_g + c_m & -c_m \\ -c_m & c_g + c_m \end{bmatrix} \frac{\partial^2 \tilde{V}}{\partial t^2} \tag{4.147}$$

where $\tilde{V} = \begin{bmatrix} V_A \\ V_Q \end{bmatrix}$, V_A is the transient voltage of the active line, V_Q is the transient voltage of the quiescent line, c_g is the line to ground capacitance, c_m is the mutual capacitance, l_s is the self-inductance, and l_m is the mutual inductance between the two conductors. The eigenvectors in (4.147) can be used to decouple the partial differential equation. Using this transformation gives the following decoupled differential equations:

$$\frac{\partial^2}{\partial x^2} V_+ (x,t) = (l_s + l_m) c_g \frac{\partial^2}{\partial t^2} V_+ (x,t) + r c_g \frac{\partial}{\partial t} V_+ (x,t) \tag{4.148}$$

$$\frac{\partial^2}{\partial x^2} V_- (x,t) = (l_s - l_m)(c_g + 2c_m) \frac{\partial^2}{\partial t^2} V_- (x,t) + r(c_g + 2c_m) \frac{\partial}{\partial t} V_- (x,t) \tag{4.149}$$

where $V_+ = V_Q + V_A$ and $V_- = V_A - V_Q$.

The plus mode has the interpretation that it is the solution to the voltage of either line when they are both excited simultaneously. For this configuration, both conductors are always at the same potential, and the mutual capacitance between the two lines does not affect the transient response of either line. In the plus solution, the effective capacitance is determined only by the ground capacitance. The currents in this configuration are also in the same direction; therefore, the magnetic flux emanating from each line is in the same direction. For this configuration, the effective flux linkage of each line is increased which produces a higher effective inductance for the plus mode. This is evident in the partial differential equations for the plus mode in that the effective inductance is the self-inductance plus the mutual inductance.

The minus mode is the solution to the transient response of two coupled lines that are switching with opposite polarity: one low to high and the other high to low. The odd mode solution is the transient response for the low to high line. In this situation, the potential difference between the two lines goes through a change of twice the switching potential. This causes the mutual capacitance to be effectively twice its original value because of the Miller effect. This is evident in the PDEs for the minus mode in which the effective capacitance is the ground capacitance plus twice the mutual capacitance. In addition, the currents in this configuration are equal and opposite. This means that the magnetic flux emanating from each line is in opposing directions. For this configuration, the effective flux linkage of each line is reduced which produces an effective lower inductance. This is evident in the PDEs for the minus mode in which the effective inductance is the self-inductance minus the mutual inductance.

The worst-case time delay occurs when the two coupled lines are switching with opposite polarity. As mentioned, this produces a Miller effect in the mutual capacitance between the wires. Using the partial differential equations with the appropriate boundary conditions, the final transient response for the worst-case time delay of the active line is given by

$$V_A\left(L,t\right) = V_{fin}\left(L,t,l=l_s-l_m,c=c_g+2c_m\right). \tag{4.150}$$

The transient response of the worst-case crosstalk occurs when both lines are initially uncharged and the active line charges to V_{dd}. Under these initial conditions, the transient response for the worst-case crosstalk is given by

$$V_Q\left(L,t\right) = \frac{1}{2}\begin{pmatrix} V_{fin}\left(L,t,l=l_s+l_m,c=c_g\right) \\ -V_{fin}\left(L,t,l=l_s-l_m,c=c_g+2c_m\right) \end{pmatrix} \tag{4.151}$$

where $V_{fin}(x,t)$ is defined in (4.133).

To verify these compact expressions, Figure 4.16-4.19 shows the comparison between the compact model results to HSPICE simulation of 1, 10, 50, and 500 lumped RLC elements. As the number of lumped elements is increased, the HSPICE simulation converges to the distributed RLC solution. Unlike HSPICE the compact expressions can be used with various searching algorithms to calculate directly without intermediate steps interconnect characteristics such as peak crosstalk voltage, time delay, and overshoot. Physical insight gained by the derivation of these compact expressions is explored in the final sections.

Three Conductor Example. In this section, the expressions for the worst-case time delay and crosstalk of three parallel lines sandwiched between two ground planes are derived. Each line has arbitrary source impedance, R_s, and an open-circuit load termination. To determine the expressions for three coupled lines, the capacitance and inductance matrices are determined using compact quasi-analytical expressions [8] combined with a quasi-TEM wave analysis [6]. The quasi-TEM wave analysis gives the following relationship between the capacitance and inductance matrix:

$$[L]\,[C] = \frac{1}{\nu^2}\,[I]. \tag{4.152}$$

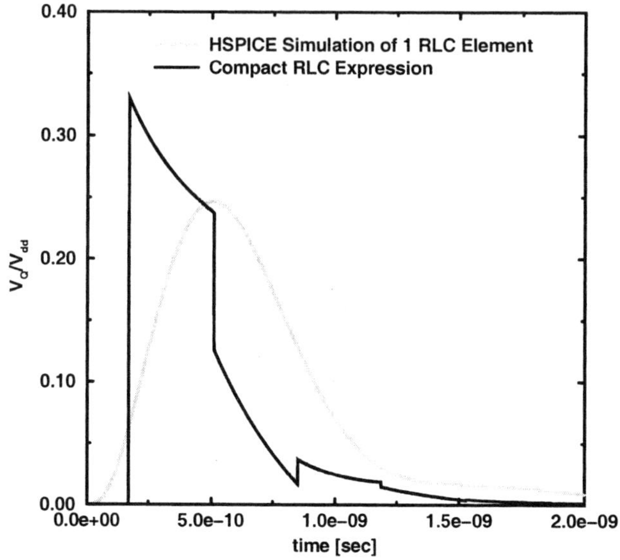

Figure 4.16. Compact distributed RLC models compared to HSPICE Simulation of 1 lumped RLC element

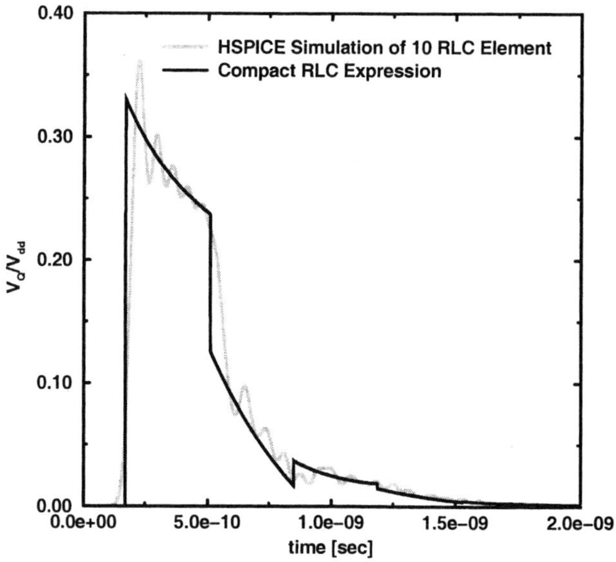

Figure 4.17. Compact distributed RLC models compared to HSPICE Simulation of 10 lumped RLC element

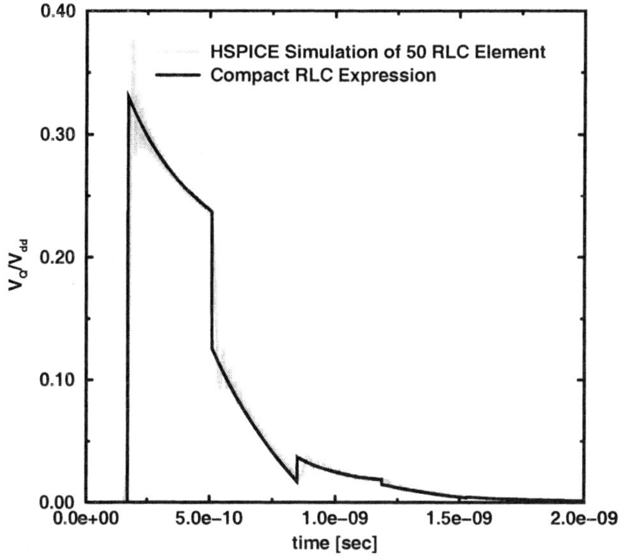

Figure 4.18. Compact distributed RLC models compared to HSPICE Simulation of 50 lumped RLC element

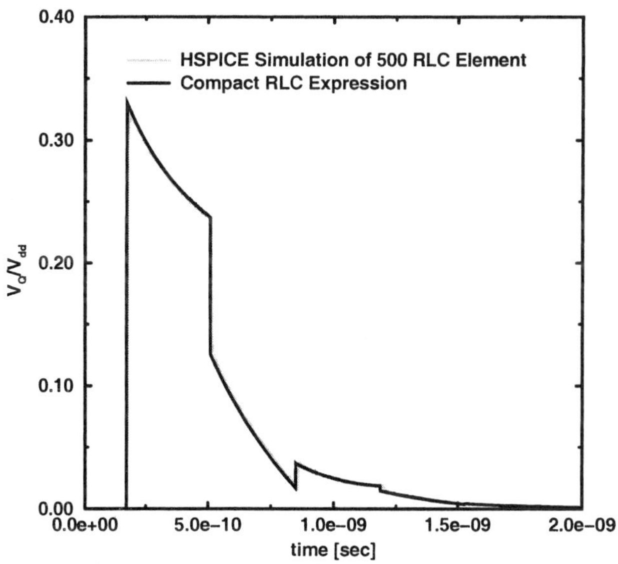

Figure 4.19. Compact distributed RLC models compared to HSPICE Simulation of 500 lumped RLC element

where ν is the speed of an electromagnetic wave in a given dielectric material, and [I] is the unity matrix with all diagonal elements equal to one and off-diagonal elements equal to zero. Physically, (4.152) assumes an ideal return path is present in the virtual ground planes that are above and below the three coupled interconnects. Because of this idealization, the models in this example project a lower limit on the crosstalk and time delay.

The partial differential equations that describe the three conductor two ground plane model are then given by

$$\frac{\partial^2 \tilde{V}}{\partial x^2} = r[C]\frac{\partial \tilde{V}}{\partial t} + \frac{1}{\nu^2}\frac{\partial^2 \tilde{V}}{\partial t^2} \tag{4.153}$$

where $\tilde{V} = \begin{bmatrix} V_1(x,t) \\ V_2(x,t) \\ V_3(x,t) \end{bmatrix}$. Once this assumption in (4.152) is made, then the transformations defined earlier in this chapter can be used. These transformation allows the single line solutions to be used to determine the three coupled interconnect solutions. The worst-case crosstalk occurs when all three lines are initially low, and the outer two lines switch to the V_{dd}. Using the transformations outlined previously, the transient expression for the worst-case crosstalk is given by

$$\begin{aligned} V_Q(L,t) &= \frac{2}{3}\left(V_{fin}\left(L,t,l=\frac{1}{2c_g\nu^2}, c=2c_g\right)\right. \\ &\quad \left. -V_{fin}\left(L,t,l=\frac{1}{(2c_g+3c_m)\nu^2}, c=2c_g+3c_m\right)\right) \end{aligned} \tag{4.154}$$

where $V_{fin}(x,t)$ is defined in (4.133).

The worst-case time delay occurs when the inner line switches with an opposite polarity of the outer two lines. The resulting expression for the worst-case time delay is given by

$$\begin{aligned} V_A(L,t) &= \frac{4}{3}V_{fin}\left(L,t,l=\frac{1}{(2c_g+3c_m)\nu^2}, c=2c_g+3c_m\right) \\ &\quad -\frac{1}{3}V_{fin}\left(L,t,l=\frac{1}{2c_g\nu^2}, c=2c_g\right) \end{aligned} \tag{4.155}$$

where $V_{fin}(x,t)$ is defined in (4.133).

Peak Crosstalk Expression. The compact solutions in (4.151) provide the basis for efficient computation of inductive effects for distributed RLC coupled interconnects. Further simplification, however, is necessary to provide physical understanding of RLC crosstalk as well as rapid calculation.

To find an approximate peak crosstalk expression, the plus and minus modes for two or three coupled semi-infinite interconnects are approximated as low-loss transmission line solutions. The transient solutions at a position x, therefore, are

$$\frac{V_+(x,t)}{V_{dd}} = \frac{Z_o^+}{Z_o^+ + R_s} e^{-\frac{rx}{Z_o^+}} u(t - \frac{x}{\nu}) \tag{4.156}$$

and

$$\frac{V_-(x,t)}{V_{dd}} = \frac{Z_o^-}{Z_o^- + R_s} e^{-\frac{rx}{Z_o^-}} u(t - \frac{x}{\nu}) \tag{4.157}$$

where $Z_o^+ = \sqrt{\frac{l_s+l_m}{c_g}}$ and $Z_o^- = \sqrt{\frac{l_s-l_m}{c_g+2c_m}}$ are the lossless characteristic impedances of the plus and minus modes for two coupled lines over a single ground plane and $Z_o^+ = \frac{1}{\nu 2c_g}$ and $Z_o^- = \frac{1}{\upsilon(2c_g+3c_m)}$ are the lossless characteristic impedances of the sum and difference modes for three coupled lines between two ground planes.

Equations (4.156) and (4.157) are plotted as a function of length in Figure 4.20. One key feature of Figure 4.20 is that there is an interconnect length at which the difference between these two modes is a maximum. This is significant because the difference of the plus and minus modes is proportional to the quiet line voltage.

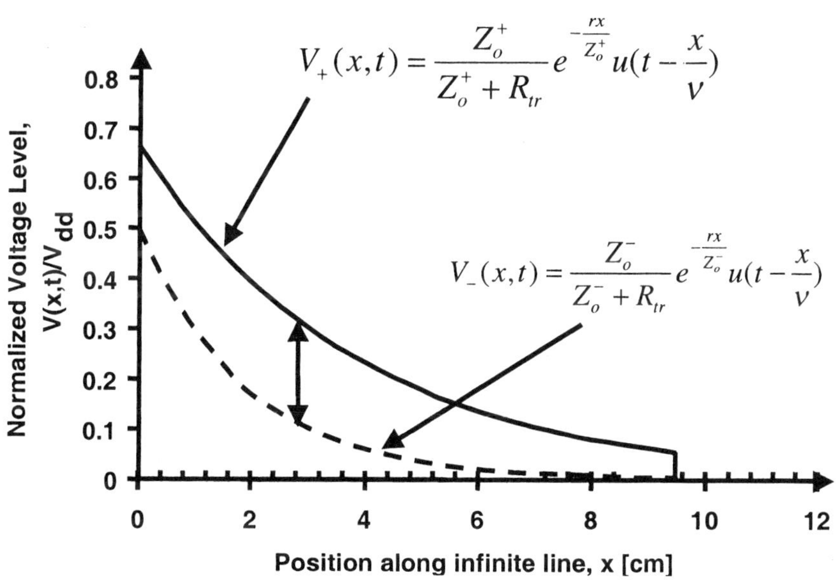

Figure 4.20. Plot of plus and minus modes traveling down a semi-infinite low-loss transmission line

For a finite line, it is assumed that the peak noise voltage occurs during the first reflection at $t = L\sqrt{lc}$ when inductive effects are dominant. Therefore, the low-loss solutions for the semi-infinite line are doubled because the finite line has open circuit terminations. Likewise, when the interconnects are long $(L >> Z_o/r)$, it is assumed that the interconnect resistance is so great that inductive effects are suppressed. In this region, it is assumed that the distributed RC model after [2] is valid. Therefore, an approximate closed-form expression for the peak crosstalk voltage for distributed RLC lines is piecewise defined as

Region I: $L \leq L_{crit}$

$$V_{peak} = \chi \left(\frac{Z_o^+}{Z_o^+ + R_s} e^{-\frac{rL}{Z_o^+}} - \frac{Z_o^-}{Z_o^- + R_s} e^{-\frac{rL}{Z_o^-}} \right) \tag{4.158}$$

Region II: $L > L_{crit}$

$$V_{peak} = \frac{1}{2} \frac{c_m}{c + c_m} \tag{4.159}$$

where $\chi = 1$, $Z_o^+ = \sqrt{\frac{l+l_m}{c_g}} = \frac{1}{vc_g}$, and $Z_o^- = \sqrt{\frac{l-l_m}{c_g+2c_m}} = \frac{1}{v(c_g+2c_m)}$ for two coupled lines and $\chi = \frac{4}{3}$, $Z_o^+ = \frac{1}{v2c_g}$, and $Z_o^- = \frac{1}{v(2c_g+3c_m)}$ for three coupled lines, and L_{crit} is determined from the solution of the following equation

$$\chi \left(\frac{Z_o^+}{Z_o^+ + R_s} e^{-\frac{rL_{crit}}{Z_o^+}} - \frac{Z_o^-}{Z_o^- + R_s} e^{-\frac{rL_{crit}}{Z_o^-}} \right) = \frac{1}{2} \frac{c_m}{c + c_m} \tag{4.160}$$

This new peak crosstalk expression is compared to the results from the compact RLC expression in Figure 4.21. Agreement is close in the non-linear region near maximum V_n/V_{dd} and again in the flat region. In the transition between these two regions, as defined in (4.158) and (4.159), the complete compact model is needed to get an accurate estimation of the peak crosstalk.

An approximate expression for L_{crit} comes from the observation that Z_o^+ is always greater than Z_o^-. With significant coupling capacitance the second exponential term in (4.160) is ignored and L_{crit} is derived to be

$$L_{crit} \approx \frac{2Z_o^+}{r} ln \left[\frac{Z_o^+ 2\chi(c_g + c_m)}{(Z_o^+ + R_s)c_m} \right] \tag{4.161}$$

This critical length demarks when inductance is important for crosstalk estimation on an interconnect. In fact, inductive effects can be *ignored* to first order if the following conditions hold:

$$R > 2Z_o^+ ln \left[\frac{Z_o^+ 2\chi(c_g + c_m)}{(Z_o^+ + R_s)c_m} \right] \tag{4.162}$$

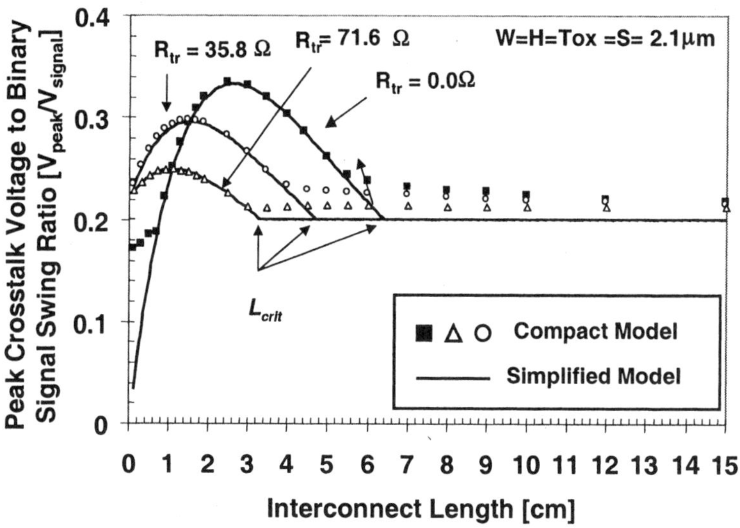

Figure 4.21. Length dependence of crosstalk revealed using new compact models for three values of the driver source impedance

or

$$R_s > Z_o^+ \left(2\chi \frac{c_m}{c_g} + (2\chi - 1) \right). \tag{4.163}$$

Both conditions in (4.162) and (4.163) come directly from (4.161). The latter condition is a result of insuring that the argument of the logarithm in (4.161) is greater than one.

Upper Bound of Peak Crosstalk. The interconnect length at which the peak noise voltage occurs is calculated from setting the derivative of (4.158) equal to zero,

$$\frac{dV_{peak}}{dL} = \left(-\frac{Z_o^+}{Z_o^+ + R_s} \frac{r}{Z_o^+} e^{-\frac{rL}{Z_o^+}} + \frac{Z_o^-}{Z_o^- + R_s} \frac{r}{Z_o^-} e^{-\frac{rL}{Z_o^-}} \right) = 0. \tag{4.164}$$

Solving for the length at which this peak occurs gives

$$L_{\max} = \frac{Z_o^- Z_o^+}{r\left(Z_o^- - Z_o^+\right)} \ln \left[\frac{(Z_o^- + R_s)}{\left(Z_o^+ + R_s\right)} \right]. \tag{4.165}$$

The peak noise at this length is determine from substituting (4.165) into (4.164), which gives

$$\frac{V_{peak,max}}{V_{dd}} = \chi \left[\frac{Z_o^- + R_s}{Z_o^+ + R_s}\right]^{-\frac{Z_o^-}{Z_o^- - Z_o^+}} \frac{Z_o^+ - Z_o^-}{Z_o^+ + R_s}. \tag{4.166}$$

This is the peak noise voltage that is possible on a given interconnect tier assuming that parallel interconnects of length $L_{\max}$ are routed on that tier.

Assuming that there are two parallel lines and $R_s = 0$, expressions (4.165) and (4.166) simplify to

$$L_{max} = \frac{1}{vrc_m} ln \left[\frac{c_g + 2c_m}{c_g}\right] \tag{4.167}$$

and

$$\frac{V_{peak,max}}{V_{dd}} = \left(\frac{c_g}{c_g + 2c_m}\right)^{\frac{c_g}{2c_m}} \left(\frac{c_g}{c_g + 2c_m}\right) \frac{2c_m}{c_g} \approx \frac{\pi}{4} \frac{c_m}{c_g + c_m}, \tag{4.168}$$

respectively. (4.168) compares directly to Sakurai's full crosstalk expression in [3] and shows that the peak crosstalk at L_{crit} is exactly $\frac{2\pi}{4} = 1.57$ times larger than predicted by a distributed RC model with zero source impedance.

For three coupled lines, the upper bound on the crosstalk voltage for ideal return paths is given by

$$\frac{V_{peak,max}}{V_{dd}} \approx \frac{\pi}{3} \frac{c_m}{\frac{4}{3}c_g + c_m} \tag{4.169}$$

4.4 Non-Ideal Return Paths

The external inductance and capacitance of a normal transmission line with ideal return path currents as seen in Figure 4.22 have a reciprocity relation independent of the geometry of the line [6]. The implications of this property is that the wave propagation speed in all transmission lines is constant and it is equal to the speed of the light in the dielectric. For these conventional transmission lines, any change in geometry which results in decreasing the capacitance, increases the inductance and vise versa. There are, however, some structures in which, inductance and capacitance can be changed independently. For example, in Figure 4.23 some extensions are periodically added along the line. Since current in x direction cannot pass through these extensions, they do not change the inductance of the structure.

However, as a wave travels along the line it has to charge the additional capacitance corresponding to these fingers, which means the equivalent capacitance of the line is increased. The structure in Figure 4.23 is a periodic

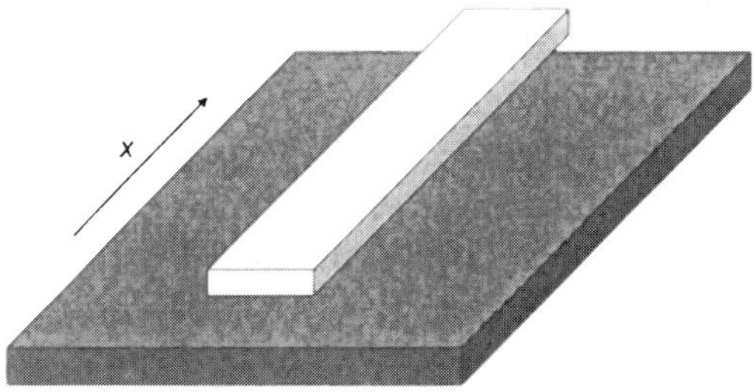

Figure 4.22. Conventional transmission line over ideal ground plane

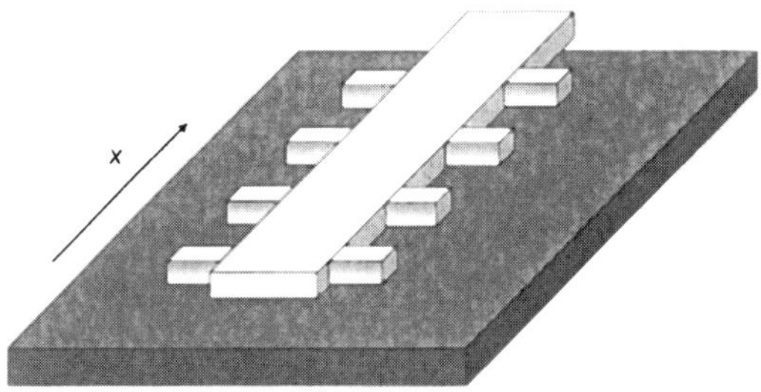

Figure 4.23. Transmission line with periodic extensions that slow wave propagation speed

structure and is widely used in microwave circuits [9] and is well characterized. Rigorous analysis shows that if the distance between the extensions is on the order of the signal wavelength, because of the reflections from discontinuities, the periodic structure behaves as a bandpass filter For the cases that the wavelength is much larger than the distance between the discontinuities, the structure behaves as a smooth transmission line with a slower propagation speed

$$\nu = \frac{1}{\sqrt{l(c + \frac{C_F}{d})}} = \frac{1}{1 + \frac{C_F}{cd}} \frac{c_o}{\sqrt{\epsilon_r}} \tag{4.170}$$

where l and c are the inductance and capacitance per unit length of the ideal transmission line, C_F is the capacitance of each finger, d is the distance between the fingers, c_o is the speed of light and ϵ_r is the dielectric constant. The additional capacitance also affects the characteristic impedance and results in a larger attenuation.

On-chip layers are very expensive, and therefore, in most of the cases there are no on-chip ground planes. Instead some power and ground lines are usually inserted along the lines to provide adequate return paths. Let's consider a coplanar transmission line structure as shown in Figure 4.24. Such a structure may be used for a critical line such as clock line. Wide ground lines can shield

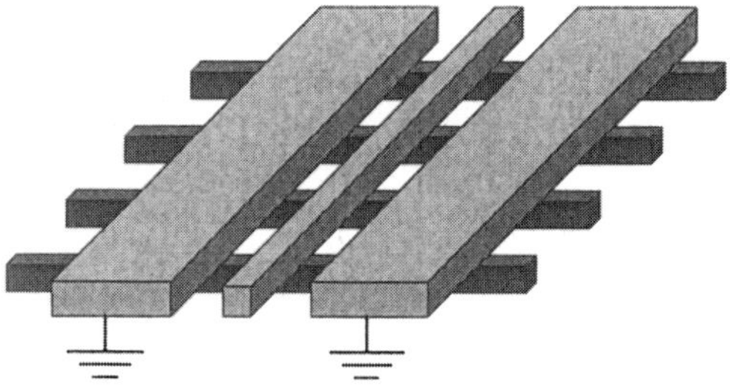

Figure 4.24. Transmission line structure with coplanar ground lines for a VLSI multilevel interconnect network

the line from far lines and they can provide a nearby return path. If the orthogonal lines didn't exist, the structure would be a transmission line and the signal would propagate with the speed of light in the dielectric. The orthogonal lines, however, make the coplanar transmission line have similar characteristics of the periodic structure in Figure 4.23.

The orthogonal lines' pitch is in the order of a few microns, which is much smaller than the wavelength of a signal that we may have. For instance, for 10GHz, the wavelength is about 1.5cm. Hence, the structure can be modeled by a smooth transmission line with an excessive capacitance per unit length. Figure 4.25 compares the voltage at the end of the line in two cases, first a signal line is over an ideal ground plane and the second is a coplanar transmission line above orthogonal lines. It can be seen that the propagation speed of the latter would be larger and due to smaller characteristic impedance, the attenuation would also be larger. In order to have a larger wiring density, the number of signal lines between the ground lines might be increased. In this case the set of differential equations should be solved. It is important that (4.170) is no longer valid. This means that the propagation speed would be a function of boundary conditions or in other words the pattern of excitation.

For the case that there are two lines between the ground lines, each signal line and its nearby ground line forms a transmission line. If there were no orthogonal lines we had

$$[L]\,[C] = \frac{\epsilon_r}{c_o} \tag{4.171}$$

and therefore, the propagation speed for the common and differential modes were equal. The orthogonal lines, however, cause additional capacitance and since the equivalent capacitance of the differential mode is larger, the percent increase in the capacitance would be smaller as compared to the common mode. Hence, the propagation speed of the differential mode is larger. The noise voltage on the quiet line consists of both common and differential modes

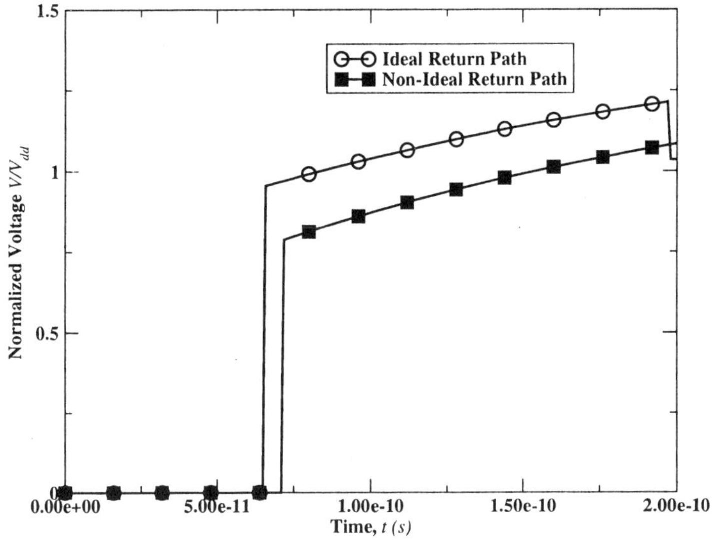

Figure 4.25. Comparison to ideal and non-ideal return path for single line transients

as

$$V_n = \frac{1}{2}\left(V_+ - V_-\right), \tag{4.172}$$

which means for a time proportional to the difference between the two speeds, an out-of-phase noise appears at the end of the line, which can be very large if the line's attenuation is very small. This is unlike the ideal return path case, where very small attenuation resulted in a small noise as seen in Figure 4.26.

4.5 Summary

Both distributed RC and RLC models are needed to describe the electrical characteristics of high-speed wires in a GSI multilevel network. In this chapter, distributed RC models that include finite rise-time effects for high-speed interconnects are rigorously derived for the worst-case time delay and crosstalk transient response of one, two, and three coupled distributed RC interconnects circuits. From the transient response, a new general expression for the peak noise voltage between two and three coupled interconnects is derived. Simplified forms of this expression reveal fundamental properties of the crosstalk.

In addition, compact expressions for a distributed RLC transmission line with step-input and finite-rise-time excitation voltages have been rigorously derived. The resulting compact expressions efficiently describe interconnect circuit behavior when neither inductance nor resistance can be treated as a perturbation. In addition, these new expressions provide physical insight into

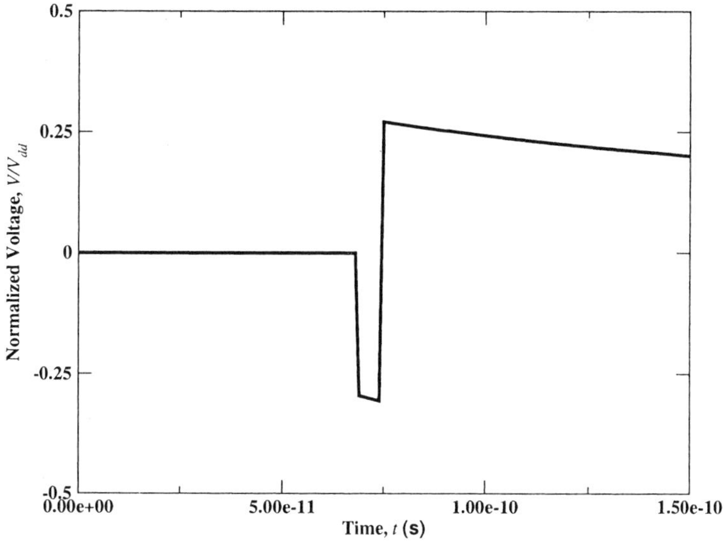

Figure 4.26. Crosstalk transients on a quiescent line with non-ideal return paths

GSI inductive effects through the derivation of simple closed-form expressions for time delay, overshoot, and crosstalk. Most notable, the inclusion of inductance has been shown to introduce a new non-linear length dependence of peak crosstalk as seen in Figure 4.21.

Acknowledgments

The authors wish to express their appreciation to DARPA, contract F33615-97-C-1132, MARCO, contract MDA 972-99-1-002, and the SRC, contract 448:048, NSF 0009240 and 0098227, for their generous support.

References

[1] Y. Ismail and E. Friedman, "Effects of inductance on the propagation delay and repeater insertion in vlsi circuits," *IEEE Transactions on VLSI Systems*, vol. 8, no. 2, pp. 195–206, 2000.

[2] H. Bakoglu, *Circuits, Interconnections, and Packaging for VLSI*. Reading, Mass.: Addison-Wesley Company, 1990.

[3] T. Sakurai, "Closed-form expression for interconnection delay, coupling, and crosstalk in vlsi's," *IEEE Transactions on Electron Devices*, vol. 40, pp. 118–124, Jan. 1993.

[4] K. Lee, "On-chip interconnects - giga hertz and beyond," *International Interconnect Technology Conference*, pp. 15–17, 1998.

[5] W.H. Beyer, *CRC Standard Mathematical Tables and Formulae.* Boca Raton, Fl: CRC Press, Inc, 1991.

[6] C. Paul, *Analysis of Multi-conductor Transmission Lines.* New York: John Wiley and Sons, 1994.

[7] O. Heaviside, *Electromagnetic Theory.* New York: Chelsea Publishing, 1971.

[8] J.Chern, J. Jurang, L. Arledge, P. Li, and P. Yang, "Multilevel metal capacitance models for cad design synthesis systems," *IEEE Electron Device Letters*, vol. 13, pp. 32–34, Jan. 1992.

[9] R.E. Collin, *Foundations of microwave engineering.* New York: IEEE Press, 1966.

Chapter 5

POWER, CLOCK, AND GLOBAL SIGNAL DISTRIBUTION

P. Zarkesh-Ha
LSI Logic Corporation

5.1 Introduction

Global interconnects are commonly considered a key potential bottleneck to the advancing performance of future integrated systems. The complete global interconnect architecture of a digital system implemented on a single chip consists of signal, clock, and power-supply distribution networks, as shown in Figure 5.1. Since the signal, clock, and power-supply networks utilize the same interconnect stacks (Figure 5.1), it is imperative to integrate the complete interconnect architecture to design and optimize global interconnects for a gigascale system-on-a-chip (GSoC). The complete interconnect models for signal, clock, and power distribution networks and the integrated architecture for global interconnects are described in Sections 5.2, 5.3, 5.4, and 5.5 respectively.

5.2 Global Signal Interconnect Modeling

Wiring area requirement, signal delay, and crosstalk noise limit are the main issues for a global signal wiring network. A stochastic wire-length distribution model for global interconnects in a heterogeneous system-on-a-chip is presented in this chapter. The model will be used to estimate the wiring area requirement of global signal interconnects in a gigascale system-on-a-chip. Also, based on the models for signal delay and crosstalk noise in a distributed RLC network derived in [1], the criteria will be defined for the interconnect bandwidth and noise limit for global wires.

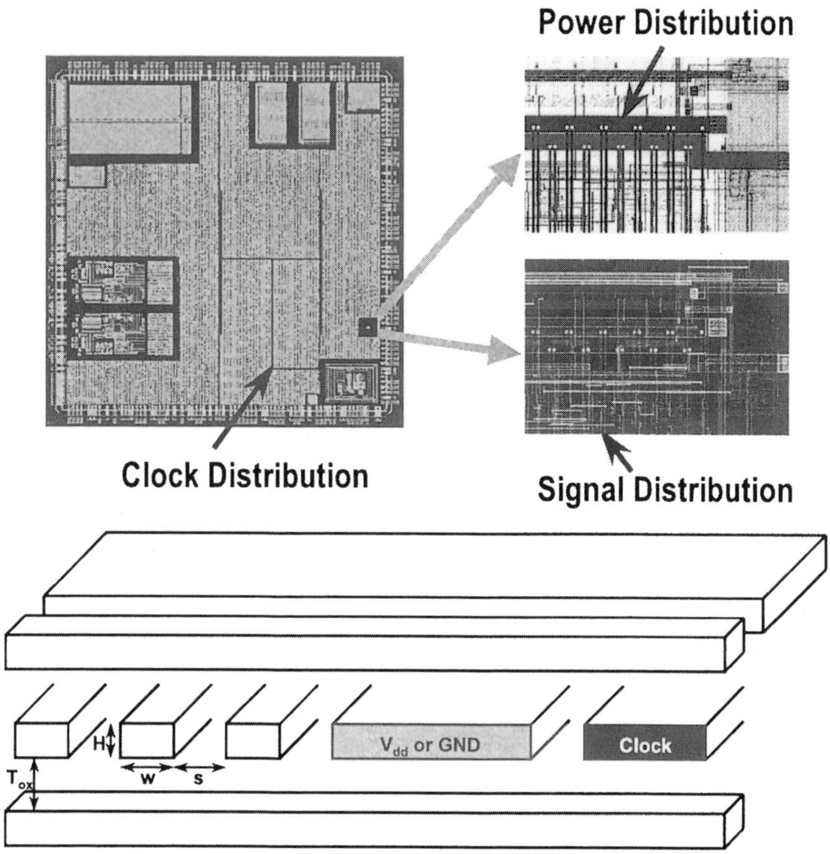

Figure 5.1. Signal, clock, and power-supply wires in a real chip utilizing the same orthogonal interconnect architecture

5.2.1 Stochastic Global Net-Length Distribution in a Heterogeneous SoC

It has been observed that the overall wire-length distribution of a heterogeneous system-on-a-chip (SoC) at the system level has a bimodal behavior [2,3], as shown in Figure 5.2.

The distribution has two peaks, the first peak representing internal megacell interconnects and the second representing global interconnects that provide communication between megacells. The reason for the bimodal behavior in wire-length distribution is that the placement and routing of the lower levels (within megacells) are done independently of the higher level (between megacells) in SoC designs. The network within megacells can be modeled as a

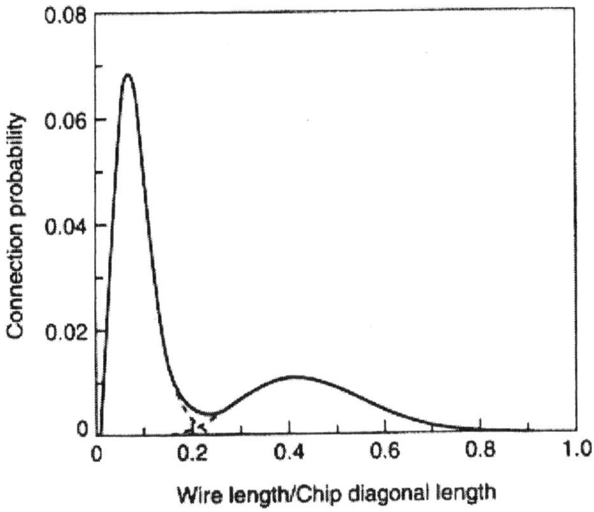

Figure 5.2. A typical distribution of wire lengths in an integrated system after [2,3]

homogeneous system whereas the global network between megacells needs to be modeled as a heterogeneous system. Unlike wire-length distribution in a homogeneous system, the prediction of global wiring demand in a nonhomogeneous system requires more information about the system, such as netlist, placement, and routing.

Modeling of Global Netlist Information. The netlist information defines the connectivity between megacells in a SoC. To predict the number of nets between megacells, a well-established empirical relationship known as Rent's rule is used. This relationship correlates the number of I/O terminals, T, to the number of gates, N, in a homogeneous logic network. This correlation is given by a simple power-law expression as

$$T = kN^p, \tag{5.1}$$

where k and p are the Rent's coefficient and exponent, respectively. Since a SoC is a heterogeneous network containing different varieties of megacells, a new definition of Rent's rule for such heterogeneous systems is required. The heterogeneous Rent's rule is rigorously derived in [4]. It is shown that the same power-law expression as (5.1) is also valid for heterogeneous networks with equivalent k and p parameters

$$\begin{cases} k_{eq} = \sqrt[N_{Geq}]{\left(\prod_{i=1}^{n} k_i^{N_{Gi}}\right)} \\ p_{eq} = \frac{\sum_{i=1}^{n} p_i N_{Gi}}{N_{Geq}} \end{cases} \tag{5.2}$$

where k_i and p_i are the usual Rent's rule parameters, N_{Gi} is the equivalent number of gates in the ith megacell, and

$$N_{G_{eq}} = \sum_{i=1}^{n} N_{Gi}.$$

Using the heterogeneous Rent's rule, the number of nets between megacells is computed. For instance, suppose that there is a block of two megacells shown in Figure 5.3(a), where T_1 and T_2 represent the number of terminals of megacell#1 and megacell#2, respectively. Using the Venn diagram of Figure 5.3(b) and de Morgan's law of set theory, it can be proven that

$$T_{Ext(1,2)} = T_1 + T_2 - T_{Int(1,2)}. \tag{5.3}$$

Here, $T_{Int(1,2)}$ is the number of internal terminals that are shared between the two megacells and $T_{Ext(1,2)}$ is the number of external terminals of the whole collection of gates, which is calculated by heterogeneous Rent's rule as

$$T_{Ext(1,2)} = k_{1,2} \left(N_1 + N_2\right)^{p_{1,2}}. \tag{5.4}$$

where N_1 and N_2 are the number of gates in megacell#1 and megacell#2, respectively, and $k_{1,2}$ and $p_{1,2}$ are the heterogeneous Rent's parameters given by

$$\begin{cases} k_{1,2} = \left(k_1^{N_1} \cdot k_2^{N_2}\right)^{\frac{1}{N_1+N_2}} \\ p_{1,2} = \frac{p_1 N_1 + p_2 N_2}{N_1+N_2} \end{cases} \tag{5.5}$$

Equation (5.3) is illustrated by examining the number of pins of Figure 5.3(a). Equations (5.4) and (5.5) can also be validated by considering Rent's rule in a homogeneous system. Substituting Rent's rule in (5.3) gives

$$k_{1,2} \left(N_1 + N_2\right)^{p_{1,2}} = k_1 N_1^{p_1} + k_2 N_2^{p_2} - T_{Int(1,2)} \tag{5.6}$$

where k_1, p_1, k_2, and p_2 are the Rent's rule parameters for megacell#1 and megacell#2, respectively, and $k_{1,2}$ and $p_{1,2}$ are the heterogeneous Rent's parameters defined in (5.5). Since there are only two megacells in the block, all shared terminals are used for two terminal nets. Note that in this study, a multiterminal net is defined in the global level, where the connections of terminals

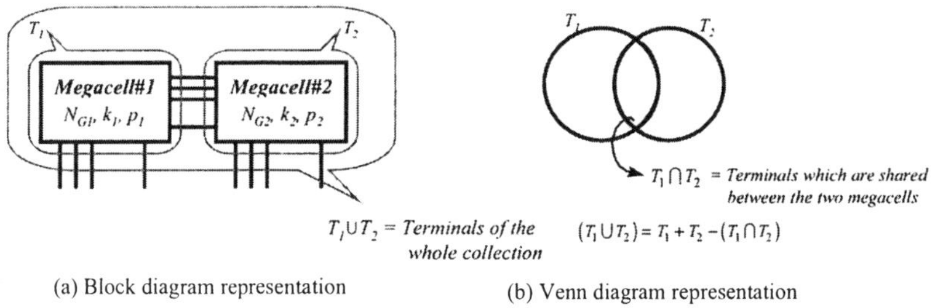

Figure 5.3. An example of a collection of two megacells

inside a megacell are not of concern. Thus, the number of two-terminal nets, $N_{net(2)}$, is given by

$$N_{net}(2) = \frac{T_{Int(1,2)}}{2} \tag{5.7}$$

Using (5.7) for all combinations of two megacells in the whole system gives the total number of two-terminal nets (fan-out of 1). Similarly, in the general case with N_{Meg} megacells, the number of m-terminal nets is computed recursively by examining all possible combinations of 2, 3,..., m out of N_{Meg} megacells. Assuming that N_{Meg}>10, which is satisfied in current and future microprocessor designs, then a closed-form expression for the number of m-terminal nets, $N_{Net(m)}$, can be approximated by a fan-out distribution model [5]

$$N_{net}(m) \approx \frac{\tilde{K} N_{Meg}\left((m-1)^{\tilde{P}-1} - m^{\tilde{P}-1}\right)}{m} \tag{5.8}$$

where $\tilde{k}$ and $\tilde{p}$ are the equivalent megacell Rent's rule parameters given by

$$\begin{cases} \tilde{k} = \sqrt[N_{Meg}]{\left(\prod_{i=1}^{N_{Meg}} k_i N_i^{P_i}\right)} \\ \tilde{p} = \dfrac{\sum_{i=1}^{N_{Meg}} p_i N_i}{N_{Meg}} \end{cases} \tag{5.9}$$

Note that the underlying assumption for derivation of (5.9) is that the heterogeneous Rent's rule is valid throughout the whole design, from the gate level to the megacell level. However, since Rent's parameters at the higher level of the hierarchy, defined as region II [6], may not be the same as the Rent's parameters at the lower levels, more accurate netlist information results from equating $\tilde{k}$ and $\tilde{p}$ to the Rent's parameters in region II if they are available. In the absence of the topology of the global Rent's parameters in region II, (5.9) gives the first-order approximation for the Rent's coefficient and exponent of the system.

Modeling of Global Placement Information. The placement information describes the average bounding area dimensions of a net connecting a group of megacells. On average, a group of m megacells is bounded by a square with an edge of length [7]

$$\hat{e} = \sqrt{\bar{A}_{Meg}\left[m\eta_p + N_m\left(1 - \eta_p\right)\right]} \tag{5.10}$$

where $\bar{A}_{Meg}$ is the average area of a megacell, N_m is the total number of megacells in the design, and η_p is the placement efficiency given by [7]

$$\eta_p = \frac{1 - \dfrac{Block \quad Bounding \quad Area}{Total \quad Chip \quad Area}}{1 - \frac{m}{N_m}} \tag{5.11}$$

The block bounding area of m megacells is the minimum rectangular area encompassing all m megacells. Likewise, the rectangular area of an m-terminal net is the minimum rectangular area covering all m terminals. Figure 5.4 illustrates the block bounding area and the net bounding area of a 4-terminal net connecting 4 megacells with a placement efficiency of 60%. As shown, the net bounding area is a fraction of block bounding area. Assuming a random placement of terminals with uniform distribution, the edge length of the net-bounding area, e, is related to the edge length of the block bounding area, $\hat{e}$ as below [7]

$$e = \frac{m-1}{m+1} \cdot \hat{e} \tag{5.12}$$

Modeling of Global Routing Information. The routing information provides the length of a net for a given net bounding area assuming a minimum rectilinear Steiner tree (MRST) construction for the net [7]. Figure 5.5(a) shows an example of a six-terminal net based on MRST. A rule of thumb to calculate the netlength is known as the half-perimeter model [8] that approximates the net length by the half perimeter of the net bounding area. For example, in the net shown in Figure 5.5, the length is estimated as

$$L_{HP} = a + b, \tag{5.13}$$

where a and b are the net bounding area dimensions and L_{HP} is the estimated net length using the half-perimeter model. It can be easily shown that the

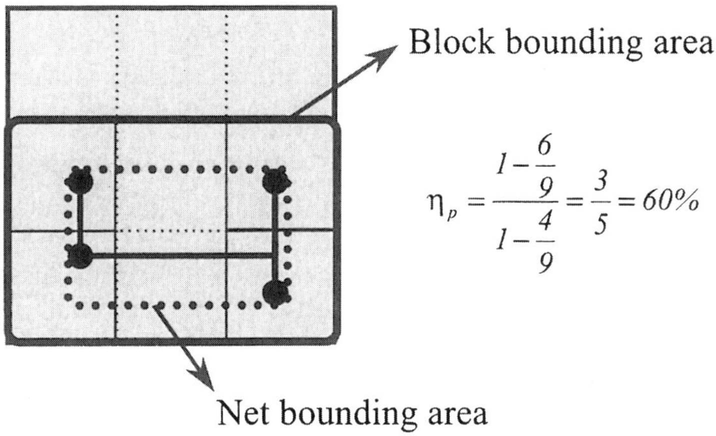

Figure 5.4. Block bounding area and net bounding area in a system with 9 megacells

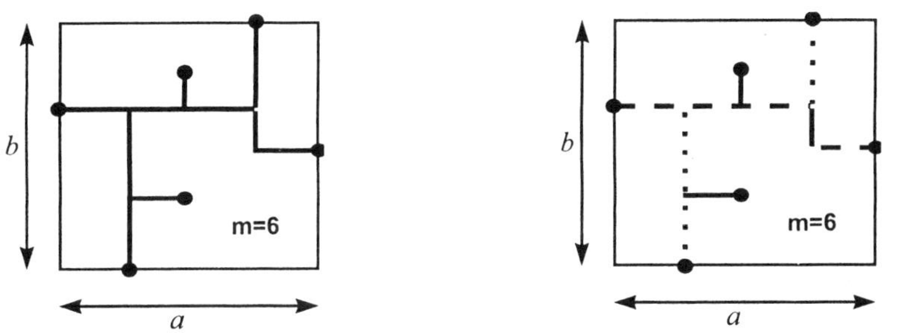

(a) The complete net structure using MRST

(b) The segments representing the net bounding area dimensions

Figure 5.5. An example of a six-terminal net based on MRST

half-perimeter model always underestimates the net length for m-terminal nets when $m > 3$. Suppose that the net in Figure 5.5(a) is divided into segments, as shown in Figure 5.5(b). The sum of the length of dashed line and dotted line segments in the net bounding area is equal to the length of a and b, respectively. Therefore, the actual wire length of the MRST net is the half-perimeter length $(a + b)$ plus the solid line segments,

$$L_{MRST} = \delta L + (a + b) \tag{5.14}$$

where δL represents the length of the additional segments, which is a function of the number of terminals and the bounding area dimensions. Also, δL depends on the placement of the terminals. Now, we assume that the terminals are placed randomly with uniform probability density function over the entire bounding area. The probability density function of net length can be computed by the random-walk technique. In this technique, first, MRST nets are constructed for several layouts with randomly placed terminals. Then, the probability density function is constructed by normalizing the plot of number of occurrence versus net length. An example of the net-length pdf for the case of a 30-terminal net over the block bounding area of $10 \times 20mm^2$ is shown in Figure 5.6. As shown, the net-length pdf has an average of 61 mm and a standard deviation of 3.98 mm.

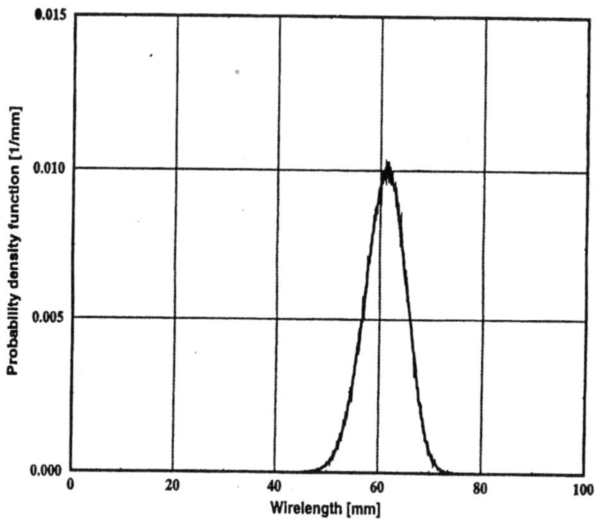

Figure 5.6. An example of net-length probability density function

To find a closed-form expression for the average MRST net length, suppose that there is an m-terminal MRST net bounded with a square of size d (a=b=d). The net length is proportional to d because scaling the bounding dimension would scale the net length with the same weight of scaling. Thus, the net length can be represented by

$$L_{MRST} = f(m) \cdot d \tag{5.15}$$

where f is a function of the number of net terminals, m, d is the net bounding area dimension, and L_{MRST} is the net length of the MRST net. Now, suppose that the area is divided into four equal pieces with dimensions of $\frac{d}{2}$ by $\frac{d}{2}$, as shown in Figure 5.7.

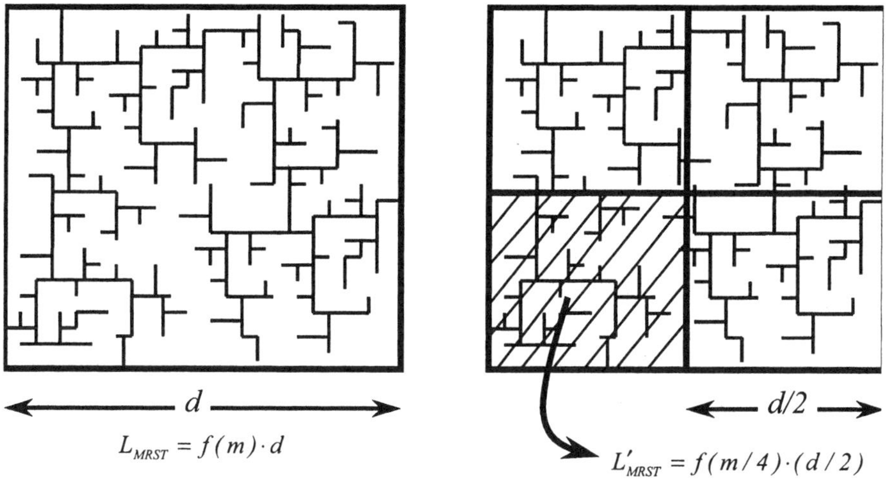

Figure 5.7. Derivation of the form of MRST net-length function based on net partitioning

Assuming uniform distribution of terminals, there are $m/4$ terminals in each piece, which means that the length of the net in each piece is $L'_{MRST} = f(\frac{m}{4}) \cdot (\frac{d}{2})$. For a very large value of m, the length of broken segments, resulting from this dividing, can be ignored compared with the total net length. Therefore, $L_{MRST} = 4L'_{MRST}$ or $f(m) \cdot d = 4 \cdot f(\frac{m}{4}) \cdot (\frac{d}{2})$ which gives

$$\frac{f(m)}{2} = f(\frac{m}{4}) \tag{5.16}$$

By examining power-law functions, the only solution for (5.16) is $f(m) = \alpha \cdot m^{\gamma}$, where $\gamma = 0.5$ precisely, and α can be any number. In the general case, when m is large, by considering the length of broken segments, the general solution of the same equation as described above is

$$f(m) = \alpha \cdot m^{\gamma} - \beta \tag{5.17}$$

where α, β, and γ are computed by curve fitting. For the case of a rectangular bounding area with the edge length of e, the form of the equation for the average net length of an MRST net is given by

$$L_{av} \approx (\alpha \cdot m^{\gamma} - \beta)\, e + (2e) \tag{5.18}$$

where α, β, and γ are the fitting parameters that have been computed as $\alpha = 0.55$, $\beta = 1.0$, and $\gamma = 0.5$. The first term, $(\alpha \cdot m^{\gamma} - \beta)\, e$, is a representation of δL in (5.14), m is the number of terminals of the net, and e is the edge length of the net bounding area.

Complete Stochastic Global Net-Length Distribution. The complete global net-length distribution is derived by combining all three distinct information bases: netlist, placement, and routing. The netlist information calculates the number of nets for each fan-out by using heterogeneous Rent's rule. The placement information gives the pdf of the net bounding area dimensions computed by the random-walk technique. Using the routing information, the net bounding area dimensions are translated to the net length, assuming MRST construction for the net.

Megacell's Name	k	N	p	Megacell's Name	k	N	p
Instruction Cache	4.12	380,000	0.20	Instr. Fetch Address	3.20	16,500	0.60
Instruction Cache Tags	3.80	18,000	0.47	Instr. Fetch Data Path	3.20	13,800	0.60
Data Cache	4.12	350,000	0.20	Instr. Fetch Control	3.20	9,500	0.60
Data Cache Tags	3.80	25,500	0.47	Address Queue	3.20	22,000	0.60
TLB	3.80	22,400	0.35	Inst. Decode & Reg. Ren.	3.20	45,300	0.60
Secondary Cache Ctrl.	3.20	15,700	0.60	Integer Data Path	3.20	43,800	0.60
External Interface	3.20	18,400	0.60	Integer Queue	3.20	19,700	0.60
Sys. Interface Buffers	3.20	22,600	0.60	Floating Point Data Path	3.20	32,600	0.60
Free List	3.20	9,800	0.60	Floating Point Queue	3.20	51,000	0.60
Graduation unit	3.20	26,300	0.60	Floating Point Multiplier	3.20	19,300	0.60

Figure 5.8. Rent's parameters of megacells

The complete global net-length distribution is the summation of net-length pdf multiplied by the corresponding number of nets for all fan-outs. Equations (5.10), (5.12), and (5.18) can be used to compute the average of net-length pdf for each fan-out. A real RISC microprocessor [9] has been modeled as an example to verify the new model derived here. The chip size is 16.6×17.8 mm^2 with 20 different megacells ranging in size from 9,500 gates to 380,000 gates. Figure 5.9 shows the Rent's parameters and the number of gates in the megacells. Equations (5.10), (5.12), and (5.18) give the average net length for each fan-out, which is depicted in Figure 5.9, assuming a placement efficiency of 80%. In the absence of detailed placement data to compute the placement efficiency in this example, the value of 80% is selected for the placement efficiency because it gives the best fit to the data.

The second column of Figure 5.9 gives the total number of nets for the fan-out specified in column one. Column three gives the average net bounding area from the placement information equation and the fourth column gives the av-

Global Fan out	Netlist Info. Total No. of Nets	Placement Info. Average Net Bounding Area	Routing Info. Average Net Length	Total Net Length
1	3632	3.03×3.03 mm^2	5.173 mm	18788 mm
2	1184	4.86×4.86 mm^2	9.068 mm	10736 mm
3	561	6.18×6.18 mm^2	12.36 mm	6933.9 mm
4	318	7.24×7.24 mm^2	15.33 mm	4874.9 mm
5	200	8.14×8.14 mm^2	18.10 mm	3620.0 mm
6	134	8.93×8.93 mm^2	20.74 mm	2779.1 mm
7	94	9.64×9.64 mm^2	23.27 mm	2187.3 mm
8	68	10.3×10.3 mm^2	25.75 mm	1751.0 mm
9	50	10.9×10.9 mm^2	28.13 mm	1406.5 mm
10	38	11.4×11.4 mm^2	30.30 mm	1151.4 mm
11	28	12.0×12.0 mm^2	32.78 mm	917.96 mm
12	22	12.5×12.5 mm^2	35.03 mm	770.66 mm
13	16	13.0×13.0 mm^2	37.32 mm	597.12 mm
14	12	13.4×13.4 mm^2	39.34 mm	527.15 mm
15	9	13.9×13.9 mm^2	41.70 mm	375.30 mm
16	6	14.3×14.3 mm^2	43.78 mm	262.68 mm
17	4	14.7×14.7 mm^2	45.88 mm	183.52 mm
18	3	15.1×15.1 mm^2	48.00 mm	144.00 mm
19	2	15.5×15.5 mm^2	50.15 mm	100.30 mm
Total	6381	-	-	58106.8 mm

Figure 5.9. Netlist, placement, and routing information

erage net length from the routing information using the average net bounding computed in the second column. The total net length is computed by multiplying the average net length by the number of nets for each fan-out. Figure 5.9 shows that the total number of global nets is 6381, the average global net length is 9.11 mm, and the total global net length is 58.1 m. Note that, based on [9], the total number of global nets is about 6000, which supports the prediction of the netlist information presented here. The complete global net-length distribution is derived by combining the three models of netlist, placement, and routing information. In summary, the netlist information is computed based on the heterogeneous Rent's rule. The placement and routing information for every fan-out is obtained from the random-walk technique described in Section 2.1.3. The global net density as a function of net length is compared to actual data from [9] in Figure 5.10.

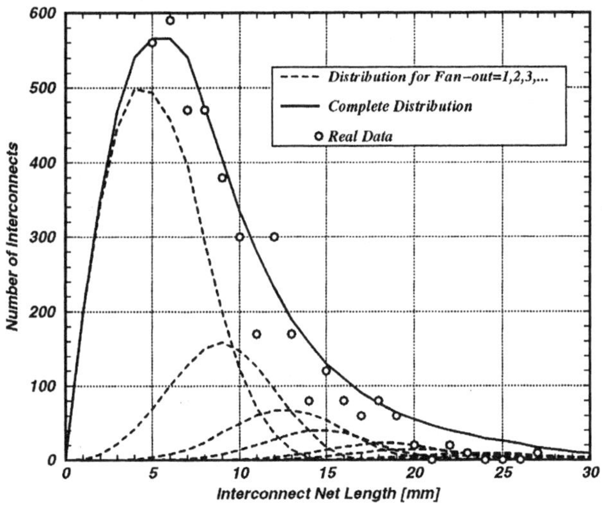

Figure 5.10. The complete global net-length distribution compared with real data

5.2.2 Signal Interconnect Bandwidth Model (RC and RLC)

The resistance and capacitance of on-chip wires create a limit on the maximum wire length for a given frequency because of the interconnect bandwidth. In this section, some models for interconnect bandwidth are presented. These models are lumped RC, distributed RC, and distributed RLC models.

Lumped RC Model for Interconnect Bandwidth. In the simplest case, an interconnect can be modeled as an RC low-pass filter with the bandwidth

$$f_{-3dB} = \frac{1}{2\pi\left(r_{int}c_{int}\right)\cdot l^2} \tag{5.19}$$

where r_{int} and c_{int} are the distributed resistance and capacitance per unit length of the interconnect, and l is the length of the line, as shown in the Figure 5.11. Although the lumped RC model underestimates the actual bandwidth of a distributed RC model, it gives the simplest expressions for the first-order approximation of the interconnect bandwidth.

Distributed RC and RLC Model for Interconnect Bandwidth. To derive a more accurate model for the interconnect bandwidth, the interconnect must be modeled as a distributed RLC line. A closed-form expression for the frequency response of a distributed RLC network with arbitrary source

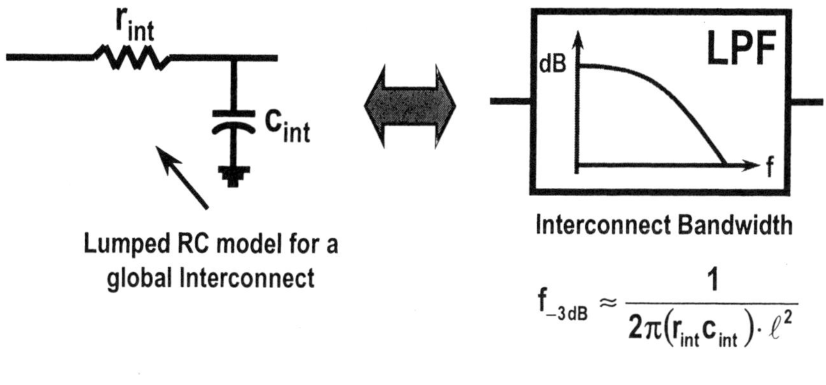

Figure 5.11. A lumped RC model for a global interconnect

and load terminations is derived in this section, by using Z-domain analysis in control theory. It has been shown that when a traveling wave reaches the end of a finite length transmission line, it reflects back to the line depending on the termination load. By using telegrapher's equation [10], the reflected wave, V_r, is computed by

$$V_r = V_i \cdot \Gamma_L = V_i \frac{Z_L - Z_o}{Z_L + Z_o} \tag{5.20}$$

where V_i is the incident wave, Γ is the reflection coefficient at the load, Z_L is the load impedance, and Z_o is the line characteristic impedance. In the case of lossy transmission line, Z_o is a complex and frequency dependent function

$$Z_o = \sqrt{\frac{r + jwl}{jwc}}, \tag{5.21}$$

where r is the distributed resistance per unit length, l is the distributed inductance per unit length, and c is the distributed capacitance per unit length. Moreover, in the general case, Z_L can be a complex and frequency dependent function. For example, when the load is an input capacitance of a CMOS buffer, C_L, the load impedance is given by

$$Z_L = \frac{1}{jwC_L} \tag{5.22}$$

Likewise, the reflected wave will be an incident wave when it reaches the beginning of the line. The reflection coefficient, in this case at the source side, is

$$\Gamma_S = \frac{Z_S - Z_o}{Z_S + Z_o} \tag{5.23}$$

where Z_S is the source impedance and Z_o is the line characteristic impedance given in (21). The electrical circuit model of a transmission line with arbitrary termination is illustrated in Figure 5.12. As shown in Figure 5.12, since the reflected wave at the load is $-\Gamma_L \cdot V_i$, the transferred wave that comes out of the line at the load, because of the energy conservation theory, will be $(1 + \Gamma_L) \cdot V_i$. Moreover, from voltage division in circuit theory, the incoming wave from the source side, V_{inc}, as a function of the input source voltage, V_S, is

$$V_{inc} = V_S \frac{Z_o}{Z_S + Z_o} = V_S \frac{1 - \Gamma_S}{2} \tag{5.24}$$

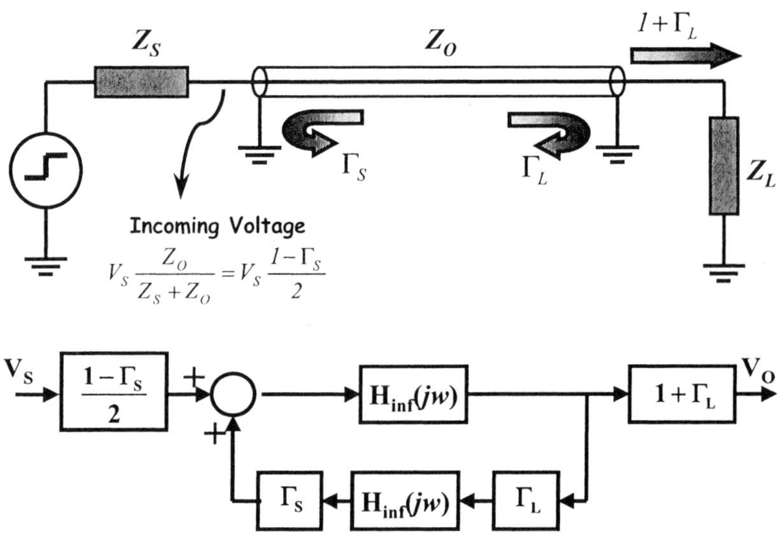

Figure 5.12. Electrical circuit model for a transmission line with arbitrary termination

Utilizing the circuit model of Figure 5.12 along with control theory, the block diagram of the system of transmission line with arbitrary termination is illustrated in Figure 5.12, where $H_{inf}(jw)$ is the transfer function of an infinite length transmission line and Γ_L and Γ_S are the reflection coefficients at the load and source, as given in (5.20) and (5.23), respectively. From control theory, the overall transfer function of the system is

$$H(jw) = \frac{V_O}{V_S} = \frac{(1 - \Gamma_S)(1 + \Gamma_L)}{2} \cdot \frac{H_{\text{inf}}(jw)}{1 - \Gamma_S \Gamma_L H_{\text{inf}}(jw)^2}. \tag{5.25}$$

To derive the transfer function of an infinite transmission line, $H_{inf}(jw)$, the partial differential equation that describes the distributed transmission line is used as

$$\frac{\partial^2}{\partial x^2}V(x,t) = lc\frac{\partial^2}{\partial t^2}V(x,t) + rc\frac{\partial}{\partial t}V(x,t), \tag{5.26}$$

where r is the distributed resistance per unit length, l is the distributed inductance per unit length, and c is the distributed capacitance per unit length. Using a single-sided Laplace transform of $V(x,t)$, the partial differential equation (5.26) becomes an ordinary differential equation. It is assumed that the initial values of the voltage and the current on the transmission line are zero, which gives

$$\frac{\partial^2}{\partial x^2}V(x,s) = lcs^2V(x,s) + rcsV(x,s) \tag{5.27}$$

The general solution to this expression in the Laplace domain is

$$V(x,s) = Ae^{-xscZ_o} + Be^{xscZ_o}. \tag{5.28}$$

The coefficient B must be zero so that the solution of (5.28) is well-behaved and finite at infinity. Likewise, the coefficient A is determined from the boundary condition at $x = 0$, i.e., $V(0,s)$. Therefore, the voltage at point of $x = L$, $V(x = L, s)$, is

$$V(x = L, s) = V(0,s)e^{-sLcZ_o} \tag{5.29}$$

Therefore, from (5.29) and considering the fact that $s = jw$, the transfer function of an infinite length transmission line, $H_{inf}(jw)$, is given by

$$H_{inf}(jw) = e^{-LZ_ojwc}. \tag{5.30}$$

Substituting (5.30) into (25) gives

$$H(jw) = \frac{(1-\Gamma_S)(1+\Gamma_L)}{2} \cdot \frac{e^{-LZ_ojwc}}{1-\Gamma_S\Gamma_L e^{-2LZ_ojwc}}, \tag{5.31}$$

where $\Gamma_S = Z_S - Z_o/Z_S + Z_o$ is the reflection coefficient at the source, $\Gamma_L = Z_L - Z_o/Z_L + Z_o$ is the reflection coefficient at the load, and $Z_o = \sqrt{(r + jwl)/jwc}$ is the transmission line characteristic impedance given in (5.23), (5.20), and (5.21), respectively. Equation (5.31) is a closed-form expression that gives the frequency response of a finite length transmission line with arbitrary termination. There are several applications for the closed-form expression of (5.31). An example of utilizing (31) to derive the frequency response of an interconnect is described below. Utilizing (5.31), an example of the frequency response of an interconnect with the lengths of 1 cm, 2 cm, and 3 cm is shown in Figure 5.13. As shown, the -3 dB bandwidth of the interconnect is almost the same for the distributed RC and RLC models. This phenomenon has also been observed in [11]. Figure 5.13 also shows that the -3 dB frequency of the 2 cm line, in this example, is about 300 MHz for distributed RLC model and 280 MHz for RC model. Likewise, using the lumped RC model (5.19), the -3 dB frequency of the 2 cm line, in this example, is

$$f_{-3dB} = \frac{1}{2\pi\left(r_{\text{int}}l + R_{tr}\right)\cdot c_{\text{int}}l} \approx 265 \text{ MHz} \tag{5.32}$$

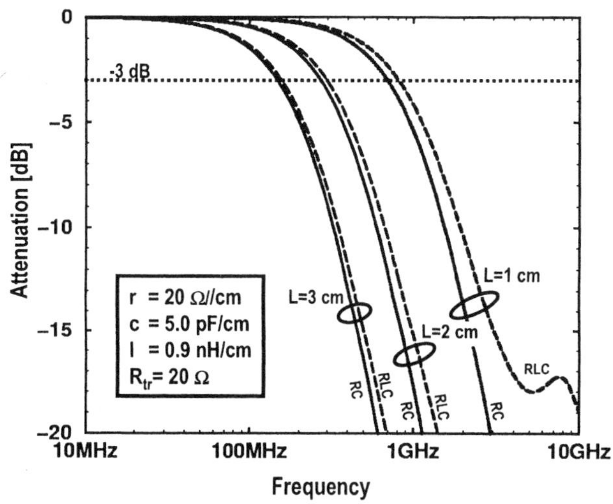

Figure 5.13. Interconnect frequency response using distributed RC and RLC models

Equation (5.32) confirms that although the lumped RC model underestimates the actual bandwidth of a distributed RC and RLC line, in the most typical cases, it gives a first-order approximation for the interconnect bandwidth.

5.2.3 Signal Interconnect Crosstalk Model (RC and RLC)

Interconnect coupling noise or crosstalk is often a major concern for global wires. The modeling of crosstalk noise is often a complex problem since the crosstalk noise is a function of input pulse rise time and width, length of the interconnect, and driver source resistance. Significant work has been done by

Sakurai [12] and Davis, *et al.* [13] on analyzing the crosstalk in distributed RC and RLC networks. In this section, the results of their work are used to model the crosstalk noise of global interconnects. Unlike the interconnect bandwidth, the peak crosstalk noise prediction usually needs an accurate distributed RLC approach. Figure 5.14 shows the HSPICE simulations for the peak crosstalk noise using distributed RC and RLC approaches. As shown, the peak crosstalk noise in an RLC line is often higher than that of an RC line. It has been shown in [14] that the worst-case peak crosstalk noise using a distributed RLC model can be simplified to

$$\frac{V_n}{V_{dd}} \approx \frac{\pi}{4} \frac{c_m}{c_{gnd} + c_m}, \tag{5.33}$$

where c_m and c_{gnd} are the mutual and ground capacitance per unit length, respectively. Also, as the length of the interconnect increases, the crosstalk noise using both RC and RLC models saturates at the same level given by [12]

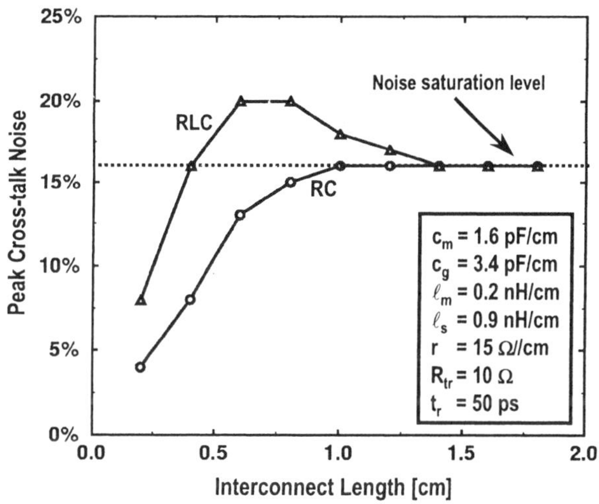

Figure 5.14. Peak crosstalk noise versus wire length using HSPICE

$$\frac{V_n}{V_{dd}} \approx \frac{1}{2} \frac{c_m}{c_{gnd} + c_m} \tag{5.34}$$

where c_m and c_{gnd} are the mutual and ground capacitance per unit length, respectively. From the comparison between (5.33) and (5.34), it is shown that the worst-case peak crosstalk noise of a distributed RLC model is about 57% more than that of an RC model.

5.3 Global Clock Distribution Modeling

In high-speed digital systems, clock distribution is a challenging problem, consuming an increasing fraction of resources such as design time, power, and wiring. There are three main issues in the design and optimization of a clock distribution network: clock skew, clock signal degradation, and clock power dissipation. A new compact model for on-chip clock skew as a function of device, interconnect, and system parameter variations is derived in this section. Unlike the previous models that describe qualitative behavior of clock skew components [15], the new model provides a closed-form expression for each clock skew component. Utilizing the interconnect bandwidth model presented in Section 5.2.2, a limit for signal degradation in the global clock distribution networks is defined.

Finally, a model for estimating the global clock distribution power dissipation is provided. Utilizing the models for clock skew and clock signal degradation, a design plane for global clock distribution network is presented.

5.3.1 Clock Skew Model

Clock skew arises mainly from unequal clock path lengths to various modules. To equalize line lengths and thus reduce the clock skew, a common practice is to use a balanced clock network. In this way, the nominal value of skew becomes zero and clock skew reduces to the variations of the clock path from the clock generator to the registers. These variations may originate from process and circuit parameter tolerances, which are codified hereafter.

Device Parameter Variations– In the IC fabrication process, all device parameters are subject to deviations from their nominal values. Statistical models have been developed for transistor parameters such as threshold voltage (ΔV_T), gate oxide thickness (Δt_{ox}), and effective channel length (ΔL_{eff}) [16].

Interconnect Parameter Variations – Interconnect width (ΔW_{int}) and thickness (ΔH_{int}) and interlevel dielectric (ILD) thickness (ΔT_{ILD}) variations are the main parameters of interest. As technology advances, the number of interconnect layers increases and the interconnect lines become more nonuniform. This nonuniformity, which is caused by manufacturing processes, produces large variations of interconnect parameter values. Chemical mechanical polishing (CMP) is a new manufacturing process for planarization of metal and ILD layers that greatly reduces the ILD nonuniformity in multilayer structures. However, the CMP process still does not eliminate interconnect parameter variations completely. Metal dishing and oxide erosion are the vexing issues in CMP that degrade the planarization efficiency [17]. The amount of metal dishing and oxide erosion, in general, is a function of metal width and pattern density [17]. Hence, interconnect and ILD thickness variations are important considerations even after using CMP processes.

System Parameter Variations – Besides process parameter variations, which are mainly the tolerances of device and interconnect physical parameters, system-level fluctuations may create clock skew. Power-supply fluctuations (IR drop, ΔV_{DD}), temperature variations (ΔT), and nonuniform distribution of clocked registers (clock driver load mismatch), (ΔC_L) are considered system-level parameter variations.

Derivation of Clock Skew Model. The goal of the derivation of a clock skew model is to understand the impact of the process and system variations in an ideally zero-skew clock network distribution. The most common strategy to ensure zero-nominal clock skew, which is often used for distributing high-frequency clock signals in digital systems, is a symmetric H-tree structure [18]. Although a model is derived here especially for the symmetric H-tree structure, the model can be easily modified for any balanced clock tree network .

Assumptions. Although the growing importance of on-chip transmission line effects has been predicted [19], the difficulty of modeling and simulating lossy and nonuniform transmission lines using existing CAD tools has prevented consideration of these effects in most cases. There are some methods, however, to reduce the inductance effects in actual design. For instance, for the clock distribution described in [9], ground return path wiring has been implemented on the two metal levels above and below the clock wire to reduce inductance effects. To simplify the derivation of a clock skew model, good return path wiring has been assumed to surround the clock-wiring network. Therefore, in this simplified study, the inductance effect is ignored. Moreover, without loss of generality, it is assumed that the clock network is a balanced H-tree structure. This model, however, can be easily modified for any balanced clock tree network.

Compact Clock Skew Model. Figure 5.15 shows a symmetric H-tree clock distribution with n=4 levels of H-tree branches. At the end of the fourth level, drivers are implemented to feed the clock signal to all registers in the subblocks. The total clock skew, by definition, is the time difference between the maximum and minimum delays, as illustrated in Figure 5.15.

Using the equivalent circuit diagram shown in Figure 5.16, the delay of the entire clock network of Figure 5.15 is sorted into three parts: i) H-tree network, ii) clock driver, and iii) subblock routing. Derivation of the above delay components is described below.

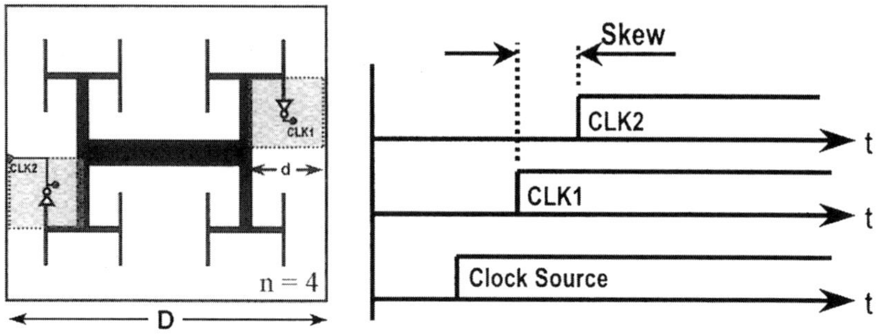

Figure 5.15. Clock skew between the two points, CLK1 and CLK2, in a symmetric H-tree structure

H-Tree Network Delay. As shown in Figure 5.16, the delay of the H-Tree network is mainly due to the interconnect delay from the clock source located at the center of the chip through the H-tree to the driver. Assuming that the H-tree network is driven by a single driver and the line capacitance of the H-tree network is much greater than the transistor input capacitance of a subblock clock driver, then the interconnect delay expression for the approximate 50% time delay of the distributed RC line using Sakurai's model is [12], [18]

$$T_{H-Tree} = 0.4\,(r_{\text{int}}c_{\text{int}}) \cdot l^2 + \frac{\sqrt{\varepsilon_r}}{c_o} \cdot l \tag{5.35}$$

where l is the length of the H-tree network, r_{int} and c_{int} are the distributed

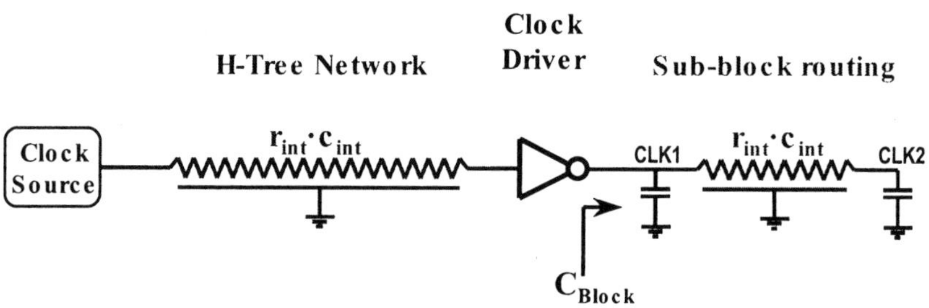

Figure 5.16. Equivalent circuit of clock network in Figure

resistance and capacitance of the line, ϵ_r is the relative dielectric constant of the ILD material, and c_o is the speed of light in free space. Since the wires in clock distributions are often much wider than the minimum wire width [19], the fringing capacitance is negligible compared with parallel plate wiring capacitance. Using the expression for the length of the H-tree versus die size, D, and the number of H-tree levels, n, defined in [18], then (5.35) becomes

$$T_{H-Tree} = 0.4\left(\frac{\rho \cdot \varepsilon_r}{H_{\text{int}} \cdot T_{ILD}}\right) \cdot D^2 \left(1 - \frac{1}{2^{n/2}}\right)^2 + \frac{\sqrt{\varepsilon_r}}{c_o} \cdot D\left(1 - \frac{1}{2^{n/2}}\right) \tag{5.36}$$

where H_{int} is the interconnect thickness, T_{ILD} is the interlevel dielectric thickness, and ρ is the line resistivity.

Clock Driver Delay. Transistor delay of the subblock clock driver defines the clock driver delay. The clocked registers within subblocks are assumed to be randomly placed and routed; therefore, the delay expression for the approximate 50% time delay of the subblock drivers is simply

$$T_{Driver} = 0.7R_{tr} \cdot C_L = 0.7\left(\frac{(L_{eff}/W)}{\mu \cdot C_{ox}\left(V_{DD} - V_T\right)}\right) \cdot C_L \tag{5.37}$$

where C_L is the total wiring and register input capacitance within the subblock; L and W are transistor channel length and width, respectively; μ is the mobility; C_{ox} is the gate oxide capacitance; and V_{DD} and V_T are supply and transistor threshold voltage, respectively.

Subblock Routing Delay. The delay of subblock routing is basically the internal wire routing delay within the subblock from the clock driver to the registers. The wiring delay inside the subblock is computed in a similar way from (5.35) except that the length of wire, l, is the Manhattan distance from center to the corner of the subblock as, described in [18].

$$T_{Sub-Blk} = 0.4\left(r_{\text{int}} c_{\text{int}}\right) d^2 + \frac{\sqrt{\varepsilon_{r,ox}}}{c_o} d = 0.4\left(r_{\text{int}} c_{\text{int}}\right) \frac{D^2}{2^n} + \frac{\sqrt{\varepsilon_{r,ox}}}{c_o} \frac{D}{2^{n/2}} \tag{5.38}$$

Since, in general, the placement of clocked registers is not uniform, the routing within the subblock is not equidistance. For instance, in Figure 5.15, the clock signal at the point CLK2 arrives later than CLK1. This delay, which is often called internal clock skew [18], in the worst case, is given by (5.38). The overall delay of the entire clock distribution network from the clock source to the clocked registers is

$$T_{Delay} = T_{H-Tree} + T_{Driver} + T_{Sub-Blk}.$$

Since the subblock size is often much less than the chip size, the wiring delay within the subblock, $T_{Sub-Blk}$, can be ignored. Therefore, the total delay is given by

$$T_{Delay} \approx \quad T_{H-Tree} + T_{Driver} = 0.4\left(\frac{\rho \cdot \varepsilon_r}{H_{\text{int}} \cdot T_{ILD}}\right) \cdot D^2\left(1 - \frac{1}{2^{n/2}}\right)^2 + \frac{\sqrt{\varepsilon_r}}{c_o} \cdot D\left(1 - \frac{1}{2^{n/2}}\right) + 0.7\left(\frac{(L_{eff}/W)}{\mu \cdot C_{ox}(V_{DD} - V_T)}\right) \cdot C_L \tag{5.39}$$

Equation (5.39) contains all device, interconnect, and system parameters described earlier. Assuming that these parameters have small variations compared with their nominal values, the clock skew, T_{CSK}, can be evaluated by

$$T_{CSK}(x) = \Delta T_{Delay} \approx \left| \frac{\partial T_{Delay}}{\partial x} \right| \Delta x \tag{5.40}$$

where T_{Delay} is the complete delay function of (5.39) and x is any variation of clock skew components such as ΔH_{int}, ΔT_{ILD}, ΔV_{DD}, ΔV_T, Δt_{ox}, ΔL_{eff}, and ΔC_L. Figure 5.17 shows the closed-form equations for each individual clock skew component by using (5.40). Therefore, the overall clock skew is given by

$$T_{CSK} = 0.4\left(r_{\text{int}}c_{\text{int}}\right)\frac{D^2}{2^n} + \frac{\sqrt{\varepsilon_{r,ox}}}{c_o}\frac{D}{2^{n/2}} + \sum_i \left|\frac{\partial T_{Delay}}{\partial x_i}\right| \cdot \Delta x_i \tag{5.41}$$

where the first and second terms, given by (5.41), are often called internal clock skew [4], which is the time difference between CLK1 and CLK2 in the subblock circuit, as shown in Figure 5.16. The third term in (5.41) is the summation of all skew components shown in Figure 5.17.

Clock Skew for Temperature (T) Variations. The clock skew resulting from temperature gradient on a chip, in general, is more complex since there are three main parameters that vary with temperature: resistivity of interconnect $\rho(T)$, threshold voltage $V_T(T)$, and mobility $\mu(T)$. Assuming that the variation of threshold voltage is greater than that of mobility and resistivity of lines, then the clock skew resulting from temperature difference is given by

$$T_{CSK}(T) = \Delta T_{Delay} \approx \left| \frac{\partial T_{Delay}}{\partial V_T} \cdot \frac{\partial V_T}{\partial T} \right| \cdot \Delta T \tag{5.42}$$

where ΔT is the temperature difference of two points in the chip. The first expression, $\partial T_{Delay}/\partial V_T$, is computed from (5.39) as

$$\frac{\partial T_{Delay}}{\partial V_T} = 0.7R_{tr} \cdot C_L \cdot \left(\frac{1}{V_{DD} - V_T}\right) \tag{5.43}$$

Also, the second expression, $\partial V_T/\partial T$, is [20]

$$\frac{\partial V_T}{\partial T} = \frac{1}{T}\left(2 - \frac{Q_B}{2C_{ox}\phi_f}\right) \cdot \left(\phi_f + \frac{E_g}{2q}\right) \tag{5.44}$$

where Q_B is the depletion region charge, C_{ox} is the gate oxide capacitance, ϕ_f is the Fermi level potential, E_g is the energy gap of Si, and q is the charge of electron. To simplify (5.44), the threshold voltage can be written as $V_T \approx \phi_{Si} - Q_B/C_{ox}$ [20]. Moreover, assuming that the substrate doping concentration is relatively high, then the surface potential of MOSFET transistor is given by $\phi_{Si} = 2\phi_f \approx 2\,(E_g/2)$. Therefore, the first-order approximation of (5.44) is given by

Physical Parameter and Derivation used		***Clock Skew Compact Model***
ILD Thickness Variation	$\left\|\frac{\partial T_{Delay}}{\partial T_{ILD}}\right\|$	$T_{CSK}(T_{ILD}) = 0.4(r_{int}c_{int}) \cdot D^2\left(1-\frac{1}{2^{n/2}}\right)^2 \cdot \left(\frac{\Delta T_{ILD}}{T_{ILD}}\right)$
Wire Thickness Variation	$\left\|\frac{\partial T_{Delay}}{\partial H_{int}}\right\|$	$T_{CSK}(H_{int}) = 0.4(r_{int}c_{int}) \cdot D^2\left(1-\frac{1}{2^{n/2}}\right)^2 \cdot \left(\frac{\Delta H_{int}}{H_{int}}\right)$
Threshold Voltage Fluctuation	$\left\|\frac{\partial T_{Delay}}{\partial V_T}\right\|$	$T_{CSK}(V_T) = 0.7R_{tr}C_L\left(\frac{V_T}{V_{DD}-V_T}\right) \cdot \left(\frac{\Delta V_T}{V_T}\right)$
Transistor Channel Length Tolerance	$\left\|\frac{\partial T_{Delay}}{\partial L_{eff}}\right\|$	$T_{CSK}(L_{eff}) = 0.7R_{tr}C_L \cdot \left(\frac{\Delta L_{eff}}{L_{eff}}\right)$
Gate Oxide Thickness Tolerance	$\left\|\frac{\partial T_{Delay}}{\partial t_{ox}}\right\|$	$T_{CSK}(t_{ox}) = 0.7R_{tr}C_L \cdot \left(\frac{\Delta t_{ox}}{t_{ox}}\right)$
IR Drop	$\left\|\frac{\partial T_{Delay}}{\partial V_{DD}}\right\|$	$T_{CSK}(V_{DD}) = 0.7R_{tr}C_L\left(\frac{V_{DD}}{V_{DD}-V_T}\right) \cdot \left(\frac{\Delta V_{DD}}{V_{DD}}\right)$
Non-uniform Register Distribution	$\left\|\frac{\partial T_{Delay}}{\partial C_L}\right\|$	$T_{CSK}(C_L) = 0.7R_{tr}C_L \cdot \left(\frac{\Delta C_L}{C_L}\right)$
Temperature Gradient	$\left\|\frac{\partial T_{Delay}}{\partial T}\right\|$	$T_{CSK}(T) = 0.7R_{tr}C_L\left(\frac{E_g/q+V_T}{V_{DD}-V_T}\right) \cdot \left(\frac{\Delta T}{T}\right)$
Internal Clock Skew	–	$T_{CSK}(internal) = 0.4(r_{int}c_{int}) \cdot \frac{D^2}{2^n} + \frac{\sqrt{\varepsilon_{r,ox}}}{c_o} \cdot \frac{D}{2^{n/2}}$

Figure 5.17. Clock skew components

$$\frac{\partial V_T}{\partial T} \approx \frac{E_g/q + V_T}{T} \tag{5.45}$$

where $E_g/q = 1.12$ V is the energy gap of Si in volts and T is the temperature in degrees Kelvin. Equation (42) along with the results of (43) and (45) give

$$T_{CSK}(T) \approx 0.7R_{tr} \cdot C_L \cdot \left(\frac{E_g/q + V_T}{V_{DD} - V_T}\right) \cdot \left(\frac{\Delta T}{T}\right) \tag{5.46}$$

An Example of a Design in 0.18μm Technology. To illustrate the clock skew model, a design example for 0.18 μm technology has been studied using the design parameters [16] illustrated in Figure 5.18. The H-tree clock distribution is assumed to be routed by the fourth metal level shielded with the third and fifth metal levels, as depicted in Figure 5.18. This wiring structure ensures minimal inductive effect. Equation (5.35) is therefore valid for interconnect propagation delay with $r_{int}c_{int} = 115$ ps/cm^2, assuming that Cu/SiO$_2$ materials are used in the wiring network. In this example, the total number of clocked registers is set to 20,000. With an input register capacitance of 5 fF, the total capacitance of each subblock, on average, is computed as $C_L = 6.25$ pF. A driver with $R_{tr} = 12.0\Omega$ output resistance is selected to ensure fast rise time for the total subblock loading capacitance. Using the expressions of Figure 5.17, the complete set of clock skew components is evaluated, as shown in Figure 5.19. In this table, the second column is expressed as the sensitivity of clock skew in picoseconds per 1% for the given parameters. The third column contains percentage of variations, which are technology and design dependent.

Parameters		Values
Process Parameter	L_{eff}	0.18 [μm]
	V_T	0.32 [V]
	V_{DD}	1.8 [V]
	$r_{int}c_{int}$	115 [ps/cm^2]
Design Parameter	R_{tr}	12.0 [Ω]
	C_L	6.25 [pF]
	D	2.0 [cm]
	n	4

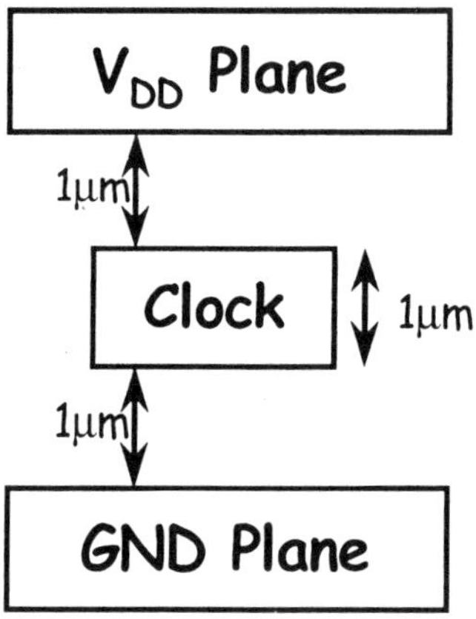

Figure 5.18. Process and design parameters as well as clock distribution wiring structure

The tolerances of interconnect and ILD thickness are roughly 3% for a well-controlled CMP process described in [17]. Statistical modeling in [16] extracted from the measurement data of 0.18 μm technology shows that the fluctuations of threshold voltage, MOS effective channel length, and gate oxide thickness are about 5%, 5%, and 1.2%, respectively. The tolerance of supply voltage is usually limited to 10% of total supply voltage [21]. Moreover, the thermal image of the Alpha microprocessor in [21] shows a 30°C temperature gradient over the entire chip, which gives a temperature variation of about 8%. The variations of loading capacitance of clock drivers highly depend on the

Parameters	Clock Skew Coeff. [ps/1%]	%of Variations	Clock Skew Comp. [ps]
T_{ILD}	1.04	3%	3.12
H_{int}	1.04	3%	3.12
V_T	0.11	5%	0.55
L_{eff}	0.53	5%	2.65
t_{ox}	0.53	1.2%	0.64
V_{DD}	0.64	10%	6.40
C_L	0.53	20%	10.6
T	0.51	8%	4.08
Internal Clock Skew [ps]			61.7
Total Clock Skew [ps]			92.9

Figure 5.19. Clock skew components for a design in 0.18 μm technology assuming n=4

uniformity of architecture. A specific investigation of a microprocessor design shows that about 20% variation exists on the subblock loading capacitance [22].

The clock skew components are evaluated based on the number of variations in the third column of Figure 5.19. Figure 5.20 illustrates a graphical view of clock skew components. A skew of 10.6 ps is created just by clock driver load mismatch . Also, IR-drop, temperature gradient, interconnect and ILD thickness variations, and MOS channel length tolerance create 6.4 ps, 4.08 ps, 3.12 ps, and 2.65 ps, respectively.

Utilizing the closed-form and compact clock skew model given by (5.41) and assuming that the clock skew is limited to 10% of clock period, the maximum clock frequency versus the number of H-tree levels is illustrated in Figure 5.21. As shown in the plot, the global clock frequency of an on-chip clock distribution network with a lower number of H-tree levels is limited by internal clock skew. This is because of the fact that fewer number of H-tree levels results in a larger subblock size and therefore more internal clock skew. On the other hand, as the number of H-tree levels increases, the clock frequency increases and finally saturates to a value that is limited by the process variations. As shown in Figure 5.21, for the example of Figure 5.18, the maximum achievable global clock frequency limited by process variations, in this example, is limited to about 2.8 GHz.

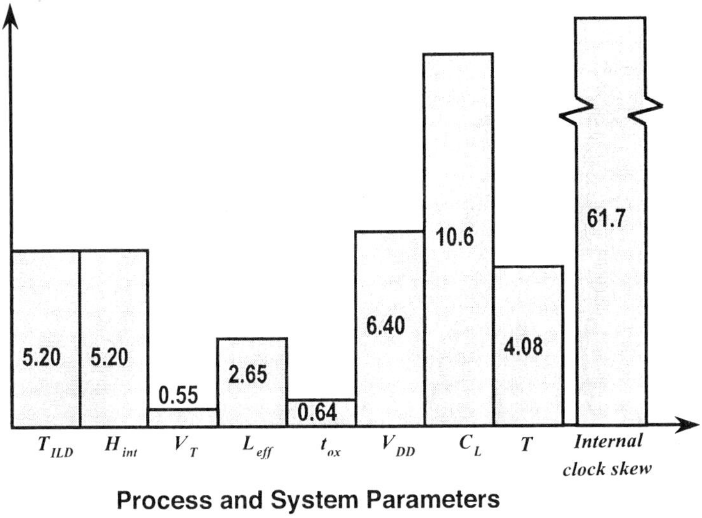

Figure 5.20. Clock skew components for a design in 0.18 μm technology assuming $n = 4$

5.3.2 Clock Signal Degradation

In a high-frequency clock distribution design, the interconnect bandwidth is often a more serious problem than interconnect delay. Figure 5.22 illustrates the importance of interconnect bandwidth in the clock distribution network design. In Figure 5.22(a), the delay is considered a simple propagation. However, as shown in Figure 5.22(b), thinking of delay as a simple propagation is not correct. In reality, the input signal is degraded as it passes through the interconnect. In the example shown in Figure 5.22, the output signal level is only 24% of the input signal level. This phenomenon is due to the fact that the interconnect has a limited bandwidth.

To compensate for the interconnect bandwidth, it is necessary to place active repeaters within the interconnect. However, utilizing active repeaters in the global clock distribution network introduces more clock skew because of the device parameter variations. Therefore, interconnect bandwidth is often a limiting factor for global clock distribution network design.

As described in Section 2.2, the resistance and capacitance of on-chip wires create a limit on the maximum wire length for a given frequency because of the interconnect bandwidth. Therefore, the global clock distribution network has a limited bandwidth that can be determined by

$$f_{-3dB} = \frac{1}{2\pi\left(r_{\text{int}}c_{\text{int}}\right) \cdot l^2} \tag{5.47}$$

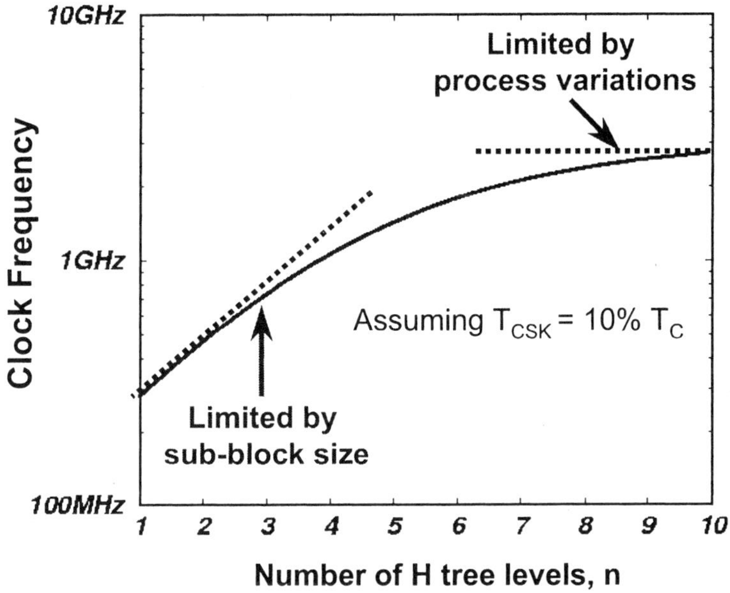

Figure 5.21. Skew-limited clock frequency versus the number of H-tree levels using equation (5.41) and the example of Figure 5.18

where r_{int} and c_{int} are the distributed resistance and capacitance per unit length of the global clock interconnect network and l is the length of the line. In this analysis, a lumped RC model approximation is used to simplify the optimal design of the maximum clock frequency described in Section 2.2.1. Using the expression for the length of the H-tree versus die size, D, and the number of H-tree levels, n, defined in [18], then (5.47) becomes

$$f_{-3dB} = \frac{1}{2\pi\left(r_{\mathrm{int}}c_{\mathrm{int}}\right) \cdot D^2 \left(1 - \frac{1}{2^{n/2}}\right)^2} \tag{5.48}$$

To illustrate the bandwidth limit of global clock distribution networks, a plot of the bandwidth-limited clock frequency versus number of H-tree levels is shown in Figure 5.23. As the number of H-tree levels increases, the global clock distribution becomes longer and therefore the bandwidth of the global clock distribution is diminished. Consequently, the maximum allowable global clock frequency will be decreased, as illustrated in Figure 5.23.

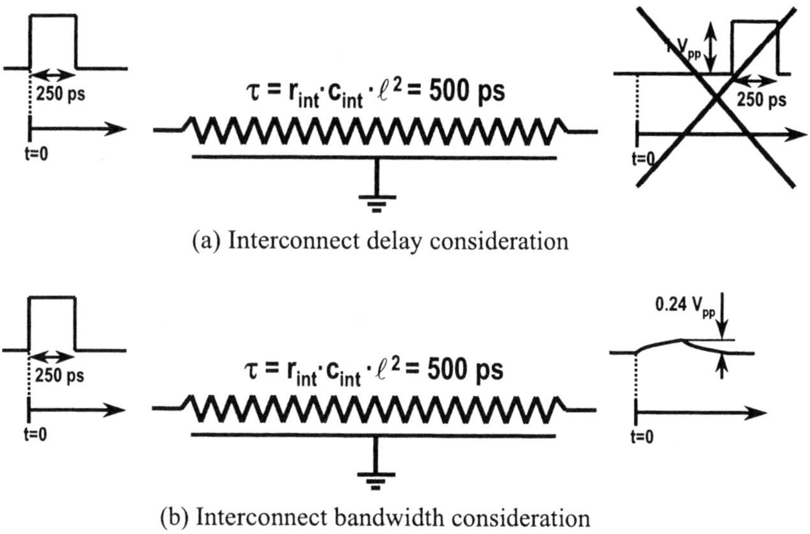

Figure 5.22. Interconnect delay versus bandwidth consideration

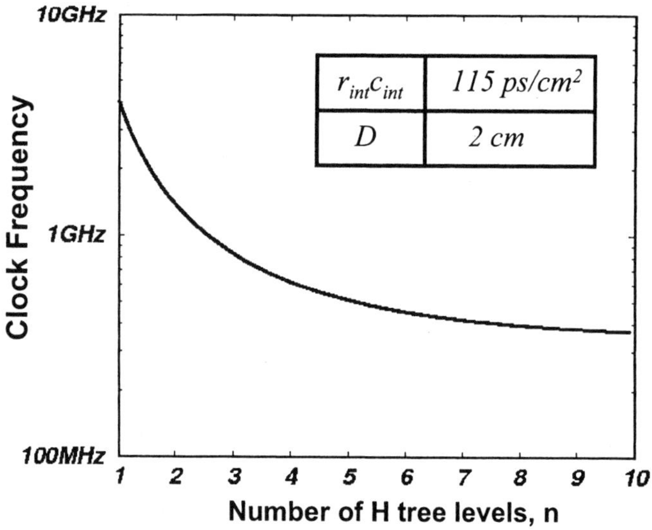

Figure 5.23. Bandwidth-limited clock frequency versus the number of H-tree levels

5.3.3 Clock Power Dissipation Model

The total power dissipation of a clock distribution network is divided into three parts associated with: i) clock wiring capacitance, ii) clock driver capacitance, and iii) clock loading capacitance. The simplified expression for clock power dissipation is

$$P_{clock} \approx f_c V_{DD}^2 \left(\frac{\varepsilon_o \cdot \varepsilon_r}{T_{ox}} W_{leaf}\, D \cdot \left(2^{n+1} + \sqrt{N_{FF}}\right) + 2^n C_{Driver} + N_{FF} C_{FF} \right) \tag{5.49}$$

where W_{leaf} is the width of the last branch of the H-tree, n is the number of branch levels of the H-tree, C_{Driver} is the clock driver input capacitance, N_{FF} is the number of clocked registers (or Flip-Flops), and C_{FF} is the input capacitance of a clocked register. In this expression, the first term, $(\varepsilon_o \cdot \varepsilon_r / T_{ox}) \cdot W_{leaf} D \cdot (2^{n+1} + \sqrt{N_{FF}})$, is the total wiring capacitance of the global clock distribution and clock wiring within the subblocks; the second term, $2^n C_{Driver}$, is the total capacitance of the clock drivers; and the third term, $N_{FF} C_{FF}$, is the total clock loading capacitance. Figure 5.24 illustrates an example of the clock distribution power dissipation versus the number of H-tree levels. As depicted in Figure 5.24, the power dissipation of clock distribution networks increases as the number of H-tree levels increases.

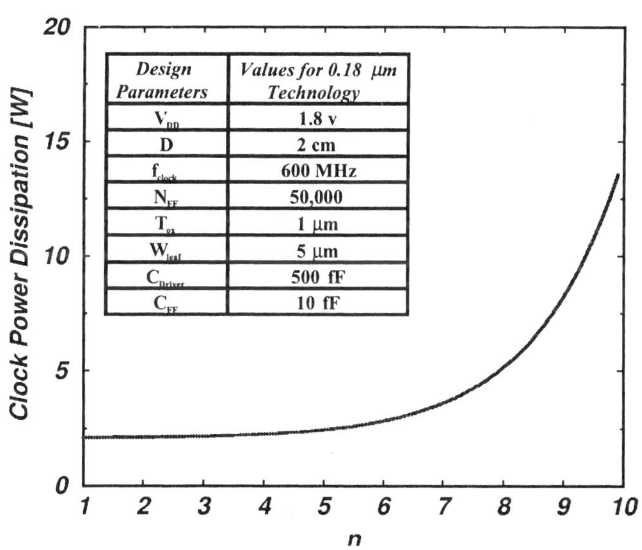

Figure 5.24. Clock power dissipation versus the number of H-tree levels

5.3.4 Optimum Clock Distribution Design

It is shown in the example of Section 5.3.1 that from the clock skew point of view, as the number of H-tree levels increases, the clock frequency increases and finally saturates to a value that is limited by the process variations. On the other hand, it is shown in the example of Section 5.3.2 that from the wiring bandwidth point of view, as the number of H-tree levels increases, the global clock distribution becomes longer and, consequently, the maximum allowable global clock frequency will be decreased. Therefore, there must be an optimum number of H-tree levels that gives the maximum achievable on-chip global clock frequency. Figure 5.21 and 5.23 are combined here to optimally design the global clock distribution network as described below.

Assume that an optimal clock distribution network is required for a design specified in Figure 5.18 with the variations given in Figure 5.19. Also, assuming that clock skew is 10% of clock cycle time, the clock frequency can be computed from $f_c = 0.1/T_{CSK}$, where T_{CSK} is given in (5.41). Using the data of a typical 0.18 μm CMOS technology given in Figure 5.18 and 5.19, a plot of clock frequency versus number of H-tree levels for the clock skew criterion is depicted in Figure 5.25. Likewise, using (5.48), the clock signal degradation criterion is represented in the same graph. Then, the optimum number of H-tree levels, $n = 4$, and the maximum achievable clock frequency, $f_c = 610$ MHz, is given by the intersection of the two curves in Figure 5.25. To have a symmetric clock distribution network, only the even number of H-tree levels is allowable.

Figure 5.25 also shows that the ultimate clock frequency is limited by process and design variations. This ultimate clock frequency can be achieved by reducing the interconnect attenuation, for instance through utilizing a low loss wires such as off-chip wiring network for global clock distribution. Note that the most binding limit for clock distribution network in today's technology is the signal bandwidth limit because for the range of $4 < n < 6$, which is being used presently, the clock frequency is limited by the signal degradation limit as depicted in Figure 5.25. Therefore, only the bandwidth limit will be used in Section 5 to identify the limits for global clock distribution network.

5.4 Global Power Distribution Modeling

Design of an optimum on-chip power distribution for high-performance chips is a challenging problem. Reliable power distribution includes consideration of IR-drop, simultaneous switching noise (SSN), and electromigration issue or chip lifetime. Moreover, the additional conflicting constraint is the minimizing the area dedicated to power routing. The problem faced by power distribution designers continuously increases in complexity because of a combination of factors: increasing design size, increasing chip current drain,

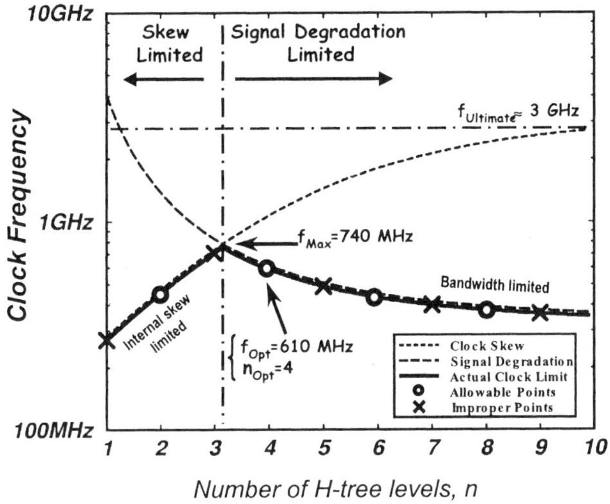

Figure 5.25. Design plane for clock distribution networks using (48) for the bandwidth limit and (41) for skew limit

shrinking supply voltage, and shrinking line widths. Therefore, it is imperative to gain a thorough understanding of the requirements for power distribution networks of present and projected gigascale integration systems.

Some early works on on-chip power distribution networks [23]-[24] consider only the IR-drop constraint for the design. However, as technology improves, the faster rise time requires more serious consideration for SSN in an on-chip power distribution network [25]. In this section, a methodology is presented to design an optimum power distribution network considering IR-drop, SSN, and the area dedicated to power distribution.

5.4.1 Modeling of On-Chip IR-Drop

In this section, an analytical model for on-chip IR-drop in a mesh structure power distribution network is derived. The model is first derived for a peripheral wire-bond system, and is then modified for chips with an area-array-bonding pad.

IR-drop in Peripheral Wire-Bond System. To simplify the model for IR drop, it is assumed that the on-chip power distribution is a mesh network connected to an equipotential outer ring. This assumption is satisfied if enough pads and pins are assigned for chip power and ground. Also, it is assumed that the power dissipation is uniform along the chip. This means that the current at every grid point of the mesh network is equal, as illustrated in Figure 5.26. The worst case of IR-drop, as shown in Figure 5.26, is the voltage drop from the center of the chip to the equipotential outer ring periphery.

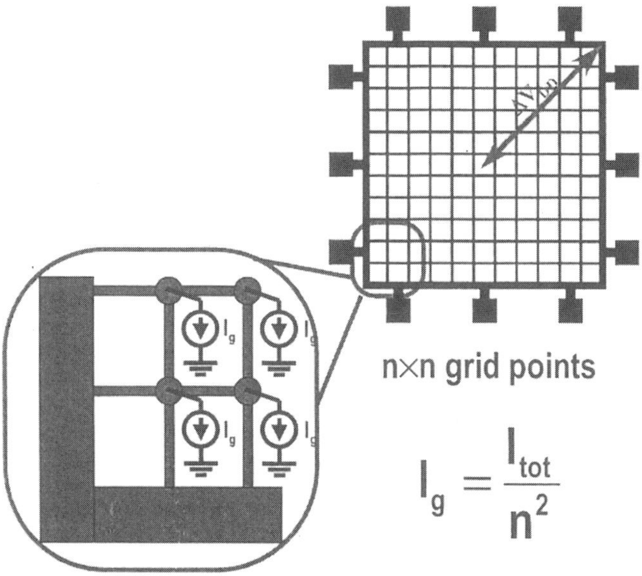

Figure 5.26. Peripheral wire-bond system with uniform power density in a mesh network

To find the potential of the center of the chip, assume that there is a square-shaped boundary cutting the grids of the mesh network at the center of the chip, as shown in Figure 5.27. In Figure 5.27, there are $3 \times 3 = 9$ nodes of the grids inside the cutting boundary that provide the current to the system. Assuming that every node supplies a current of I_g to the system, the total current needed within the cutting boundary is $9I_g$. Likewise, $4 \times 3 = 12$ grids (resistors) are intersected by the cutting boundary. Assuming that the current density on the chip is uniform, the current that passes through every sliced resistor or branch is $9I_g/12$. Similarly, the current of the sliced resistor from the center of the chip to the boundary is determined by examining different sizes of the cutting boundary, as shown in Figure 5.27.

Summing the voltage of the resistors from the center of the chip to the equipotential outer ring, using data of Figure 5.27, gives the overall IR-drop as

$$\Delta V_{DD} = \frac{I_g}{4} n \cdot r + \frac{I_g}{4}(n-2) \cdot r + \frac{I_g}{4}(n-4) \cdot r + \cdots + \frac{I_g}{4} \cdot r \quad (5.50)$$

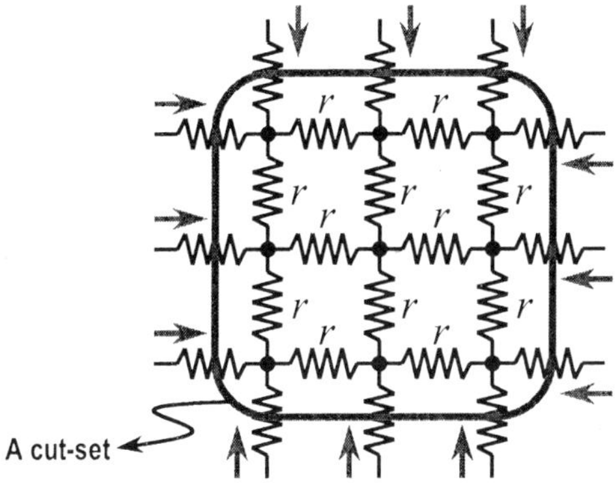

No. of Branches	Total Current	Current / Branch
$4n$	$n^2 \cdot I_g$	$n \cdot (I_g/4)$
$4(n\text{-}2)$	$(n\text{-}2)^2 \cdot I_g$	$(n\text{-}2) \cdot (I_g/4)$
$4(n\text{-}4)$	$(n\text{-}4)^2 \cdot I_g$	$(n\text{-}4) \cdot (I_g/4)$
......	......	......
12	$9I_g$	$3(I_g/4)$
4	I_g	$(I_g/4)$

Figure 5.27. Calculation of the mesh current using cutting boundary method

where ΔV_{DD} is the worst case of IR-drop in an $n \times n$ grid network, I_g is the current of a grid point, and r is the resistance of a segment of the grid. Using series theory [26], (5.50) can be simplified as

$$\Delta V_{DD} = \frac{I_g \cdot r}{4} \left[n + (n-2) + (n-4) + \cdots + 1 \right] = \frac{I_g \cdot r}{4} \left(\frac{(n+1)^2}{4} \right) \tag{5.51}$$

Substituting I_g with I_{tot}/n^2 gives

$$\Delta V_{DD} = \frac{I_{tot} \cdot r}{16} \left(\frac{n+1}{n} \right)^2 \tag{5.52}$$

Assuming that the number of grid points, n, is very large (i.e., $n > 10$),

$$\Delta V_{DD} \approx \frac{I_{tot} \cdot r}{16} \tag{5.53}$$

Equation (5.53) shows that the equivalent resistance of a mesh network is about $r/16$, where r is the resistance of a segment of the grid. Now, to relate the IR-drop to the physical characteristics of the power distribution network, a new variable - percentage of metal coverage - is defined by considering the mesh construction, as shown in Figure 5.28.

The percentage of metal coverage, by definition, is the ratio of power distribution metal area to the chip area, which is

$$\%Cov = \frac{A_{Power}}{A_{Chip}}. \tag{5.54}$$

Assuming that the length of each mesh segment is m squares, as shown in Figure 5.28, and considering the mesh element, shown by the hash lines, the

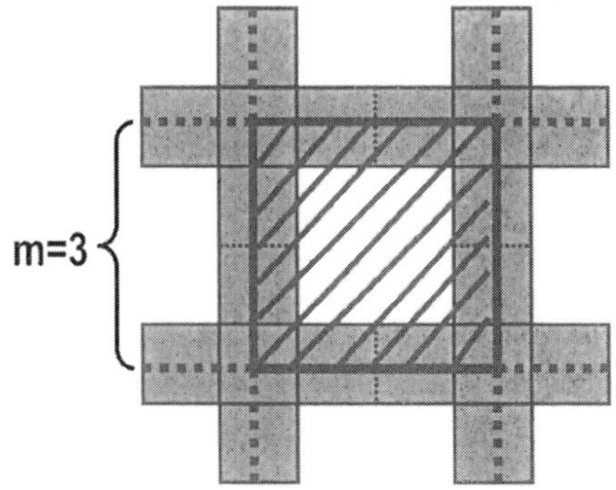

Figure 5.28. A cutting boundary at the center of the mesh network

total area of the mesh element is $m \times m = m^2$. Likewise, the metal area inside the mesh element is the total mesh element area, $m \times m = m^2$, minus the clear area inside the mesh element, $m^2 - (m-1)^2 = 2m - 1$. Therefore, (5.54) can be rewritten as

$$\%Cov = \frac{2m - 1}{m^2} \tag{5.55}$$

The value of m, the number of squares in each segment, is given by

$$m = \frac{r}{R_s} \tag{5.56}$$

where R_s is the metal sheet resistance, $R_s = \rho_w / H$, ρ_w is the metal resistivity, H is the metal thickness, and r is the resistance of the segment. The value of r is determined from the IR-drop requirement, (5.53), as

$$r = \frac{16 \Delta V_{DD}}{I_{tot}} = \frac{16 \Delta V_{DD}}{P_{tot} / V_{DD}} \tag{5.57}$$

where P_{tot} is the total chip power dissipation and V_{DD} is the supply voltage. Defining a new variable δ, normalized IR-drop, as

$$\delta = \Delta V_{DD} / V_{DD}$$

and replacing in (5.57) gives

$$r = \frac{16 \delta V_{DD}^2}{P_{tot}} \tag{5.58}$$

Substituting (5.58) in (5.56) and considering that

$$R_s = \rho_w / H$$

gives

$$m = \frac{16\delta V_{DD}^2 H}{P_{tot}\rho_w} \tag{5.59}$$

Expressions (5.59) and (5.55) relate the IR-drop requirement to the on-chip power distribution area in a peripheral wire-bond system as

$$\begin{cases} \%Cov = \frac{2m_p - 1}{m_p^2} \\ m_p = \frac{16\delta V_{DD}^2 H}{P_{tot}\rho_w} \end{cases} \tag{5.60}$$

where the subscript p in m_p stands for peripheral wire-bond system. For instance, assuming that $P_{tot} = 50W$, $V_{DD} = 1.8V$, $\rho_w = 1.72\mu\Omega$-cm for copper, $H = 2\mu$m, and $\delta = 5\%$, the required area for the power distribution network satisfying the IR-drop requirement will be % Cov=30.4%. Likewise, there is a need to have 30.4% of metal area for the ground distribution network to fulfill the requirements for 5% ground rise. Note that in the case of $m_p \gg 1$, and from (5.60), the area of the power distribution network can be approximated by

$$A_{Power} \approx \frac{P_{tot}\rho_w}{8\delta V_{DD}^2 H} \cdot A_{Chip} \tag{5.61}$$

Likewise, the same expression is valid for the required area for the ground distribution network.

IR-Drop in Area-Array-Bonding Pads. The modeling of IR-drop in the area-array-bonding technology is similar to that of the peripheral bonding pad system. Figure 5.29 illustrates the pad placements of both technologies.

To derive the model for IR-drop in an area-array-bonding pad system, assume that there is a pad at the center of the cutting boundary shown in Figure 5.27. If the total number of pads associated to the power distribution network is n_{pad}, then the equivalent current delivered by every pad in the area-array-bonding system is equal to the total current, I_{tot}, divided by the number of pads, n_{pad}. Therefore, from Figure 5.27 and expression (5.53), the value of IR-drop in area-array-bonding pads, ΔV_{DD}, is given by

$$\Delta V_{DD} \approx \frac{I_{tot} \cdot r}{16 n_{pad}} \tag{5.62}$$

where n_{pad} is the number of pads in the area-array-bonding system. Utilizing (5.62) instead of (5.53) in the derivation of IR-drop for a peripheral wire-bonding system leads to

$$\begin{cases} \%Cov = \frac{2m_a - 1}{m_a^2} \\ m_a = \frac{16\delta V_{DD}^2 H n_{pad}}{P_{tot}\rho_w} \end{cases} \tag{5.63}$$

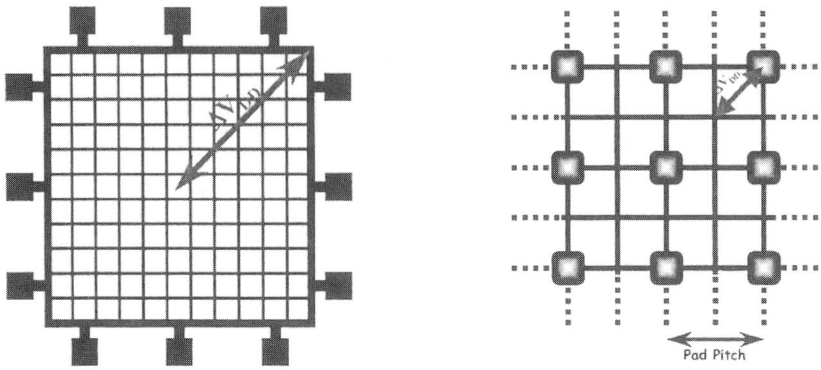

(a) peripheral bonding pads (b) area-array-bonding pads

$$m_p = \frac{16\delta V_{DD}^2 H}{P_{tot}\rho_w} \qquad m_a = \frac{16\delta V_{DD}^2 H n_{pad}}{P_{tot}\rho_w}$$

Figure 5.29. Power-supply distribution models

where the subscript a in m_a stands for area-array-bonding system. Note that in the case of $m_a >> 1$, and from (5.63), the area of the power distribution network can be computed by

$$A_{Power} \approx \frac{P_{tot}\rho_w}{8\delta V_{DD}^2 H n_{pad}} \cdot A_{Chip}. \tag{5.64}$$

Based on the IR-drop models in peripheral wire-bond and area-array-bonding systems, (5.60) and (5.63), respectively, the normalized power distribution wiring areas to chip area versus total chip power dissipation are depicted in Figure 5.30. As shown in Figure 5.30, the power-supply wiring area reduces significantly by using area-array-bonding pads instead of peripheral wire bonds. In this example, in a peripheral wire-bond system, a complete solid plane is required to provide the maximum possible power dissipation of 45 W. However, by using area-array-bonding pads with 50 pads, for instance, only 10% of the chip area will be consumed by the power distribution network.

IR-drop in Nonuniform Systems (Nonuniform Current Density). The power density is often highly nonuniform in a high-performance system-on-a-chip. Therefore, the models for IR-drop assuming uniform power dissipation are not accurate. Unlike the case of uniform power dissipation, in a hetero-

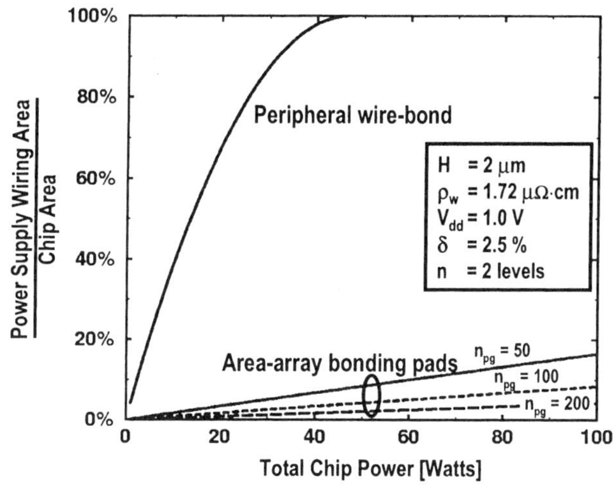

Figure 5.30. Normalized power distribution area versus total chip power

geneous system, generally, the value and location of the maximum IR-drop depend on the way of placement, power dissipation, and the area of megacells.

Although calculation of IR-drop in a peripheral wire-bond system requires a numerical simulation, a closed-form model for IR-drop in a heterogeneous system can be found for area-array-bonding system. From (5.62) and assuming that the system in Figure 5.31 is a uniform system with the average power dissipation, the value of the IR-drop is given by

$$\Delta V_{DD}|_{chip} = \frac{P_{chip}}{n_{chip}}\left(\frac{r}{16V_{DD}}\right) \tag{5.65}$$

where $\Delta V_{DD}|_{chip}$ is the value of the IR-drop, assuming uniform power density, P_{chip} is the total chip power dissipation, and n_{chip} is the total number of area-array-bonding pads. Now, assuming a localized IR-drop profile in the non-uniform system shown in Figure 5.31, the highest power density block can be modeled as an independent block with its own area-array-bonding pads. Therefore, considering that the block has the highest power density, then from (5.62), the value of the worst case of the IR-drop will be given by

$$\Delta V_{DD}|_{block} = \frac{P_{block}}{n_{block}}\left(\frac{r}{16V_{DD}}\right) \tag{5.66}$$

where $\Delta V_{DD}|_{block}$ is the value of the heterogeneous IR-drop, P_{block} is the block power, and n_{block} is the number of area-array-bonding pads over the

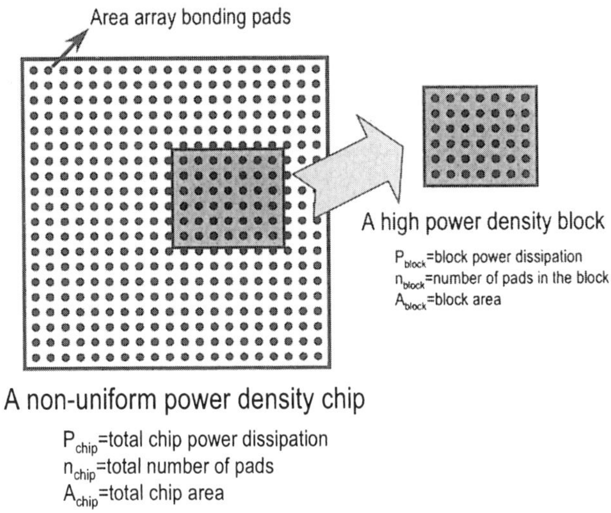

Figure 5.31. A high-power density block in a nonuniform system can be modeled as an independent block with its own area-array-bonding pads

block. Comparing (5.65) with (5.66) gives

$$\frac{\Delta V_{DD}|_{block}}{\Delta V_{DD}|_{chip}} = \frac{P_{block}}{P_{chip}} \cdot \frac{n_{chip}}{n_{block}} \tag{5.67}$$

Assuming that the area-array-bonding pads are uniformly distributed on the chip, then the following relationship is valid:

$$\frac{n_{chip}}{n_{block}} = \frac{A_{chip}}{A_{block}} \tag{5.68}$$

where A_{chip} is the total chip area and A_{block} is the block area. Substituting (5.68) into (5.67) leads to

$$\Delta V_{DD}|_{block} = k \cdot \Delta V_{DD}|_{chip} = \frac{\left(\frac{P_{block}}{A_{block}}\right)}{\left(\frac{P_{chip}}{A_{chip}}\right)} \cdot \Delta V_{DD}|_{chip} \tag{5.69}$$

where k, defined as the heterogeneity factor, is the ratio of the maximum to average power density in a non-uniform system. The heterogeneity factor, k, shows the level of nonuniformity in a heterogeneous system, as defined in (5.69). In high performance microprocessors, often the main clock driver has the maximum power density. For instance, in the example of Alpha processor 21164 [21] the ratio of clock driver, which has the maximum power density, to the average power density is about 10. Therefore, in this case, the actual value of the IR-drop, considering a heterogeneous system, is 10 times more than the value of the IR-drop, considering a uniform power density.

5.4.2 Modeling of Electromigration

Electromigration is also an essential limit on power-supply distribution networks. In power distribution networks, the power-supply pads carry the whole current to the entire chip and the size of the power-supply pad stub is limited by the supply pad size. Moreover, the width of the metals for an on-chip power distribution network can be designed wide enough to reduce the current density in the mesh. Therefore, the maximum current density is usually in the power or ground pad stubs.

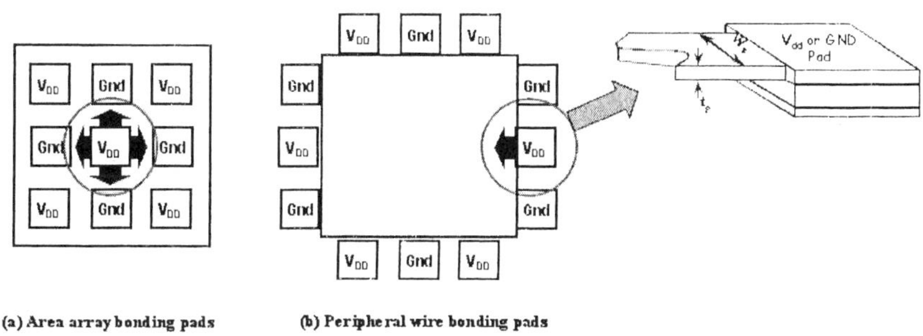

Figure 5.32. Pad placement for (a) area-array-bonding pads and (b) peripheral wire-bonding pads

Like IR-drop modeling, electromigration modeling will be discussed in this section for both a peripheral wire-bond system and area-array-bonding pads.

Electromigration in a Peripheral Wire-Bond System. Figure 5.32(b) illustrates a structure of a power-supply pad in a peripheral wire-bond system. From the limit on the maximum current density and Figure 5.32(b), the maximum current that can be supplied by every power pad, I_{Max}, is determined by

$$I_{Max} = J_{Max}\left(W_P \cdot t_P\right) \tag{5.70}$$

where W_p and t_p are the width and thickness of the pad stub, respectively, and J_{Max} is the maximum current density of the metal limited by electromigration [23]. Using (5.70) and assuming that P_{tot} is the total chip power dissipation and V_{DD} is the supply voltage, then in order to have a reliable power distribution network, the number of required power-supply and ground pads, n_{PG}, is given by

$$n_{PG} = \frac{2P_{tot}}{W_P t_P J_{Max} V_{DD}} \tag{5.71}$$

For instance, assuming that $P_{tot} = 90$ W, $V_{DD} = 1.8$ V, $W_p = 50$ μm, $t_p = 1$ μm, and from the electomigration limit of Aluminum, $J_{max} = 1$ mA/μm^2 [27], then the required number of power-supply/ground pads will be 2000 (i.e., 1000 for power and 1000 for ground). Note that using copper technology can significantly improve the electromigration limit on the power distribution networks and therefore can reduce the minimum number of power-supply pads limited by electromigration [27].

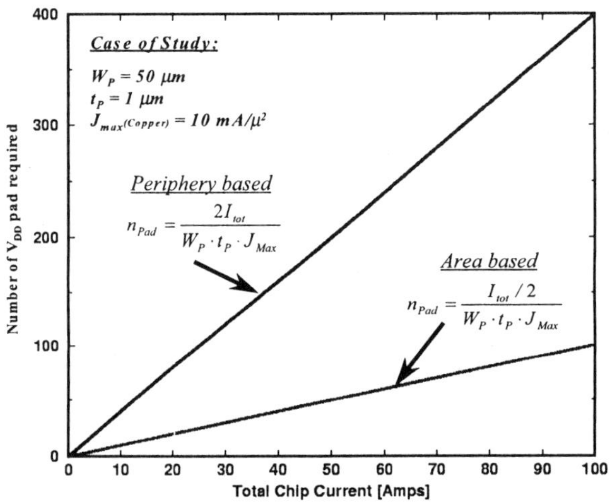

Figure 5.33. Minimum required number of pads based on electromigration limit versus total chip current

Electromigration in Area-Array-Bonding Systems. The modeling of electromigration in area-array-bonding technology is similar to that of a peripheral bonding pad system. Figure 5.32(a) illustrates the pad placements of both technologies. As shown, when the power-supply pads are placed at the periphery of the chip, the only path for the current to enter the chip is through one side of the pad that is toward the chip. In the area-array-bonding pad, however, the current passes through all four sides of the pad. Therefore, the electromigration limit for the number of power and ground pads is simply four times better than that of a peripheral wire-bond system, i.e., in this case, (5.71) can be rewritten as

$$n_{PG} = \frac{P_{tot}/2}{W_P t_P J_{Max} V_{DD}} \tag{5.72}$$

Based on electromigration modeling in power distribution networks, the minimum required number of power-supply pads from (5.71) and (5.72) is depicted in Figure 5.33. In this figure, it is assumed that $W_p = 50\ \mu$m, $t_p = 1\ \mu$m, and from the electomigration limit of AlCu alloy, $J_{max} = 10$ mA/μm^2 [27]. As seen in Figure 5.33, for the case of this study, using area-array-bonding pads requires four times fewer pads for a power-supply distribution network than using a peripheral wire-bond system.

5.4.3 Modeling of On-Chip Simultaneous Switching Noise

Simultaneous switching noise (SSN) is already a critical issue for on-board and on-package power distribution designs [28]. Likewise, the importance of SSN for the global on-chip power distribution networks for future generations of technology has been reported in [29]. The power-supply switching noise is often modeled as

$$V_{SSN} = L_{eff}\frac{di}{dt} \tag{5.73}$$

where L_{eff} is the effective inductance of the power-supply distribution network and di/dt is the peak derivative of switching current. In the conventional wire-bonding technology, L_{eff} is often approximated by the wire-bond inductance. As technology improves, the peak switching current increases and the signal rise time decreases, both resulting in an increase in the di/dt ratio. Moreover, because of the supply voltage scaling of future generations of technology, the value of SSN will be a large percentage of supply voltage if the power distribution network is not carefully designed.

Advanced electronic packaging with an area array of solder bumps instead of peripheral wire bonds and solder balls instead of peripheral leads is required to manage the SSN problem in future high-performance chips. Therefore, a simple and generic model for estimation of on-chip SSN in conventional and advanced bonding technologies is imperative for the designs of future generations of technology.

Simultaneous Switching Noise in Peripheral Wire-Bond Systems. The main source of SSN in conventional wire-bonding technology is the inductance of the wire bonds, as shown in Figure 5.34 [24]. Here, it is assumed that the supply voltage at the pin (V_1) is clean, as is shown [24]. The on-chip supply voltage (V_2), however, is noisy as a result of the switching current drawn by the drivers. The on-chip noise voltage can be simply estimated by

$$V_{SSN} = NL_p\frac{di}{dt} \tag{5.74}$$

where L_p is the effective inductance of the power-supply network, N is the

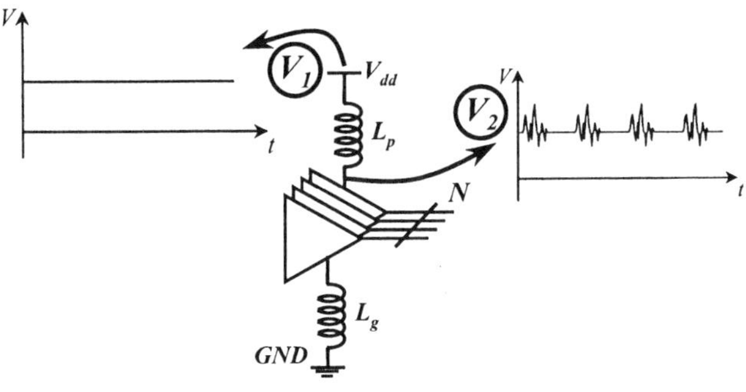

Figure 5.34. Equivalent circuit of a conventional wire-bond system

number of drivers switching simultaneously, and di/dt is the peak derivative of switching current per driver. In conventional wire-bond technology, the value of SSN can be controlled by the following methods:

Decoupling Capacitors – On-chip decoupling capacitors reduce the peak current demand on the power distribution network. The models for on-chip decoupling capacitors include both the built-in (intrinsic) capacitors such as the n-well capacitor and the circuit capacitor, and the add-on capacitors such as the thin oxide capacitor and trench capacitor. Often, in high-performance chips, the intrinsic decoupling capacitance is not enough. In this case, a large on- chip area may be needed for the additional decoupling capacitor [30].

Number of Power/Ground Pins – As the number of power/ground pins in the wire-bond technology increases, the effective inductance decreases. However, it is important to note that there exists a limit of the number of power/ground pins beyond which the effective inductance value is clamped and will not decrease with an increase in the number of power/ground pins any further. In the plastic quad flat pack (PQFP) packages, for instance, it has been reported [31] that the effective inductance of a pin is about 10 nH. It is also shown in [31] that the saturated value for the effective inductance of PQFP packages is about 0.15 nH for power/ground pin pairs in excess of 20.

Architectural Solutions – The value of SSN can be managed by a careful design of architecture. As presented in [32], the value of SSN can be reduced by controlling the time of transition from "sleep mode" to "active mode" of logic blocks. A detailed description of this method has been shown in [32].

Because of severe IR-drop and switching noise problems, it has been shown that conventional wire-bonding technology, which constrains the power pads to be located at the chip periphery, is not a scalable concept for deep sub-micron technology [24]. On the other hand, utilizing area-array-bonding pads, such as flip-chip technology, chip-scale packaging (CSP), and sea of leads [33] (SoL) can almost eliminate the parasitic inductance of the input/output leads. Moreover, it allows for the power pads to be distributed anywhere on the chip using solder bumps. In the next section, a new analytical model for SSN in the area-array-bonding pad technology is derived.

Simultaneous Switching Noise in Area-Array-Bonding Systems. In today's technology, electronic packaging has progressed to packages that have power and ground planes. Wire bonds have been replaced by solder bumps and peripheral leads have been replaced by solder balls. The new structures may have less than 1/100 of the parasitic inductance of the conventional wire-bond electronic packages [28].

Therefore, the major issue with modern on-chip SSN is managing the return current that occurs in the on-chip power distribution network. As shown in Figure 5.35(a), when N drivers switch simultaneously, the return current, $i(t)$, that comes out of the driver to the load capacitor should return through the power distribution network. Since the return path has some parasitic inductance and resistance, a noise voltage, V_{noise}, will be created on the receiver ground line. Similarly, the same process may happen on the power line.

Assumptions. To derive an analytical model for the switching noise voltage, it is assumed that the parasitic inductance of the solder bumps is negligible. To support this assumption, one can eliminate the effect of the parasitic inductance of the solder bumps by using proper on-chip decoupling capacitors placed under every power-supply solder bump [28]. Also, to simplify the analysis, the power-supply return path is modeled by a lumped RL circuit, as shown in Figure 5.35(b).

Derivation. In Figure 5.35(b), if the signal rise time at the load side is t_r, the voltage at the load capacitor can be written as

$$V_o(t) = V_{dd}(1 - e^{-t/t_r}) \tag{5.75}$$

Then, the return current is

$$i(t) = NC_l \frac{dV_o(t)}{dt} = -\frac{NC_l V_{dd}}{t_r} e^{-t/t_r} \tag{5.76}$$

where C_l is the total wire and load capacitance per driver and N is the number of drivers that switch simultaneously. In this analysis, instead of the driver

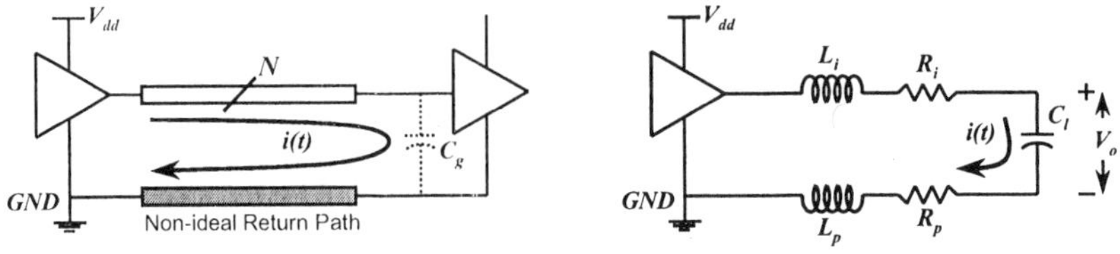

Figure 5.35. Simultaneous switching noise resulting from a nonideal return path and a simplified lumped model for derivation of SSN

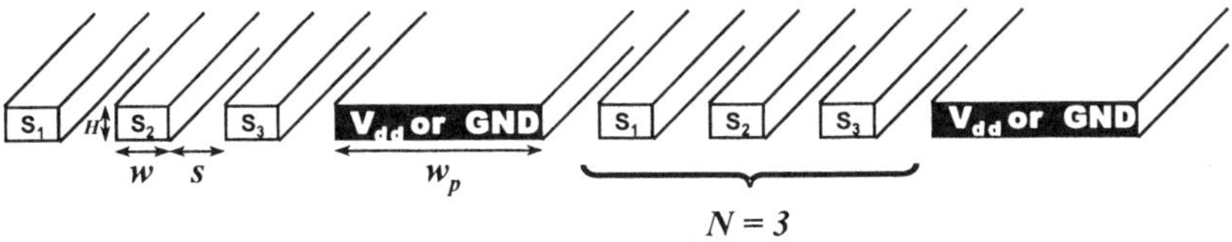

Figure 5.36. Interconnect architecture for simultaneous switching noise analysis

resistance, the value of rise time is considered as a known parameter since often the value of the driver resistance is set such that the desired rise time is achieved. Assuming that the value of noise voltage, V_{noise}, is the sum of resistance and self- and mutual inductance noises, then

$$\begin{aligned} V_{noise}(t) &= |R_p i(t)| + \left|L_p \frac{di(t)}{dt}\right| + \left|M_{pi}\frac{di(t)}{dt}\right| \\ &= \frac{NC_l V_{dd}}{t_r}\left(R_p + \frac{L_p}{t_r} + \frac{M_{pi}}{t_r}\right) e^{-t/t_r} \end{aligned} \tag{5.77}$$

where R_p and L_p, are the resistance, self-inductance of the power distribution network and M_{pi} is the mutual-inductance between the signal interconnects and the power network. The peak value of noise in (5.77) happens at $t = 0$. Considering the interconnect architecture shown in Figure 5.36, and assuming that $s = w$, the normalized peak value of SSN versus interconnect physical dimensions is given by

$$\%SSN = \frac{V_{noise}}{V_{dd}} = \frac{NC_l}{t_r}\left[\frac{\rho \cdot \bar{\ell}}{Hw_p} + \frac{1}{t_r}\left(L_{self} + \frac{2}{N}\sum_{i=1}^{N/2} M_i\right)\right] \tag{5.78}$$

It is shown in Chapter 3 and [34] that assuming $\bar{\ell} \gg w, w_p, H$, the closed-form expressions for self- and mutual inductance of the interconnect architecture shown in Figure 5.36 are given by

$$L_{self} \approx \frac{\mu_o \cdot \bar{\ell}}{2\pi} \left[Ln \left(\frac{2\bar{\ell}}{w_p + H} \right) + \frac{1}{2} \right] \tag{5.79}$$

$$M_i \approx \frac{\mu_o \cdot \bar{\ell}}{2\pi} \left[Ln \left(\frac{2\bar{\ell}}{(2i-1)\, w} \right) - 1 \right] \tag{5.80}$$

where w, w_p, and H are the signal line width, power line width, and the interconnect thickness, respectively, and $2\bar{\ell}$ is the average distance between the two points with clean supply voltages. In our analysis, $2\bar{\ell}$ is the power-supply pad pitch.

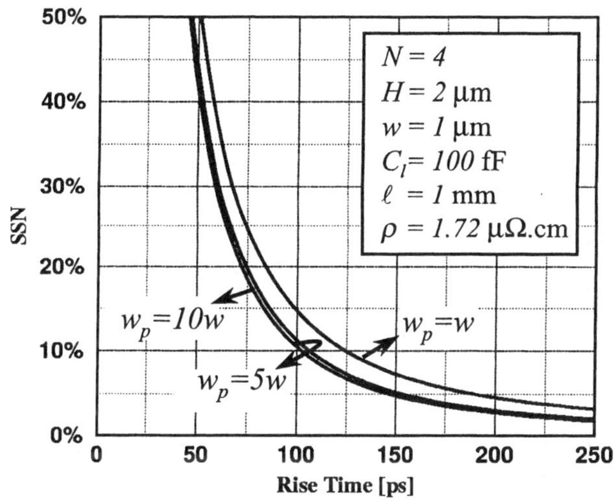

Figure 5.37. Simultaneous switching noise versus signal rise time

Figure 5.37 illustrates the plot of simultaneous switching noise versus signal rise time by using (5.78). As shown in Figure 5.37, the faster rise time results in more SSN. It is also shown that to reduce SSN, one can widen the power and ground lines. However, there exists a limit of the width of the power and ground lines beyond which SSN will not decrease with widening of power and ground lines any further. Figure 5.37 shows that the value of SSN is reduced by selecting $w_p = 5w$. However, SSN does not change noticeably when $w_p = 10w$ is selected instead of $w_p = 5w$.

5.4.4 Optimum On-Chip Power Distribution Network

It was shown in previous sections that because of severe IR-drop and switching noise problems, conventional wire-bonding technology is not a scalable concept for deep submicron technology [24]. Therefore, the use of area-array-bonding pads, such as flip-chip technology, chip-scale packaging (CSP), and sea of leads [35] (SoL) for future gigascale integration (GSI) systems is unavoidable. Since GSI systems require an area-array-bonding pad technology, all the models used in this section are for area-array-bonding pad technology.

An optimal design for the power distribution network is the one that satisfies both requirements of IR-drop and simultaneous switching noise with the minimum power-supply area. It has been shown in Section 5.4.1 that the area of power distribution network, A_{Power}, in a heterogeneous system is a function of the chip IR-drop requirement given by

$$A_{Power} \approx \frac{kP_{tot}\rho_w}{8\delta V_{DD}^2 H n_{pad}} \cdot A_{Chip} \tag{5.81}$$

where P_{tot} is the total chip power dissipation, n_{pad} is the number of power-supply pads, δ is the normalized maximum allowable IR-drop, H is metal thickness, ρ_w is metal resistivity, A_{Chip} is the total chip area, V_{DD} is the power-supply voltage, and k is the heterogeneity factor defined in (5.69). Equation (5.81) shows that the area dedicated to the power distribution network is proportional to the chip power dissipation and inversely proportional to the square of supply voltage. It also shows that the lower IR-drop requires a larger area dedicated to the power distribution network. Note that the same expression can be used for the area of the ground distribution network.

Although the on-chip power distribution area can be determined from the IR-drop requirement, as given in (5.81), no further information such as power line width and number of power line grids can be estimated from (5.81). To design and optimize the on-chip power distribution network, it is necessary to consider the SSN requirement, as well. From the analytical model presented in Section 5.4.3, it has been shown that the value of SSN is given by (5.78). Making the substitution of (5.79) and (5.80) into (5.78) and simplifying gives the final expression for SSN, which is

$$\begin{aligned} \%SSN &= \frac{V_{noise}}{V_{dd}} \qquad (5.82) \\ &= \frac{NC_l}{t_r}\left\{\frac{\rho \cdot \bar{\ell}}{Hw_p} + \frac{1}{t_r}\frac{\mu_o \cdot \bar{\ell}}{2\pi}\left[\begin{array}{c} Ln\left(\frac{2\bar{\ell}}{w_p+H}\right) - \frac{1}{2} \\ +Ln\left(\frac{\bar{\ell}}{w}\right) - \frac{2}{N}Ln\left(\frac{\Gamma\left(\frac{N+1}{2}\right)}{\sqrt{\pi}}\right)\end{array}\right]\right\}, \end{aligned}$$

where $\Gamma(x)$ is gamma function defined as

$$\Gamma(x) = \int_0^\infty t^{x-1} e^{-t} dt \tag{5.83}$$

Instead of the summation in (5.78), the gamma function, $\Gamma(x)$, is used here to make the model continuous as a function of N. A new relationship between the power distribution area and power line width is required to relate the two models of IR-drop, (5.81), and SSN, (5.82). Figure 5.36 helps to relate the two models, as

$$\frac{A_{Power}}{A_{Chip}} = \frac{w_p + w}{2Nw + w_p + w}. \tag{5.84}$$

Substituting (5.81) into (5.84) gives

$$\frac{w_p + w}{2Nw + w_p + w} = \frac{kP_{tot}\rho_w}{8\delta V_{DD}^2 H n_{pad}} \tag{5.85}$$

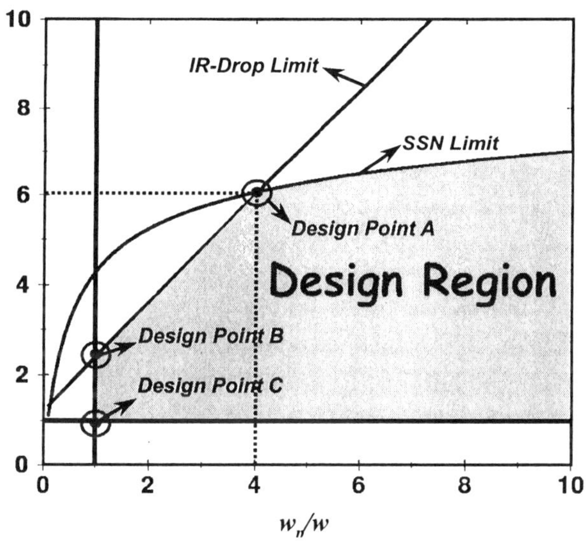

Figure 5.38. Design parameters for the design plane shown in Figure 5.39

Utilizing (5.82) and (5.85) in a single plot, as shown in Figure 5.39, an optimum design for a power distribution network can be found. The design parameters for Figure 5.39 are summarized in Figure 5.38.

As depicted in Figure 5.39, the design region is the shaded area under the two curves. The region in which $N < 1$ is a forbidden region since at least one signal line must be routable between the power lines. Likewise, the region in which $w_p/w < 1$ is forbidden since the power line width cannot be less than the minimum signal line width.

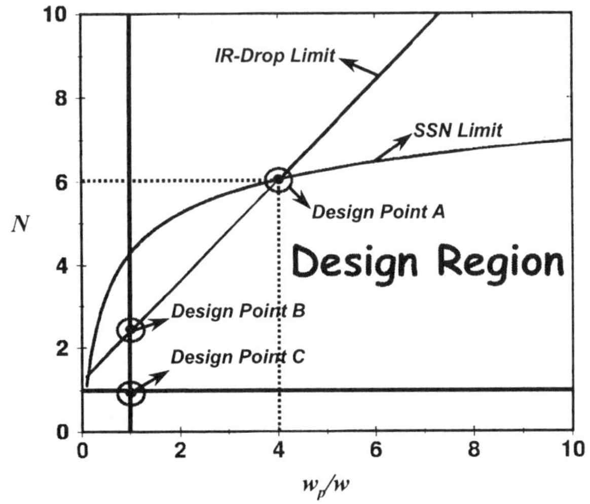

Figure 5.39. Design plane for optimum on-chip power distribution networks, IR-drop in (5.85) and SSN in (5.82)

There are three interesting design points, A, B, and C, in the design plane of Figure 5.39 that can be used as optimum points. Design point A consists of a design in which there are six signal lines with w in width between power/ground lines with $w_p = 4w$ in width. This design point satisfies both IR-drop and SSN requirements with minimum power distribution area. Design point B, on the other hand, consists of a design in which there are two signal lines with w in width between power/ground lines with $w_p = w$ in width. This design point satisfies IR-drop with minimum power distribution area. Although this design point provides a better SSN feature, as illustrated in Figure 5.39, it requires more routing efforts since for every two signal lines a power line has to be routed. All other points in the shaded area (design region) satisfy the requirements for IR-drop and SSN; however, they consume more area dedicated to power distribution network. For instance, design point C consists of a design in which every signal line is sandwiched between power/ground lines with minimum width, $w_p = w$. As shown in Figure 5.39, this design consumes more area dedicated to power distribution network.

5.5 An Integrated Architecture for Global Interconnects

In this section, utilizing the new set of models for the global signal, clock, and power distribution networks, an integrated architecture for the global interconnects is defined. The integrated architecture for global interconnects in a GSoC is based on the models for wiring resource demand, wiring noise limit, and wiring bandwidth requirement, as briefly described below.

> **Wiring Resource Requirement –** The required area for global wiring networks is often limited by the GSoC area. Using the models for wiring requirements of signal, clock, and power-supply distribution networks, a limit on the global interconnect physical dimensions is defined.
>
> **Wiring Bandwidth Requirement –** Since clock is the fastest signal in a GSoC, it requires a high-bandwidth interconnect so that the rise time is preserved when the clock signal passes through the interconnect. Therefore, the bandwidth of global clock distribution often defines the speed limit of the interconnect architecture.
>
> **Wiring Noise Limit –** Interconnect coupling noise or crosstalk is often a major concern for global wires. Unlike the interconnect bandwidth, the peak crosstalk noise prediction usually needs an accurate distributed RLC approach.

5.5.1 Wiring Resource Requirement

Figure 5.40 illustrates the wiring breakdown for two different cases: 1) a low-power design and 2) a high-performance chip. In this figure, the area for the clock distribution network is always negligible (about 1%). The area for the power distribution network depends on the power dissipation of the chip. In a high-performance design, about 32% of the chip area is dedicated to power distribution, whereas in a low-power design, only 7% of the chip area is dedicated to power distribution. The rest of the chip area is used for signal wires. However, depending on the routing efficiency and via blockage, only a part of the chip area will be used by signal interconnects, as shown in Figure 5.40.

As shown in Figure 5.40, the wiring requirement for signal, power, and clock is highly design dependent. Therefore, accurate models for wiring requirements of signal, clock, and power-supply networks are required to define a limit on the global interconnect physical dimensions. The total required area for the wiring networks is often limited by the SoC area, A_{SoC},

$$A_{Signal} + A_{Clock} + A_{Power} \leq A_{SoC} \tag{5.86}$$

where A_{Signal}, A_{Clock}, and A_{Power} are the area of signal, clock, and power-supply networks, respectively.

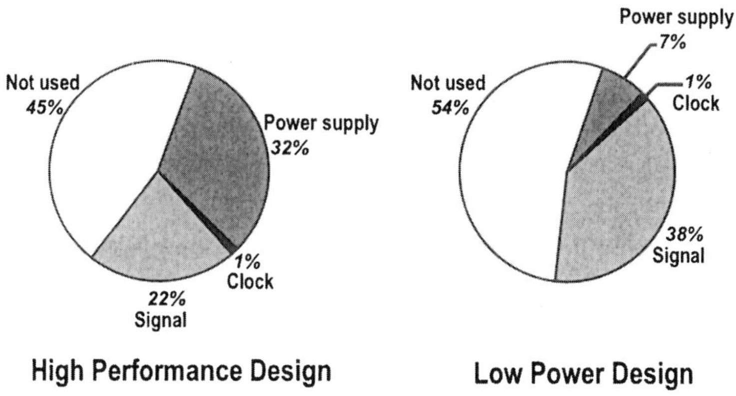

Figure 5.40. Wiring area breakdown in a low-power design and a high-performance chip

The required area for a signal wiring network, by definition, is given by

$$A_{Signal} = \frac{(w+s)L_{tot}}{\eta_r} \tag{5.87}$$

where L_{tot} is the total global wire length for signal interconnects, η_r is the router efficiency, and w and s are the width and spacing of signal lines, as shown in Figure 5.41, respectively. The global signal wiring requirements can be estimated by utilizing the stochastic net-length distribution model for global interconnects in a heterogeneous SoC described in Section 5.2.1. It was shown in Section 5.2.1 that the total global wire length, L_{tot}, is estimated from

$$L_{tot} = \sum_{m=2}^{N_m} N_{net}(m) \cdot L_{av}(m) \tag{5.88}$$

where $N_{net}(m)$ is the number of m-terminal nets, $L_{av}(m)$ is the average length of m-terminal nets, and N_m is the total number of megacells in the SoC. From the netlist information model presented in Section 5.2.1, $N_{net}(m)$ can be estimated from (8) by

$$N_{Net}(m) \approx \frac{K_{eq}N_m\left(m^{p_{eq}-1} - (m+1)^{p_{eq}-1}\right)}{m+1} \tag{5.89}$$

where K_{eq} and p_{eq} are the equivalent megacell Rent's rule for the heterogeneous system defined by (9). Moreover, from placement and routing information models, presented in Section 5.2.1, $L_{av}(m)$ can be estimated from (5.10), (5.12), and (5.18) by

$$L_{av}(m) \approx (0.5\sqrt{m}+1)\frac{m-1}{m+1}\sqrt{A_{SoC}\left(\eta_p + \frac{N_m}{m}(1-\eta_p)\right)} \tag{5.90}$$

where η_p is the placement efficiency.

The area for global clock distribution, on the other hand, is often negligible compared with the chip area, as shown in Figure 5.40. As shown in this figure, the area for clock distribution networks is usually about only 1% of the total chip area.

Finally, utilizing the design plane for optimum power distribution networks presented in Section 5.4, the area for a power distribution network in an optimum design is computed so that it satisfies both requirements of IR-drop and simultaneous switching noise with the minimum power-supply area. As presented in Section 5.4 and by utilizing the design plane of Figure 5.39 and (5.81), the minimum wiring area dedicated to the power distribution network in a heterogeneous system is given by IR-drop limit, i.e.

$$A_{Power} \approx \frac{kP_{tot}\rho_w}{8\delta V_{DD}^2 H n_{pad}} \cdot A_{Chip} \tag{5.91}$$

where P_{tot} is the total chip power dissipation, n_{pad} is the number of power-supply pads, δ is the normalized maximum allowable IR-drop, H is metal thickness as shown in Figure 5.41, ρ_w is metal resistivity, A_{Chip} is the total chip area, V_{DD} is the power-supply voltage, and k is the heterogeneity factor defined in (5.69).

Substituting (5.87) and (5.91) into (5.86) and simplifying gives the expression for the wiring resource requirement as

$$(w+s)\sum_{m=2}^{N_m} N_{net}(m) \cdot L_{av}(m) \leq \eta_r \left(1 - \frac{kP_{tot}\rho_w}{8\delta V_{DD}^2 H n_{pad}}\right) A_{SoC} \tag{5.92}$$

where $N_{net}(m)$ and $L_{av}(m)$ are defined in (5.89) and (5.90), respectively.

5.5.2 Wiring Bandwidth Requirement

Since the clock signal is the fastest signal in a GSoC, it requires a high-bandwidth interconnect so that the rise time is preserved when the clock signal passes through the interconnect. Therefore, the bandwidth of a global clock distribution often defines the speed limit of the interconnect architecture.

It was shown through an example that the -3 dB bandwidth of an on-chip interconnect is almost the same for the distributed RC and RLC models. Moreover, it was shown that although the lumped RC model underestimates the actual bandwidth of a distributed RC and RLC line, in most typical cases, it gives a first-order approximation for the interconnect bandwidth. Therefore, a simplified lumped RC model of (5.47) is used here to express the bandwidth requirement for an on-chip global interconnect.

Utilizing the expression for the length of the H-tree clock distribution network versus die size, D, and the number of H-tree levels, n, described in Section 5.3.1, (5.47) becomes

$$f_{-3dB} = \frac{1}{2\pi\left(r_{\text{int}}c_{\text{int}}\right) \cdot D^2 \left(1 - \frac{1}{2^{n/2}}\right)^2} \tag{5.93}$$

Assuming a large number of H-tree levels, $n \gg 1$, and considering that $A_{SoC} = D^2$, then (5.93) is simplified to

$$f_{-3dB} = \frac{1}{2\pi\left(r_{\text{int}}c_{\text{int}}\right) \cdot A_{SoC}} \tag{5.94}$$

where r_{int} and c_{int} are the distributed resistance and capacitance per unit length of the interconnect, respectively, and A_{SoC} is the SoC area.

5.5.3 Wiring Noise Limit

Interconnect coupling noise or crosstalk is often a major concern for global wires. The modeling of crosstalk is often a complex problem since the crosstalk noise is a function of input pulse rise time and width, length of the interconnect, and driver/source resistance.

It was described in Section 5.2.3 that unlike the interconnect bandwidth, the peak crosstalk noise prediction usually needs an accurate distributed RLC approach. It has been shown in [14] that the worst-case peak of crosstalk noise using a distributed RLC model can be simplified to

$$\frac{V_n}{V_{dd}} \approx \frac{\pi}{4}\frac{c_m}{c_{gnd} + c_m} \tag{5.95}$$

where c_m and c_{gnd} are the mutual and ground capacitance per unit length, respectively. In this section, the simplified expression of (5.95) for the worst-case peak of crosstalk noise is used to express the noise limit for an on-chip global interconnect. Equation (5.95) assumes a high quality nearby return path.

5.5.4 A Complete Global Interconnect Design Plane

Assuming the interconnect architecture of Figure 5.41 for global signal, clock, and power distribution networks, the global interconnect parameters can be described by the interconnect physical dimensions as expressed below

$$r_{\text{int}} \approx \frac{\rho_w}{Hw}, c_{gnd} \approx \frac{\varepsilon_o\varepsilon_r w}{T_{ox}}, c_m \approx \frac{\varepsilon_o\varepsilon_r H}{s}, c_{\text{int}} \approx 2\left(c_{gnd} + c_m\right) \tag{5.96}$$

where r_{int}, c_{gnd}, c_m, and c_{int} are distributed resistance, ground capacitance, mutual capacitance, and total capacitance per unit length of the interconnect, respectively.

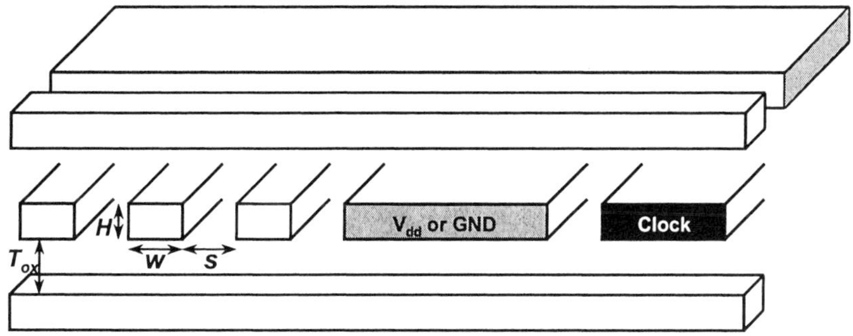

Figure 5.41. Global interconnect architecture including the physical dimensions

Substituting (5.96) into the expressions for the wiring resource requirement, (5.92), wiring bandwidth demand, (5.94), and wiring noise limit, (5.95) gives the complete closed-form model for the integrated architecture of global signal, clock, and power distribution networks. The simplified expressions are given in Figure 5.42.

Assuming that $w = s$ and $T_{ox} = H$, and utilizing the models for the wiring resource requirement, wiring bandwidth demand, and wiring noise limit, given in Figure 5.42, a new design plane for global interconnects is defined.

The global interconnect design plane includes the restrictions for global signal, clock, and power distribution networks all at once. Figure 5.43 illustrates a plot of interconnect thickness, H, versus interconnect width, w, which defines the design plane. The design parameters shown in Figure 5.43 are for a design in 180 nm technology generation with two global wiring levels, assuming that the routing efficiency, η_r, is 50% [36]. As shown in Figure 5.43, for the design in 180 nm technology generation with two global wiring levels, the forbidden zone of operation lies external to the bounded region defined by the three wiring requirements. Moreover, the interconnect physical dimensions for the minimum pitch, minimum aspect ratio, and maximum speed design points can be obtained from the design plane shown in Figure 5.43. Finally, the design point, maximum reliability, for global interconnect physical dimensions can be defined as a point that is placed at the center of the design space, the furthest possible place to the prohibited edges, as shown in Figure 5.43. Assuming that the design region is a triangle the central point can be easily determined by calculating the center of the three design points; minimum pitch, minimum aspect ratio, and maximum speed.

Equation Description	Simplified Expressions for Global Wiring Requirements in Terms of w, s, H, and T_{ox}
Wiring resource requirement	$(w+s)\sum_{m=2}^{N_m} N_{Net}(m)\cdot L_{av}(m) \le \eta_r\left(1-\frac{kP_{tot}\rho_w}{8\delta V_{dd}^2 H n_{pg}}\right)A_{SoC}$ where: $N_{Net}(m)\approx \frac{K_{eq}N_m\left(m^{P_{eq}-1}-(m+1)^{P_{eq}-1}\right)}{m+1}$ $L_{av}(m)\approx(0.5\sqrt{m}+1)\frac{m-1}{m+1}\sqrt{A_{SoC}\left(\eta_p+\frac{N_m}{m}(1-\eta_p)\right)}$
Wiring bandwidth requirement	$f_c \le \frac{1}{4\pi\rho_w\varepsilon_o\varepsilon_r\left(1/HT_{ox}+1/ws\right)A_{SoC}}$
Wiring noise limit	$\frac{\pi}{4}\frac{1/ws}{\left(1/HT_{ox}+1/ws\right)} \le \%Noise$

Figure 5.42. The simplified expressions for global interconnects

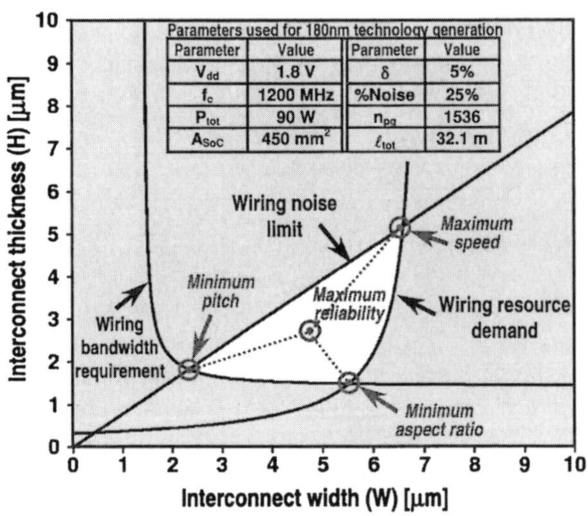

Figure 5.43. A complete interconnect design plane for a design with 20 megacells assuming that the heterogeneous Rent's parameters are

5.5.5 Applications of Global Interconnect Design Plane

The global interconnect design plane presented in Section 5.4 can be used to enhance the understanding of limits associated with the global integrated architectures.

For instance, as depicted in Figure 5.44(a), assuming two global levels, the size of the design region appears to be vanishingly small, as a result of scaling technology into deep submicron range. Figure 5.44(b) shows that the actual design space, assuming the global interconnect variations, is smaller than in the ideal case since in this case the borders defining the design space are fuzzy lines instead of solid lines. Also, utilizing on-board wiring network for part of the on-chip global clock distribution, provides a larger design space as shown in Figure 5.44(c). Likewise, utilizing on-board wiring network for part of the on-chip global signal interconnects, provides a larger design space as shown in Figure 5.44(d).

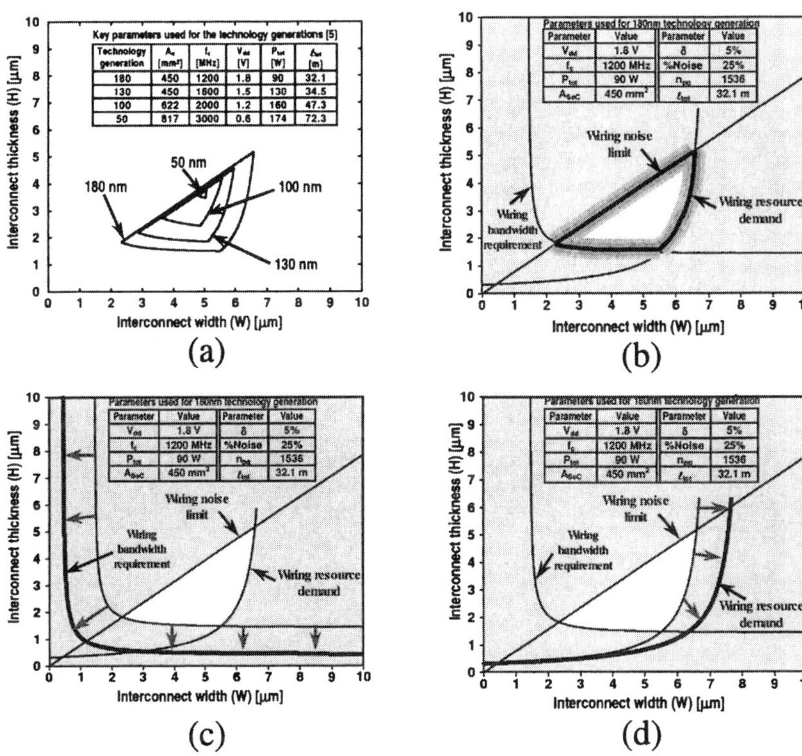

Figure 5.44 Applications of global interconnect design plane

5.6 Conclusions

The design of global interconnect architecture is a critical issue in gigascale integration. The complete global wiring network in a gigascale system-on-a-chip (GSoC) consists of signal, clock, and power distribution networks. Since the signal, clock, and power-supply networks utilize the same interconnect stacks, it is imperative to integrate the complete interconnect architecture to design and optimize the global interconnects for a GSoC.

In this chapter, an integrated architecture for global interconnects in a gigascale system-on-a-chip (GSoC) is presented by using the models for global signal, clock, and power-supply wiring networks. Based on the models for wiring resource demand, noise limit, and bandwidth requirement, an interconnect design plane is proposed. The new design plane demonstrates the limits imposed on global on-chip interconnect physical dimensions for the future generations of technology. It is shown that the size of the design region is vanishingly small as a result of scaling to deep submicron technology, assuming two global wiring levels. Also, the impact of interconnect process variations is studied using the developed design plane. Finally, the impact of utilizing on-board intrachip wiring networks on the global interconnect design space is investigated.

References

[1] J.A. Davis and J.D. Meindl, "Compact distributed RLC models for multilevel interconnect networks," *IEEE Symposium on VLSI Technology*, pp. 165-166, June 1999.

[2] S. M. Kang, "Performance-driven Layout of CMOS VLSI Circuits," *IEEE International Symposium on Circuits and Systems*, vol. 2, pp. 881-884, 1990.

[3] D. Stroobandt, "A priori wire length estimation based on Rent's rule," First International Workshop on System-Level Interconnect Prediction, April 1999.

[4] P. Zarkesh-Ha, J.A. Davis, W. Loh, and J.D. Meindl, "On a pin versus gate relationship for heterogeneous systems: heterogeneous Rent's rule," IEEE Custom Integrated Circuit Conference, pp. 93-96, May 1998.

[5] P. Zarkesh-Ha, J.A. Davis, W. Loh, and J.D. Meindl, "Prediction of interconnect fan-out distribution using Rent's rule," International Workshop on the System-Level Interconnect Prediction, pp. 107-112, April 2000.

[6] B. S. Landman and R. L. Russo, "On a pin versus block relationship for partitions of logic graphs," *IEEE Trans. Computers*, Vol. C-20, pp. 1469-1479, Dec. 1971.

[7] P. Zarkesh-Ha, J.A. Davis, and J.D. Meindl, "Prediction of net-length distribution for global interconnects in a heterogeneous system-on-a-chip," *IEEE Transaction on Very Large Scale Integration (VLSI) Systems*, vol. 8, pp. 689-692, Dec. 2000.

[8] K. Doll, F. M. Johannes, and G. Sigl, "Accurate net models for placement improvement by network flow methods," IEEE ACM International Conference on Computer Aided Design ICCAD, pp. 594-597, 1992.

[9] N. Vasseghi, K. Yeager, E. Sarto, and M. Seddighnezhad, "200-MHz superscalar RISC microprocessor," *IEEE Journal of Solid State Circuits*, vol. 31, No.11, pp. 1675-1686, Nov. 1996.

[10] J. Watt, "Pointing the way," *Electronics & Wireless World*, vol. 94, no. 1624, pp. 115-120, Feb 1988.

[11] A. Deutsch, G. Kopcsay, V. Ranieri, J. Cataldo, E. Galligan, W. Graham, R. McGouey, S. Nunes, J. Paraszczak, J. Ritsko, R. Serino, D. Shih, and J. Wilczynski, "High-speed signal propagation on lossy transmission lines," *IBM Journal of Research and Development*, Vol. 34, No. 4, pp. 601-615, July 1990.

[12] T. Sakurai, "Closed-form expressions for interconnect delay, coupling and crosstalk in VLSIs," *IEEE Transaction on Electron Devices*, Vol. 40, pp. 118-124, Jan. 1993.

[13] J. A. Davis and J. D. Meindl, "Compact distributed RLC interconnect models - part I: single line transient, time delay, and overshoot expressions," *IEEE Transaction on Electron Devices*, Vol. 47, No. 11, pp. 2068-2077, Nov. 2000.

[14] J. A. Davis and J. D. Meindl, "Compact distributed RLC interconnect models - part II: coupled line transient expressions and peak crosstalk in multilevel networks," *IEEE Transaction on Electron Devices*, Vol. 47, No. 11, pp. 2078-2087, Nov. 2000.

[15] D. C. Keezer, "Design and evaluation of wafer scale clock distribution," IEEE International Conference on Wafer Scale Integration, pp. 168-175, 1992.

[16] A. Azuma, A. Oishi, Y. Okayama, K. Kasai, and Y. Toyoshima, "Methodology of MOSFET characteristics fluctuation description using BSIM3v3 SPICE model for statistical circuit simulations," The 3rd International Workshop Statistic Metrology, pp. 14-17, June 1998.

[17] S. J. Fang, A. Barda, T. Janecko, W. Little, D. Outley, G. Hempel, S. Joshi, B. Morrison, G. Shinn, and M. Briang, "Control of dielectric CMP using an interferometry based endpoint sensor," IEEE International Interconnect Technology Conference, pp. 76-78, June 1998.

[18] H. Bakoglu, J. Walker, and J. Meindl, "A symmetric clock distribution tree and optimized high-speed interconnections for reduced clock skew in ULSI and WSI circuits," IEEE International Conference on Computer Design, pp. 118-122, Oct. 1986.

[19] P. J. Restle and A. Deutsch, "Designing the best clock distribution network," IEEE Symposium on VLSI Circuit Design, pp. 2-5, June 1998.

[20] E. S. Yang, Microelectronic Devices, McGraw-Hill, 1988.

[21] P. E. Gronowski, W. Bowhill, R. Preston, M. Gowan, and R. Allmon, "High-performance microprocessor design," *IEEE Journal of Solid-State Circuits*, pp. 676-686, May 1998.

[22] W. Loh and P. Zarkesh-Ha, private communication.

[23] W. Song and L. Glasser, " Power distribution techniques for VLSI circuits," *IEEE Journal of Solid-State Circuits*, pp. 150-156, Feb. 1986.

[24] D. Sylvester and K. Keutzer, "Getting to the bottom of deep submicron: A global wiring paradigm," *Proceedings of International Symposium on Physical Design*, pp. 193-200, 1999.

[25] M. Benoit, S. Taylor, D. Overhauser, and S. Rochel, " Power distribution in high-performance design," Proceedings of International Symposium on Low Power Electronics and Design, pp. 274-278, 1998.

[26] D. Braess, L. SchumakerA, Numerical methods in approximation theory, International series of numerical mathematics.

[27] G. Harsanyi, " Copper may destroy chip-level reliability: handle with care-mechanism and conditions for copper migrated resistive short formation," *IEEE Electron Device Letters*, vol. 20, pp. 5-8, Jan. 1999.

[28] L. Smith, " Simultaneous switching noise and power plane bounce for CMOS technology," *IEEE Electrical Performance of Electronic Packaging*, pp. 163-166, Oct. 1999.

[29] M. Cases, B. Singh, and L. Smith, "Delat-I noise avoidance methodology for high-performance chip design," *IEEE Electrical Performance of Electronic Packaging*, pp. 163-166, Oct. 1997.

[30] H. Chen and J. Neely, "Interconnect and circuit modeling techniques for full-chip power supply noise analysis," *IEEE Transaction on Components, Packaging, and Manufacturing Technology*, pp. 209-215, Aug. 1998.

[31] L. Lin and J. Prince, "SSO noise electrical performance limitations for PQFP packages," *IEEE Transaction on Components, Packaging, and Manufacturing Technology*, pp. 292-297, Aug. 1997.

[32] M. Deb-Pant, P. Pant, D. S. Wills, and V. Tiwari, " Inductive noise reduction at the architectural level," Proceedings of the 13th International Conference on VLSI Design, Calcuta, India, pp. 162-167, Jan. 2000.

[33] A. Naeemi, C. Patel, M. Bakir, P. Zarkesh-Ha, K. Martin, and J. Meindl, " Sea of Leads: A Disruptive Paradigm for a System-on-a-Chip," IEEE International Solid-State Circuits Conference, pp.280-281, Feb. 2001.

[34] X. Qi, G. Wang, Z. Yu, R. Dutton, T. Young, and N. Chang, "On-Chip Inductance Modeling and RLC Extraction of VLSI Interconnects for Circuit Simulation," Proceedings of the IEEE Custom Integrated Circuits Conference, pp. 487-490, May 2000.

[35] W. Lynch and L. Arledge, " Power supply distribution and other wiring issues for deep-submicron IC's," Proceedings of Materials Research Society Symposium, pp. 11-27, 1998.

[36] D. Sylvester and K. Keutzer, "Getting to the bottom of deep submicron II: A global wiring paradigm," Proceedings of International Symposium on Physical Design, pp. 193-200, April 1999.

Chapter 6

STOCHASTIC MULTILEVEL INTERCONNECT MODELING AND OPTIMIZATION

J. Davis, R. Venkatesan, and J. Meindl
School of Electrical and Computer Engineering
Georgia Institute of Technology

6.1 Introduction

Clock frequency, power consumption, and chip size are largely determined by the wiring requirements of a VLSI system [1, 2, 3]. It is, therefore, imperative to gain thorough understanding of wiring requirements for present and projected gigascale integrated (GSI) systems. It has been shown that optimized logic networks have certain collective properties that can be described with Rent's Rule. Using this well-established empirical relationship as a starting point, a wire length distribution is rigorously derived to enable first order estimation of the local, semi-global, and global wiring requirements for GSI random logic networks. Applications to critical path modeling, power dissipation modeling, and die size estimation are explored. Finally, this chapter will conclude with an n-tier multilevel interconnect optimization that uses these models to help guide the design of advanced multilevel interconnect architectures for gigascale integration (GSI).

6.2 Wire-Length Distribution Model

6.2.1 Assumptions

Early work on wiring distributions, especially [4], assumed that the wires emanating from a block of logic follow a Poisson distribution. The primary assumption of this wire length distribution model, however, is based upon a well-established empirical relationship commonly known as Rent's Rule [5, 6, 7, 8]. This relationship correlates the number of signal input and output (I/O) terminals, T, to the number of gates, N, in a random logic network. This correlation is given by a simple power law expression [5, 6, 7], which is

$$T = kN^p \tag{6.1}$$

where the parameters k and p are empirical constants.

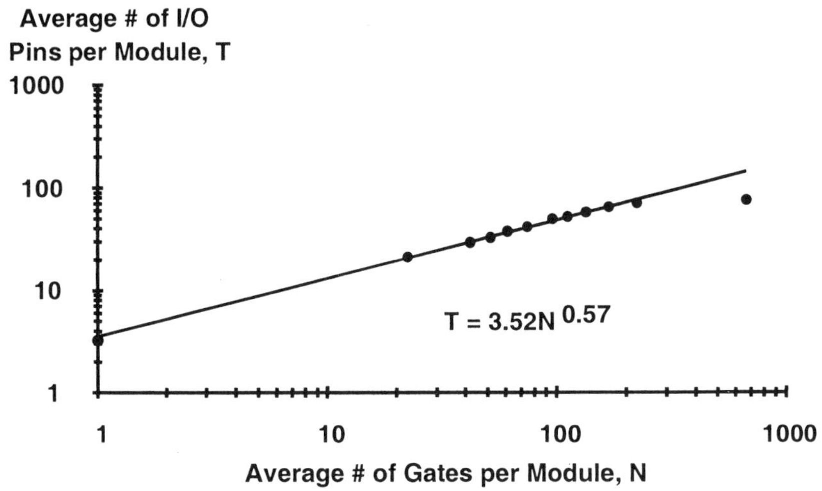

Figure 6.1. Landman and Russo experiment which verified "internal" Rent's Rule

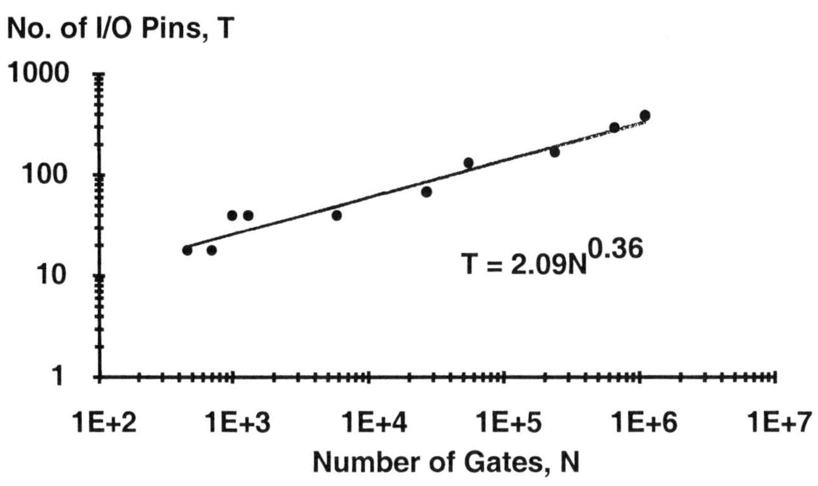

Figure 6.2. External Rent's Rule describes Intel microprocessor family over the past thirty years

Early compelling evidence of Rent's Rule comes from a study by Landmann and Russo [6], who partitioned existing scientific computers into modules and discovered an average relationship between the number of gates in a module and the number of module I/O terminals as illustrated in Figure 6.1. Bakoglu in [9] examined a variety of microprocessor, ASIC, memory, and gate array chips, and found that similar power law relationships describe the external I/O requirements of these systems. For example, the external I/O pins for the Intel microprocessor family, from the Intel 4004 in 1971 up to the Pentium Pro in 1996, are estimated by Rent's Rule as seen in Figure 6.2.

The underlying assumption of this derivation is based upon the recursive application of Rent's Rule throughout an entire monolithic system. For any arbitrary closed path within a system of N gates, Rent's Rule determines the I/O requirements of the enclosed collections of gates.

6.2.2 Predicting Point-to-Point Wiring Distributions

To predict an entire distribution for a GSI chip, we will first illustrate the estimation of a stochastic wire length distribution of a single logic gate in a square array of gates. To illustrate this calculation for a single gate without loss of generality, consider the corner element of a square array of gates as seen in Figure 6.3. The expected number of interconnects from the corner element to all gates that are a distance ℓ away is determined using Rent's Rule.

The gates in Figure 6.3 are grouped into three distinct but adjacent blocks, A, B, and C such that a single closed path can encircle one, two or all three of these blocks. The number of connections between Block A and Block C is calculated by conserving all I/O terminals for blocks A, B, and C. Conservation of I/O terminals dictates that the terminals for block A, B, and C are either inter-block connections or external system connections.

For instance, applying the principle of conservation of I/O's to the three block system in Figure 6.3 gives

$$T_A + T_B + T_C = T_{A-to-C} + T_{A-to-B} + T_{B-to-C} + T_{ABC} \tag{6.2}$$

where these variables are defined in Table 6.1. Because blocks A and B are adjacent, one closed curve is drawn to encircle A and B. From conservation of terminals, the number of I/O's between blocks A and B is written as

$$T_{A-to-B} = T_A + T_B - T_{AB}. \tag{6.3}$$

Likewise, because blocks B and C are adjacent, the number of terminals between block B and C is written as

$$T_{B-to-C} = T_B + T_C - T_{BC}. \tag{6.4}$$

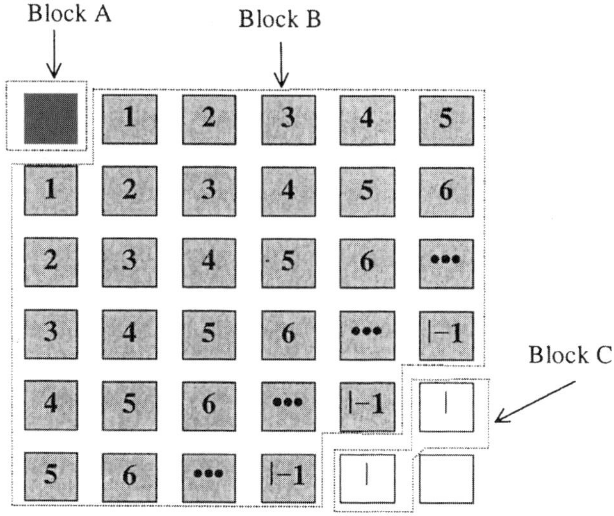

Figure 6.3. Determination of wire-Length distribution for a single gate.

Table 6.1. Variable Definitions

VARIABLE	DEFINITION
T_A	No. of I/Os for block A
T_B	No. of I/Os for block B
T_C	No. of I/Os for block C
T_{A-to-B}	No. of I/Os connecting block A to B
T_{A-to-C}	No. of I/Os connecting block A to C
T_{B-to-C}	No. of I/Os connecting block B to C
T_{AB}	No. of I/Os connecting block A+B
T_{BC}	No. of I/Os connecting block B+C
T_{ABC}	No. of I/Os connecting block A+B+C

Substituting (6.3) and (6.4) into (6.2) and simplifying gives

$$T_{A-to-C} = T_{AB} - T_B + T_{BC} - T_{ABC}. \tag{6.5}$$

The number of I/O terminals for a single block is directly calculated from Rent's Rule. Assuming, N_A, N_B, and N_C are the number of gates in block A, block B, and block C, respectively, from (6.1) the number of terminals for each block is

$$T_B = k(N_B)^p \tag{6.6}$$
$$T_{AB} = k(N_A + N_B)^p \tag{6.7}$$
$$T_{BC} = k(N_B + N_C)^p \tag{6.8}$$
$$T_{ABC} = k(N_A + N_B + N_C)^p. \tag{6.9}$$

Substituting (6.6)-(6.9) into (6.5) gives

$$T_{A-to-C} = k[(N_A + N_B)^p - (N_B)^p + (N_B + N_C)^p - (N_A + N_B + N_C)^p]. \tag{6.10}$$

To calculate the number of interconnects between blocks A and C, we define a variable α that is the fraction of terminals that are sinks (or input terminals) [8]. Therefore, the expected number of point-to-point interconnects between blocks A and C in Figure 6.3 is

$$I_{A-to-C} = \alpha k[(N_A + N_B)^p - (N_B)^p + (N_B + N_C)^p - (N_A + N_B + N_C)^p] \tag{6.11}$$

where the α factor is expressed in terms of the average fanout of the system, f.o., as $\alpha = \frac{f.o.}{f.o.+1}$.

Using (6.11) to calculate the number of interconnects for each length, ℓ, in Figure 6.3 in the range from 1 gate pitch to $2\sqrt{N}$ gate pitches, gives the complete stochastic wire-length distribution for the corner element. This is accomplished by tabulating the number of gates contained in N_A, N_B, and N_C for each length under consideration.

Once the stochastic wire-length distribution is determined for the corner element, it is removed from the system of gates for calculating the remainder of the wiring distribution in order to prevent multiple counting of interconnects. The same process is repeated for all other gates in the system. This algorithm is illustrated in Figure 6.4. The wire-length distributions for individual gates are superimposed to obtain the wire-length distribution for the entire system of N gates.

To describe this algorithm mathematically, we define a function $\Phi(i, j, \ell)$ that gives the number of gates that are a distance ℓ away from the gate in the ith row and the jth column in a square array of gates, as seen in Figure 6.4. This function excludes all gates that have been previously analyzed (i.e. the dashed gates in Figure 6.4). $\Phi(i, j, \ell)$ is defined in the next section.

Derivation of $\Phi(i, j, \ell)$. This section describes the overall strategy that is used to determine the expression, called $\Phi(i, j, \ell)$, for all possible gate pairs separated by a length ℓ for a gate in the ith row (from 1 to $\sqrt{N}$) and j^{th} column (from 1 to $\sqrt{N}$) as seen in Figure 6.4. A main assumption in this derivation is that the length connecting two gates is determined by the shortest orthogonal distance between the two elements.

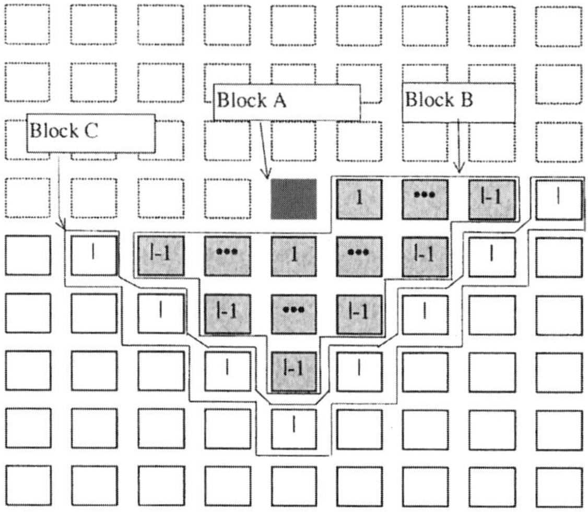

Figure 6.4. Algorithm for exact wire-length distribution calculation

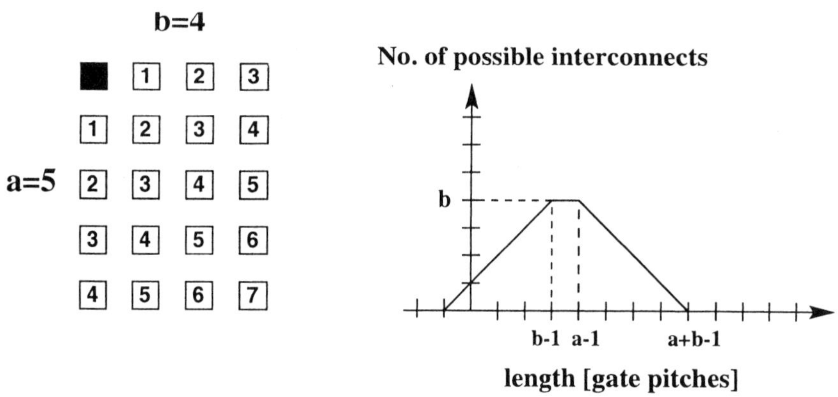

Figure 6.5. Number of interconnects of length ℓ for a corner interconnect

First consider the total number of interconnects with length ℓ, from a corner element of a rectangular $a \times b$ array of elements, this is seen graphically in Figure 6.5. From the graph in Figure 6.5, all possible connections from the corner element that extend to a length ℓ is given by

$$\begin{aligned}\delta_{ab} &= (\ell+1)u_o(\ell+1) - (\ell-b+1)u_o(\ell-b+1) \\ &\quad -(\ell-a+1)u_o(\ell-a+1) + (\ell-a-b+1)u_o(\ell-a-b+1)\end{aligned} \tag{6.12}$$

where $u_o(x)$ is a unit step function. Note that this function reaches a maximum equal to b or a depending on which is the smaller quantity.

In addition, the total number of connections of a corner gate in a rectangular $a \times b$ array to all gates *except* those in the same row and column as the corner element is observed graphically in Figure 6.6. From inspection this function

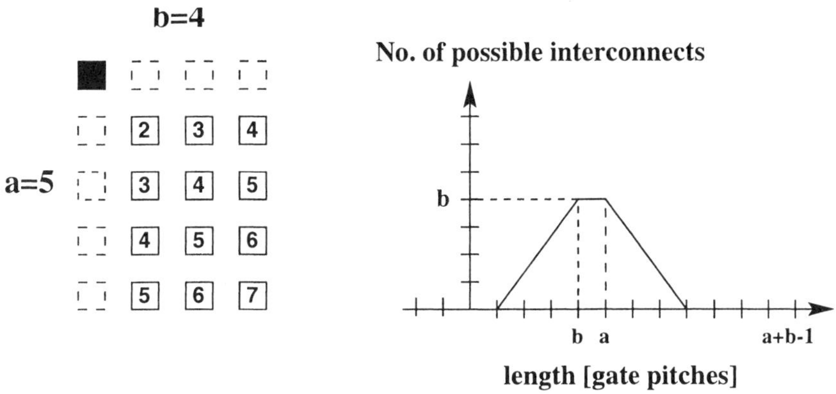

Figure 6.6. Number of gates at a length ℓ away NOT in the same row or column

illustrated in Figure 6.6 is

$$\begin{aligned} \delta'_{a'b'} &= (\ell-1)u_o(\ell-1) - (\ell-b')u_o(\ell-b') - (\ell-a')u_o(\ell-a') \\ &\quad +(\ell-a'-b'+1)u_o(\ell-a'-b'+1) \end{aligned} \tag{6.13}$$

$\Phi(i,j,\ell)$ is found in terms of (6.12) and (6.13) by letting $a = (\sqrt{N}-i+1)$ and $b = (\sqrt{N}-j+1)$ and $b' = j$ and $a' = (\sqrt{N}-i+1)$

$$\Phi(i,j,\ell) = \delta_{ab} + \delta'_{a'b'}. \tag{6.14}$$

Substituting (6.12) and (6.13) into (6.14) gives the final expression for $\Phi(i,j,\ell)$ to be

$$\Phi(i,j,\ell) = \begin{bmatrix} (\ell+1)\,u_o(\ell+1) - \left(\ell-\sqrt{N}+j\right)u_o\left(\ell-\sqrt{N}+j\right) \\ -\left(\ell-\sqrt{N}+i\right)u_o\left(\ell-\sqrt{N}+i\right) + (\ell-1)\,u_o(\ell-1) \\ +\left(\ell-2\sqrt{N}+j+i-1\right)u_o\left(\ell-2\sqrt{N}+j+i-1\right) \\ +\left(\ell-\sqrt{N}+i-j\right)u_o\left(\ell-\sqrt{N}+i-j\right) \\ -\left(\ell-\sqrt{N}-1+i\right)u_o\left(\ell-\sqrt{N}-1+i\right) \\ -(\ell-j)\,u_o(\ell-j) \end{bmatrix} \tag{6.15}$$

Exact Wire-Length Distribution. N_A, N_B and N_C in (6.11) are rigorously defined as

$$N_A = 1 \tag{6.16}$$

$$N_B = \sum_{r=1}^{r=\ell-1} \Phi(i,j,r) \tag{6.17}$$

$$N_C = \Phi(i,j,\ell) \tag{6.18}$$

Using (6.11) and (6.16)-(6.18) gives the exact formula for the discrete interconnect distribution,

$$i(\ell) = \sum_{i=1}^{\sqrt{N}} \sum_{j=1}^{\sqrt{N}} \left[\begin{array}{c} \left(1 + \sum_{r=1}^{\ell-1} \Phi(i,j,r)\right)^p - \left(\sum_{r=1}^{\ell-1} \Phi(i,j,r)\right)^p + \\ \left(\sum_{r=1}^{\ell} \Phi(i,j,r)\right)^p - \left(1 + \sum_{r=1}^{\ell} \Phi(i,j,r)\right)^p \end{array} \right] \tag{6.19}$$

Using (6.19), computer simulation of this algorithm can be performed, but for a large number of gates, simulation time can become excessive.

6.2.3 Closed-Form Wire-Length Distribution

This section outlines the derivation of a closed-form expression that approximates the exact wire-length distribution that appears in (6.19). To obtain a closed-form analytical expression for the wire-length distribution that appears in (6.19), it is assumed that on average the partitioning strategies are similar to the partial manhattan circle as seen in Figure 6.4. From geometrical inspection, a partial manhattan circle with radius ℓ in a large square array of gates is the most common configuration for gates that are at a distance greater than ℓ from the chip perimeter.

The number of gates on the periphery of the partial manhattan circle is equal to twice the radius of the circle (2ℓ). Therefore, from Figure 6.4 the expressions for N_A, N_B, and N_C are approximately given by:

$$N_A = 1 \tag{6.20}$$

$$N_B = \sum_{r=1}^{r=\ell-1} 2r = \ell(\ell-1). \tag{6.21}$$

$$N_C = 2\ell. \tag{6.22}$$

where N_A, N_B, and N_C are the number of logic gates in block A, block B, and block C in Figure 6.4. Substituting (6.20)-(6.22) into (6.11) gives the expected number of connections from the center gate of the partial manhattan circle (p.m.c) to all the periphery gates, which gives

$$I_{p.m.c.}(\ell) = \alpha k \left[(1 + \ell(\ell-1))^p - (\ell(\ell-1))^p + (\ell(\ell+1))^p - (1 + \ell(\ell+1))^p \right] \tag{6.23}$$

where α is the fraction of I/O terminals that are sink terminals in ($\alpha = \frac{f.o.}{f.o.+1}$).

The average number of interconnects connecting each gate pair separated by a length ℓ in a given partial manhattan circle is obtained by dividing the number of interconnections given by (6.23) by the number of gates on the periphery of a partial manhattan circle, ℓ. Using this partial manhattan circle approximation, the expected number of interconnects connecting gate pairs separated by a length ℓ in a given system is

$$I_{\exp}(\ell) = \frac{\alpha k}{2\ell} \left[(1 + \ell(\ell-1))^p - (\ell(l-1))^p + (\ell(\ell+1))^p - (1 + \ell(\ell+1))^p \right]. \tag{6.24}$$

To complete the derivation for the entire wire-length distribution, the number of gate pairs separated by a length ℓ, $M(\ell)$, in a square array of N gates must be determined.

Derivation of $M(\ell)$. To derive the number of gates separated by a length ℓ in a logic block as seen in Figure 6.4, the expressions in (6.12) and (6.13) are used. This first step toward this goal is to determine a function that gives the total number of interconnects with length ℓ that could be connected to *the first row* of the 2-D array as seen in Figure 6.7.

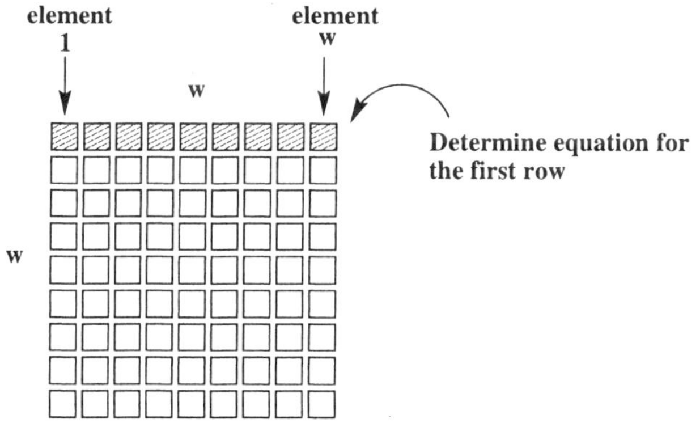

Figure 6.7. Determine gate pair distribution for the first row

In order to determine a function that gives the total number of possible interconnects connected to the first row, the b and b' elements in the δ_{ab} and $\delta'_{a'b'}$

will be summed from 1 to w, where w is the dimension of the array. Graphically this can be seen in Figure 6.7. Because of the use of step functions, the solution for each element in the first row is easily superposed

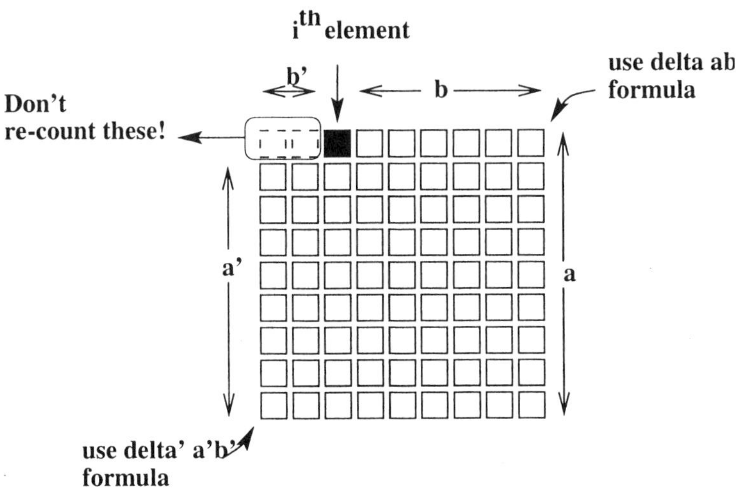

Figure 6.8. Graphically illustration of counting methodology

Because the solutions are easily superimposed, consider the summation of δ_{ab} and $\delta'_{a'b'}$ separately. The δ_{ab} function will be examined for $b = w$, $b = w-1$, $b = w-2,\ldots, b = 1$. Likewise, $b' = 1$ (when $b = w - 1$), $b' = 2$ (when $b = w-2$), $\ldots b' = w$ (when $b = 1$). Because the order that the superimposed solutions are summed for each logic gate does not matter, the final mathematical expression for the total number of interconnects that connect the first row to the rest of the array is

$$\sum_{b=1}^{b=w} \left(\delta_{ab} + \delta'_{ab}\right). \tag{6.25}$$

To obtain the number of all possible interconnects for the entire two dimensional array, this process is repeated again with another array of elements minus the top row, as seen in Figure 6.9. The resulting mathematical relation for the total number of possible interconnects in a square array is

$$\sum_{a=1}^{a=w}\sum_{b=1}^{b=w} \left(\delta_{ab} + \delta'_{ab}\right). \tag{6.26}$$

Useful Summations. Because the derivation involves discrete summation of shifted step functions, the following summations are used in the evaluation

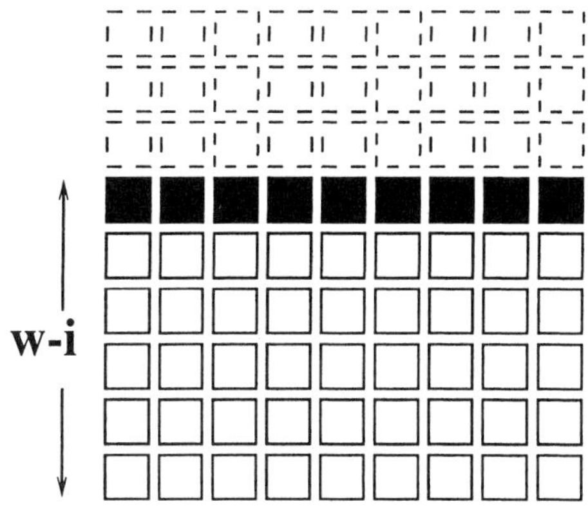

Figure 6.9. Repeat analysis on each row

of equation (6.26). The first summation that is useful in evaluating (6.26) is

$$\sum_{b=1}^{b=w} u_o(\phi - b) = u_o(\phi - 1) + u_o(\phi - 2) + u_o(\phi - 3) ... + u_o(\phi - w) \quad (6.27)$$

The solution is determined by graphically superimposing shifted step functions. The summation is then evaluated to be

$$\sum_{b=1}^{b=w} u_o(\phi - b) = \phi u_o(\phi) - (\phi - w) u_o(\phi - w). \quad (6.28)$$

The second summation that is useful in evaluating (6.26) is

$$\sum_{b=1}^{b=w} b u_o(\phi - b) = u_o(\phi - 1) + 2u_o(\phi - 2) + 3u_o(\phi - 3) ... + w u_o(\phi - w) \quad (6.29)$$

The result of (6.29) is determined by graphically superimposing these shifted step functions.

$$\sum_{b=1}^{b=w} b u_o(\phi - b) = \frac{\phi(\phi + 1)}{2} u_o(\phi) - (\frac{\phi(\phi + 1)}{2} - \frac{w(w + 1)}{2}) u_o(\phi - w) \quad (6.30)$$

Finally, the third summation that is useful in evaluating (6.26) is

$$\sum_{b=1}^{b=w} b^2 u_o(\phi - b) = u_o(\phi - 1) + 4u_o(\phi - 2) + 9u_o(\phi - 3) ... + w^2 u_o(\phi - w) \quad (6.31)$$

The solution is determined graphically by superimposing these shifted step functions. The summation evaluates to

$$\sum_{b=1}^{b=w} b^2 u_o(\phi - b) = \frac{\phi(\phi+1)(2\phi+1)}{6} u_o(\phi) - \left(\frac{\phi(\phi+1)(2\phi+1)}{6} - \frac{w(w+1)(2w+1)}{6}\right) u_o(\phi - w) \quad (6.32)$$

Final Form of $M(\ell)$. To obtain the expression for the total number of possible interconnects of length ℓ, or $M(\ell)$, the summations in (6.26) must be evaluated. Substituting (6.12) and (6.13) into (6.26) gives

$$\sum_{a=1}^{a=w}\sum_{b=1}^{b=w}\left(\begin{array}{l}(\ell+1)u_o(\ell+1) - (\ell-b+1)u_o(\ell-b+1)- \\ (\ell-a+1)u_o(\ell-a+1) + (\ell-a-b+1)u_o(\ell-a-b+1) \\ +(\ell-1)u_o(\ell-1) - (\ell-b)u_o(\ell-b)- \\ (\ell-a)u_o(\ell-a) + (\ell-a-b+1)u_o(\ell-a-b+1)\end{array}\right) \quad (6.33)$$

Using the identities in (6.27) - (6.32) to evaluate the summations in (6.33) gives

$$\begin{gathered}\left[-w\ell + w^2 - \tfrac{1}{3}\ell + w^2\ell - w\ell^2 + \tfrac{1}{3}\ell^3\right] u_o(\ell+1) \\ + \left[w^2(\ell-1)\right] u_o(\ell-1) + \left[-w\ell^2 + w\ell\right] u_o(\ell) \\ + \left[-\tfrac{2}{3}w + \ell w - w^2 + \tfrac{2}{3}\ell - 4w^2\ell + 3w\ell^2 - \tfrac{2}{3}\ell^3 + \tfrac{5}{3}w^3\right] u_o(\ell+1-w) \\ + \left[w\ell^2 - 2w^2\ell - w\ell + w^3 + w^2\right] u_o(\ell - w) \\ \left[\tfrac{2}{3}w - \tfrac{1}{3}\ell + 4w^2\ell - 2w\ell^2 + \tfrac{1}{3}\ell^3 - \tfrac{8}{3}w^3\right] u_o(\ell+1-2w)\end{gathered} \quad (6.34)$$

Re-writing (6.34) in terms of a piece-wise defined function gives

Region I : $1 \leq \ell < w$

$$M(\ell) = \frac{1}{3}\ell^3 - 2w\ell^2 + \frac{1}{3}\ell\left(6w^2 - 1\right) \quad (6.35)$$

Region II: $w \leq \ell < 2w - 1$

$$M(\ell) = -\frac{1}{3}\ell^3 + 2w\ell^2 - \frac{1}{3}\ell\left(12w^2 - 1\right) + \frac{2}{3}w(2w-1)(2w+1) \quad (6.36)$$

Region III : $\ell \geq 2w - 1$

$$M(\ell) = 0 \quad (6.37)$$

For systems with a large number of gates, where $w = \sqrt{N}$ and $\sqrt{N} >> 1$, $M(\ell)$ simplifies to its final form that is used in the rest of this chapter

Region I: $1 \leq \ell < \sqrt{N}$

$$M(\ell) = \left(\frac{\ell^3}{3} - 2\ell^2\sqrt{N} + 2\ell N\right) \quad (6.38)$$

Region II: $\sqrt{N} \leq \ell < 2\sqrt{N}$

$$M(\ell) = \frac{1}{3}\left(2\sqrt{N} - \ell\right)^3 \tag{6.39}$$

Final Closed-Form Wire-Length Distribution. The approximate expression for the discrete wire length distribution, $i(\ell)$, that gives the expected number of interconnects for a given length ℓ is

$$i\left(\ell\right) \approx M\left(\ell\right) I_{exp}\left(\ell\right). \tag{6.40}$$

$I_{exp}(\ell)$ in (6.24) is simplified using a binomial expansion and is given by

$$I_{\text{exp}}\left(\ell\right) \propto \ell^{2p-4} \tag{6.41}$$

Finally, we will insure proper normalization of this expression to make sure that the distribution predicts the total number of wires is [10]

$$I_{total} = \alpha k N(1 - N^{p-1}) \tag{6.42}$$

For a discrete random variable ℓ the following expression can be used to determine the proper normalization factor:

$$I_{total} = \Gamma \sum_{l=1}^{2\sqrt{N}} i\left(\ell\right) \tag{6.43}$$

where $i(\ell)$ is defined in (6.40).

By making the assumption that $2\sqrt{N} >> 1$, then the discrete distribution is approximated by a continuous interconnect density function. The normalizing factor is determined from

$$\Gamma = \frac{I_{total}}{\int_{\ell=1}^{2\sqrt{N}} i\left(\ell\right)}. \tag{6.44}$$

Evaluating (6.44) gives

$$\Gamma = \frac{2N\left(1 - N^{p-1}\right)}{\left(-N^p \frac{1+2p-2^{2p-1}}{p(2p-1)(p-1)(2p-3)} - \frac{1}{6p} + \frac{2\sqrt{N}}{2p-1} - \frac{N}{p-1}\right)}. \tag{6.45}$$

At p=0.5, Γ is of the form of $\frac{0}{0}$, but using L'Hospital's rule at p=0.5, Γ converges to

$$\Gamma = \frac{4N - 4\sqrt{N}}{\sqrt{N}\left(-2.0\ln N - 6.0 + 2.0\ln 4\right) + 4.0N - \frac{2}{3}}. \tag{6.46}$$

Making this substitution gives the final form of the interconnect density function, $i(\ell)$. The product $i(\ell)d\ell$ gives the number of interconnects in the infinitesimal range from ℓ to $\ell + d\ell$. Substituting (6.42) in (6.44) gives

Region I: $1 \leq \ell < \sqrt{N}$

$$i\left(l\right) = \frac{\alpha k}{2}\Gamma\left(\frac{\ell^3}{3} - 2\sqrt{N}\ell^2 + 2N\ell\right)\ell^{2p-4} \tag{6.47}$$

Region II: $\sqrt{N} \leq \ell < 2\sqrt{N}$

$$i\left(l\right) = \frac{\alpha k}{6}\Gamma\left(2\sqrt{N} - \ell\right)^3 \ell^{2p-4} \tag{6.48}$$

The interconnect density function is defined such that the number of interconnects between length $\ell = a$ and $\ell = b$ is

$$I(a < \ell < b) = \int_a^b i(\ell)d\ell. \tag{6.49}$$

The cumulative interconnect distribution function, c.i.d.f., is determined from

$$I(l) = I_{total}P(l). \tag{6.50}$$

Substituting (6.42) and (6.44) in (6.50) gives

Region I: $1 \leq \ell < \sqrt{N}$

$$I\left(\ell\right) = \frac{\alpha k}{2}\Gamma\left(\frac{\ell^{2p} - 1}{6p} + 2\sqrt{N}\frac{-\ell^{2p-1} + 1}{(2p-1)} - N\frac{-\ell^{2p-2} + 1}{(p-1)}\right) \tag{6.51}$$

Region II: $\sqrt{N} \leq \ell < 2\sqrt{N}$

$$I\left(\ell\right) = \frac{\alpha k}{2}\Gamma\begin{pmatrix} \frac{\sqrt{N}^{2p}-1}{6p} + 2\sqrt{N}\frac{-\sqrt{N}^{2p-1}+1}{(2p-1)} - N\frac{-\sqrt{N}^{2p-2}+1}{(p-1)} \\ -\frac{8}{3}N^{\frac{3}{2}}\frac{-l^{2p-3}+N^{p-\frac{3}{2}}}{(2p-3)} + 2N\frac{-l^{2p-2}+N^{p-1}}{(p-1)} \\ -2\sqrt{N}\frac{-l^{2p-1}+N^{p-\frac{1}{2}}}{(2p-1)} + \frac{-l^{2p}+N^p}{6p} \end{pmatrix} \tag{6.52}$$

The cumulative interconnect distribution function gives the total number of interconnects that have a length less than and equal to ℓ given by

$$I(\ell) = \int_1^{\ell} i\left(\zeta\right)d\zeta. \tag{6.53}$$

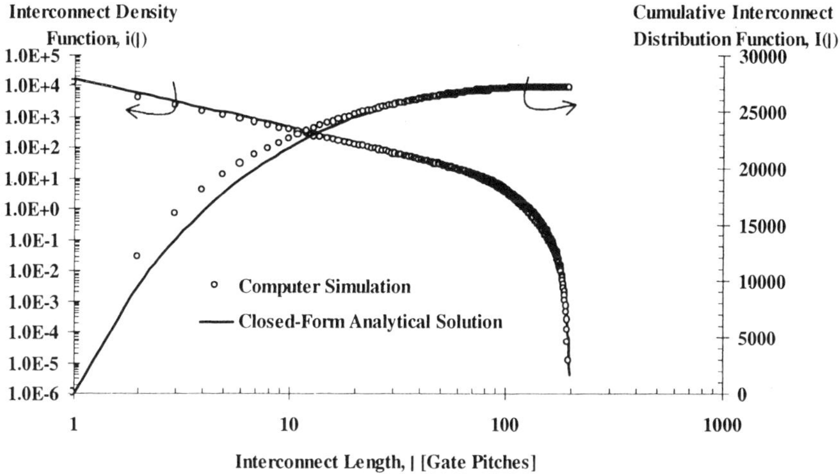

Figure 6.10. The i.d.f. and the c.i.d.f. compared to computer simulation

The closed-form analytical expressions for the i.d.f. and the c.i.d.f. are compared to computer simulation of the exact distribution in Figure 6.10.

A very useful function that is used in the following chapters is the length demand function. This function gives the total length of interconnect from a length of 1 gate pitch to ℓ gate pitches. Using this definition, the length demand function is given by

$$D(\ell) = \int_1^{\ell} \zeta i(\zeta) d\zeta \tag{6.54}$$

Evaluating (6.54) leads to

Region I: $1 \leq \ell \leq \sqrt{N}$

$$D(\ell) = \frac{\alpha k}{2}\Gamma\left(\frac{1}{3}\frac{\ell^{1+2p}-1}{1+2p} - \sqrt{N}\frac{\ell^{2p}-1}{p} + 2N\frac{\ell^{-1+2p}-1}{(-1+2p)}\right) \tag{6.55}$$

Region II: $\sqrt{N} \leq \ell \leq 2\sqrt{N}]$

$$\begin{aligned} D(\ell) &= \frac{\alpha k}{2}\Gamma\left(\frac{1}{3}\frac{N^{\frac{1}{2}+p}-1}{1+2p} - \sqrt{N}\frac{N^p-1}{p} + 2N\frac{N^{-\frac{1}{2}+p}-1}{(-1+2p)}\right) \\ &+ \frac{\alpha k}{6}\Gamma\left(-4\left(\sqrt{N}\right)^3\frac{-\ell^{2p-2}+N^{p-1}}{(p-1)} + 12N\frac{-\ell^{-1+2p}+N^{-\frac{1}{2}+p}}{(-1+2p)}\right. \\ &\left. + 3\sqrt{N}\frac{\ell^{2p}-N^p}{p} - \frac{\ell^{1+2p}-N^{\frac{1}{2}+p}}{1+2p}\right) \end{aligned} \tag{6.56}$$

6.3 Net Model Approximation

In real random logic networks, the source terminals (i.e. the output I/Os) are usually connected to multiple sink terminals (i.e. the input I/Os). The interconnect density function predicts the number of point-to-point interconnects connecting a source terminal to each of its sink terminals. Real designs, however, use wiring nets that more efficiently connect the source to its sink terminals as illustrated in Figure 6.11 [9].

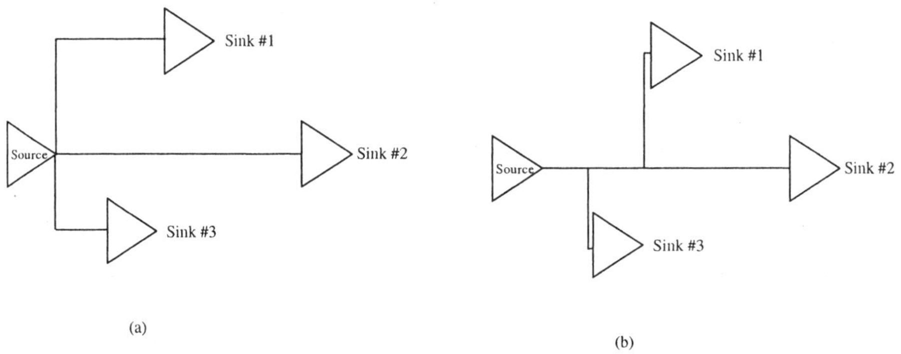

Figure 6.11. Schematic of (a) a point-to-point interconnect and (b) a real wiring net.

Wiring nets, therefore, must be considered when estimating total wiring area, total wiring capacitance, and wiring net performance [9]. In order to estimate these physical quantities, a regular linear net model is used to approximate the geometrical layout of a real wiring net. The geometrical configuration of this net model is seen in Figure 6.12, where each subsection of the net is assumed to have a length equal to s. Given a fanout, f.o., the cumulative length of the point-to-point interconnects, L_{point}, for one linear wiring net is

$$L_{point} = 2s + 3s ... + (f.o. + 1)s = \sum_{k=1}^{f.o.} (k+1)s = s\left(\frac{f.o.}{2}(f.o. + 3)\right). \tag{6.57}$$

The length of each segment, s, is

$$s = \frac{L_{net}}{2(f.o.)} \tag{6.58}$$

where L_{net} is the net length for a single wiring net with a given fanout, f.o. Therefore, the net length in terms of the cumulative length of the point-to-point interconnects, L_{point}, for the linear net model is:

$$L_{net} = \chi L_{point} \tag{6.59}$$

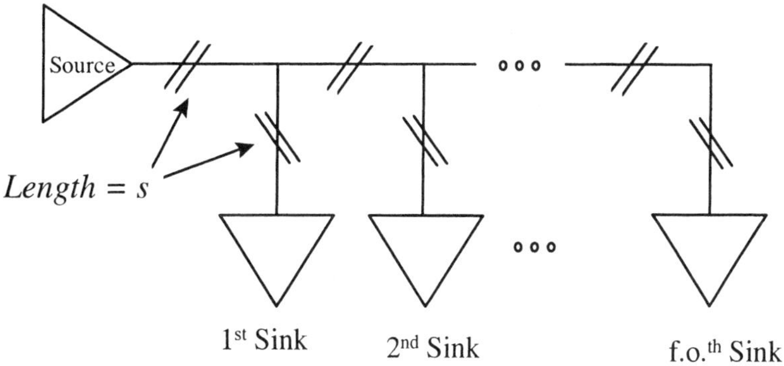

Figure 6.12. Linear Wire Net Model

where

$$\chi = \frac{4}{f.o. + 3}. \tag{6.60}$$

6.4 Comparisons with Actual Data

This interconnect density function characterizes real data with high accuracy. In Figure 6.13 it is compared to actual point-to-point interconnect data taken from a real system [10] and also to a widely referenced wire-length distribution model [10]. As seen in Figure 6.13, the new complete interconnect density function describes a real wiring histogram with higher fidelity than previous models [10, 11]. Moreover, comparisons to current data on microprocessors indicate that the new i.d.f. accurately estimates the wiring for these VLSI systems. Figure 6.14 contains a wiring distribution for a microprocessor design. The uncertainty in the actual data is due to different wiring net models that were used to convert the net list information for this microprocessor to point-to-point wiring information. A large sample set of approximate net models was used to exact point-to-point information so that a reasonable range of uncertainty could be represented. The k and p empirical parameters are chosen to obtain a best fit of the data for microprocessor A in Figure 6.14. Using the k and p parameters derived from microprocessor A, the i.d.f. is now used to *predict* the interconnect density function for subsequent generations of microprocessors from the same company as seen in Figure 6.15 and Figure 6.16. The i.d.f. provides a good first order estimation of the wiring requirements of microprocessors B and C.

Determination of the best values of k and p parameters for future generation products mandates an investigation into previous generations of a product family. As shown with microprocessor A, B, and C, the heredity of a particular

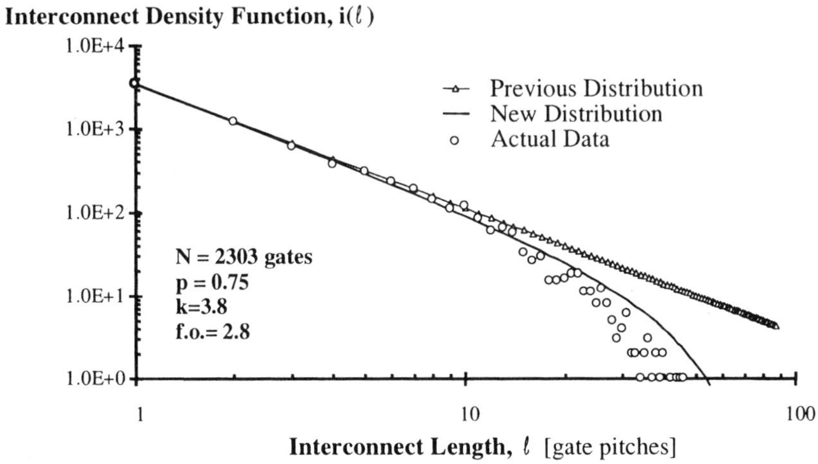

Figure 6.13. Comparison of the new stochastic model to actual data and a previous stochastic model

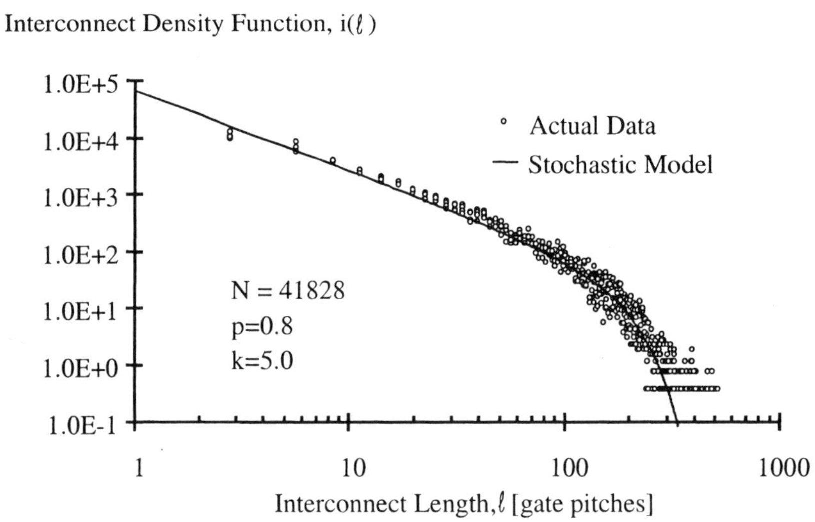

Figure 6.14. Interconnect density function compared to the interconnect histogram for microprocessor A

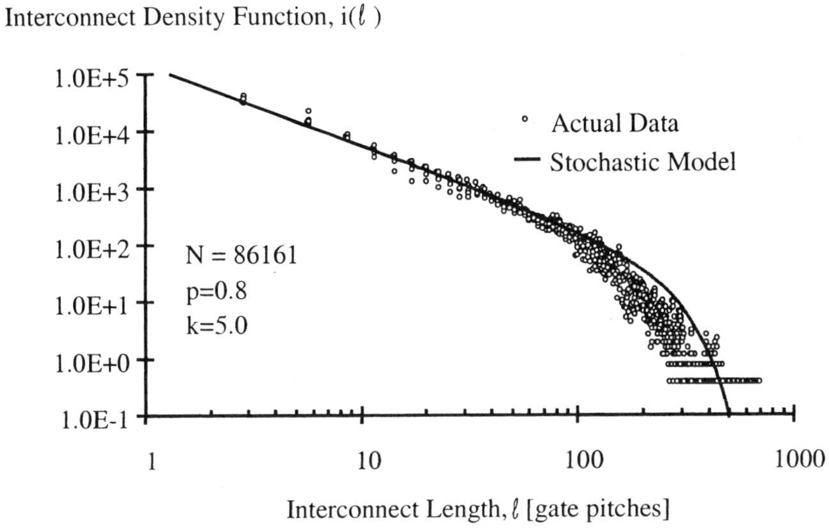

Figure 6.15. Interconnect density function compared to the interconnect histogram for microprocessor B

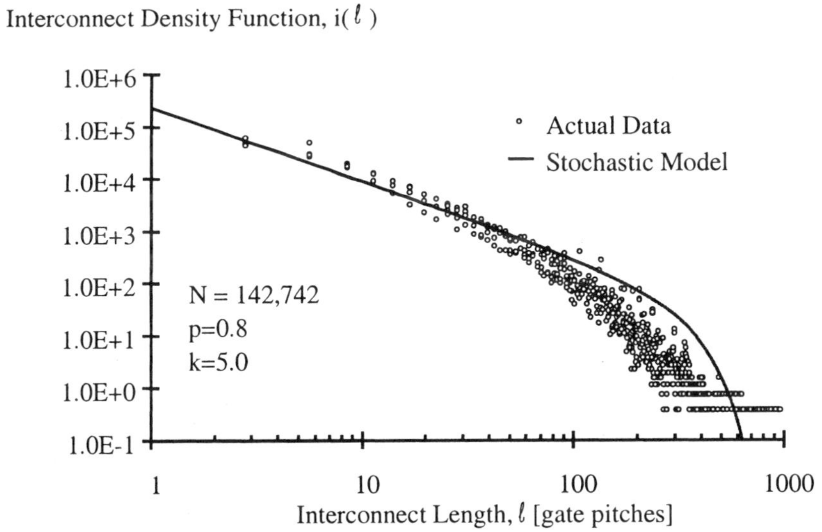

Figure 6.16. Interconnect density function compared to the interconnect histogram for microprocessor C

product generation usually has a strong influence on its k and p values. This also is particularly striking for the Intel microprocessor family as illustrated in Figure 6.2.

6.5 Critical Path Model

Various critical path models have been proposed to estimate cycle time [9, 12]. One essential component of a critical path model for CMOS circuits is the size of the wiring networks loading the critical path gates. The complete distribution enables a more accurate representation of the average wire length and estimation of the longest interconnect in the system.

An established critical path model from [12] has all but one of its gates loaded by an average interconnect net. The remaining gate is loaded by a single global interconnect [12]. From [2], the limit on the clock period is given by:

$$T_c \geq T_{cs} + n_{cp} t_d + T_{LD} \tag{6.61}$$

where T_c is the clock period, T_{cs} is the clock skew, n_{cp} is the number of gates in the critical path, t_d is the average time delay of each gate loaded by an average wiring net, and T_{LD} is the time delay of the longest global interconnect in the system.

6.5.1 Average Wire-Length Calculation

The average point-to-point interconnect length is determined directly from the interconnect density function according to the following expression

$$L_{avg} = \frac{L_{total}}{I_{total}} = \frac{\int_1^{2\sqrt{N}} \ell i(\ell)\, d\ell}{\int_1^{2\sqrt{N}} i(\ell)\, d\ell} \tag{6.62}$$

where L_{total} is the total point-to-point interconnect length and I_{total} is the total number of interconnects. Evaluating (6.62) gives the final form of the average wire length, L_{avg}.

$$L_{avg} = \frac{\left(\frac{p-0.5}{p} - \sqrt{N} - \frac{p-0.5}{6\sqrt{N}(p+0.5)} + N^p\left(\frac{-p-1+4^{p-0.5}}{2(p+0.5)p(p-1)}\right)\right)}{\left(N^{p-0.5}\frac{-2p-1+2^{2p-1}}{2p(p-1)(2p-3)} - \frac{(p-0.5)}{6p\sqrt{N}} + 1 - \frac{(p-0.5)\sqrt{N}}{(p-1)}\right)} \tag{6.63}$$

This average interconnect length is dependent only on the number of gates in the system and Rent's exponent, p. A graph of the average wire length vs. number of gates for various p values is seen in Figure 6.17. A previous

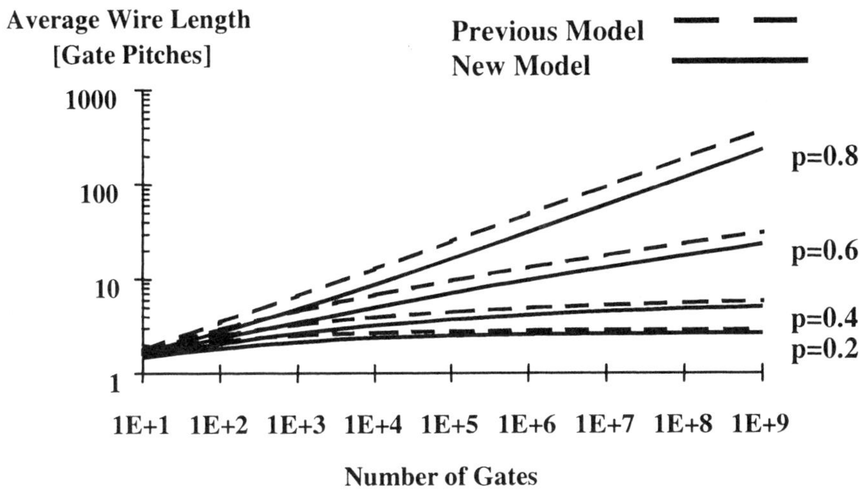

Figure 6.17. Average interconnect length versus number of gates for p=0.2,0.4,0.6,0.8

distribution gives higher average interconnect lengths due to overestimation of the number of longer interconnects as seen in Figure 6.13 [10, 11, 8].

The average wire length expression can be simplified for a system with a large number of gates. For large N the expression for the average wire length simplifies to

$$L_{avg} = \frac{-\sqrt{N} + N^p \left(\frac{-p-1+4^{p-0.5}}{2(p+0.5)(p)(p-1)}\right)}{N^{p-0.5}\frac{-2p-1+2^{2p-1}}{2p(p-1)(2p-3)} - \frac{(p-0.5)\sqrt{N}}{(p-1)}} \tag{6.64}$$

In addition, if $p > 0.5$ then the average interconnect length simplifies to

$$L_{avg} = N^{p-0.5}\left(\frac{p+1-4^{p-0.5}}{2\,(p-0.5)\,(p+0.5)\,(p)}\right) \tag{6.65}$$

If $p < 0.5$ then the average interconnect length simplifies to

$$L_{avg} = -\frac{(p-1)}{(p-0.5)}\left(-1 + N^{p-0.5}\left(\frac{-p-1+4^{p-0.5}}{2\,(p+0.5)\,(p)\,(p-1)}\right)\right) \tag{6.66}$$

As N goes to infinity, (6.66) saturates to its maximum value of

$$L_{avg} = \frac{(p-1)}{(p-0.5)} \tag{6.67}$$

The implications of (6.67) is that for $p < 0.5$ designs the average wire length does not change for larger systems. Unfortunately, the complexity of today's systems seems to be greater than 0.5, which means that average wire length also increases as seen in Figure 6.17.

Evaluating the L_{avg} at p=0.5 gives gives 0/0, therefore L'Hospital's rule is used. Taking the derivative of the numerator and the denominator with respect to p and evaluating at p=0.5 gives

$$L_{avg} = \frac{\left(\frac{1}{0.5N^{0.5}}\right) - \left(\frac{1}{6NN^{0.5}}\right) + (\ln N + 1 - 2\ln 4)}{\left(\frac{(-3+2\ln 2)}{\sqrt{N}} + \frac{\ln N}{\sqrt{N}}\frac{-1}{2(0.5)(-0.5)(-2)}\right) - \frac{1}{6(0.5)N} + \frac{1}{(0.5)}} \tag{6.68}$$

Letting N become very large gives the final expression

$$L_{avg} = 0.5\left(\ln N + 1 - 2\ln 4\right)$$

In the critical path model, each critical path gate is loaded by an average wiring net, and the length of an average wiring net is

$$L_{avg\_net} = \frac{L_{total\_net}}{I_{nets}} = \frac{\chi L_{total\_p-to-p}}{\frac{I_{total}}{f.o.}} = \chi f.o. L_{avg} \tag{6.69}$$

where χ is a correction factor that converts the point-to-point interconnect length to wiring net length (using a linear net model $\chi = \frac{4}{f.o.+3}$), L_{total_net} is the total net length, f.o. is the average fanout, and I_{nets} is the total number of nets.

6.5.2 Longest Interconnect Definition

The longest interconnect also is estimated using the interconnect density function, i.d.f., which has a range from 1 gate pitch to $2\sqrt{N}$ gate pitches. The i.d.f. predicts that one interconnect exists in the interval between a given length, ℓ_{max}, and $2\sqrt{N}$ from the following expression,

$$1 = \int_{\ell=\ell_{\max}}^{\ell=2\sqrt{N}} i(\ell)\, d\ell \tag{6.70}$$

Because the interconnect density function is monotonically decreasing, then ℓ_{max} is defined as the most probable value for the longest interconnect length. The expression for this length is also written in terms of the c.i.d.f., $I(\ell)$, as

$$I(2\sqrt{N}) - I(\ell_{\max}) = 1. \tag{6.71}$$

A graph of the longest interconnect, ℓ_{max}, normalized to twice the chip edge for $p = 0.2, p = 0.4, p = 0.8$ values appears in Figure 6.18 with $\alpha k = 3.0$.

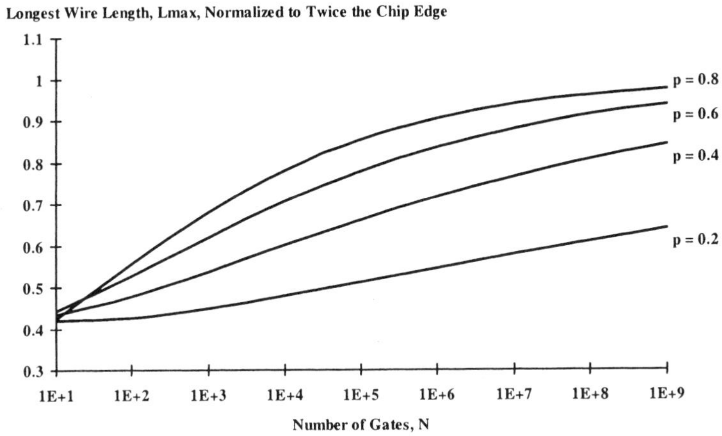

Figure 6.18. Longest interconnect length normalized to twice the chip edge versus number of gates for p=0.2, 0.4, 0.6, and 0.8.

6.6 Dynamic Power Dissipation Model

The dominant load capacitance for many VLSI CMOS circuits is the wiring capacitance [9]. The interconnect density function provides a priori information concerning the distribution of capacitive loads present in a GSI system. Assuming a constant activity factor for each capacitive node, the average dynamic power dissipation of the signal interconnects, P_{avg}, is given by [9]

$$P_{avg} = a\frac{1}{2}C_{total}V_{dd}^2 f_c, \tag{6.72}$$

where V_{dd} is the supply voltage, a is the average activity factor for each gate, C_{total} is the total capacitive load of the wiring network, and f_c is the clock frequency. The C_{total} term is estimated from the geometrical configurations of the interconnects in a multilevel wiring network and from the interconnect density function. A conventional multilevel network, for example, has two main interconnect types: local interconnects and global interconnects. The cross-sectional dimensions and the capacitive contributions for an arbitrary interconnect are illustrated Figure 6.19(a) and 6.19(b).

In general a capacitance density function is defined as

$$C(\ell)d\ell = c_{dist}(\sqrt{\frac{A_c}{N}}\ell)i(\ell)d\ell \tag{6.73}$$

where c_{dist} is the distributed capacitance per unit length as a function of interconnect length. This expression is explored in terms a two tier multilevel

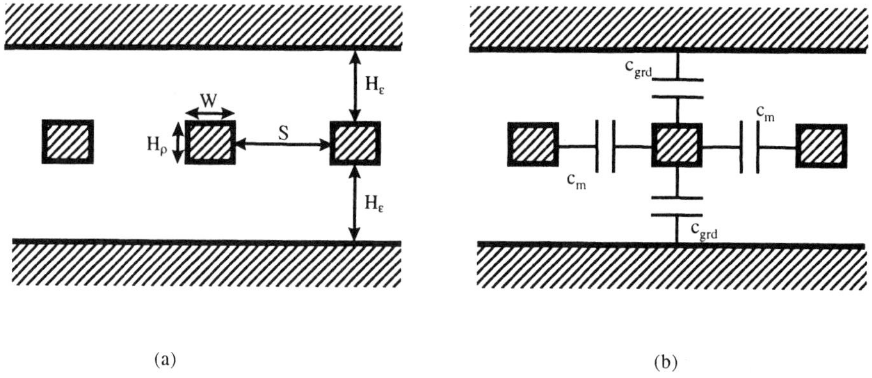

Figure 6.19. The definition of (a) the interconnect dimensions and (b) the capacitive components

interconnect architecture where the distributed capacitance per unit length is assumed to be constant across all local and global tiers. A tier is defined as a collection of metal levels.

Assuming that the neighboring wiring planes in the multilevel network provide an virtual ground plane, then the capacitance per unit length is given by

$$c_{total} = 2c_{grd} + 2c_m \tag{6.74}$$

where c_{grd} is the line-to-ground capacitance, and c_m is the line-to-line mutual capacitance. The values of c_{grd} and c_m that include fringing effects are given by [13]. There is a small typographical error [13] and is correctly given by:

$$\frac{c_{grd}}{\varepsilon} = \frac{W}{H_\varepsilon} + 1.086\left(1 + 0.685e^{-\frac{1}{1.343}\frac{H_\rho}{S}} - 0.9964e^{-\frac{1}{1.421}\frac{S}{H_\varepsilon}}\right)\left(\frac{S}{S+2H_\varepsilon}\right)^{0.0476}\left(\frac{H_\rho}{H_\varepsilon}\right)^{0.337} \tag{6.75}$$

and

$$\begin{aligned}\frac{c_m}{\varepsilon} &= \left(\frac{H_\rho}{S}\right)\left(1 - 1.897e^{\frac{-H_\varepsilon}{0.31S} + \frac{-H_\rho}{2.474S}} + 1.302e^{\frac{-H_\varepsilon}{0.082S}} - 0.1292e^{\frac{-H_\rho}{1.326S}}\right) \\ &\quad + 1.722\left(1 - 0.6548e^{\frac{-W}{0.3477H_\varepsilon}}\right)e^{\frac{-S}{0.651H_\varepsilon}}.\end{aligned} \tag{6.76}$$

where the geometrical variables are defined in Figure 6.19(a). For the general case, the capacitances per unit length for the local and global interconnects are dissimilar, and the expressions for the total capacitance of the local levels, $C_{total,loc}$, and the total capacitance of the global levels, $C_{total,glob}$, are given by

$$C_{total,loc} = c_{loc}\sqrt{\frac{A_c}{N}}\chi\int_1^{L_{loc}} \ell i(\ell)d\ell \tag{6.77}$$

$$C_{total,glob} = c_{glob}\sqrt{\frac{A_c}{N}}\chi \int_{L_{loc}}^{2\sqrt{N}} \ell i(\ell) d\ell \tag{6.78}$$

where symbol χ is the correction factor that converts the total length of point-to-point interconnects to total net length, A_c is the chip area, N is the number of gates, c_{loc} is the distributed capacitance for the local interconnects, c_{glob} is the distributed capacitance for the global interconnects, and L_{loc} is the length in gate pitches of the longest local interconnect. Assuming that the longest local interconnect, L_{loc}, is less than the chip edge, the above expressions simplify to

$$C_{total,loc} = c_{loc}\sqrt{\frac{A_c}{N}}\chi D(L_{loc}) \tag{6.79}$$

$$C_{total,glob} = c_{glob}\sqrt{\frac{A_c}{N}}\chi \left(D(2\sqrt{N}) - D(L_{loc}\right) \tag{6.80}$$

where $D(\ell)$ is defined has the demand function in (6.54).

Evaluating these two expressions and adding the total capacitance of the local and global interconnects give the final expression for the average dynamic power dissipation for a multilevel wiring network as

$$P_{avg} = a\frac{1}{2}V_{dd}^2 f_c\sqrt{\frac{A_c}{N}}\frac{\Gamma\alpha k\chi}{2}\left[c_{loc}D(L_{loc}) + c_{global}\left(D(2\sqrt{N}) - D(L_{loc}\right)\right] \tag{6.81}$$

Table 6.2. Microprocessor Application

PARAMETERS	VALUE
Number of Gates, N	8.0 million
Rent's Exponent, p	0.6
Rent's Coefficient, k	4.0
Minimum Feature Size,F	$0.18\mu m$
Supply Voltage, V_{dd}	1.5 [V]
Max number levels, n_{max}	6
Metal Resistivity, Copper	$1.673e-6[\Omega - cm]$
Dielectric Constant, Polymer	$\epsilon_r = 2.5$
Wiring Efficiency Factor, e_w	0.4

Given the two-tier multilevel network with the physical characteristics as outlined in Table 6.2 and Table 6.3, assuming that $W = H_\epsilon = H_\rho = S$ for

Table 6.3. Two-Tier Conventional Design

PARAMETERS	PITCH	# OF LEVELS
Local	$0.36\mu m$	2
Global	$1.53\mu m$	4
*****	*****	******
Clock Frequency	410Mhz	******
Wire-Limited Chip Area	4.92 cm^2	******
Transistor-Limited Chip Area	$N \times 200F^2 = 0.52cm^2$	******

the local and global tiers such that $c_{loc} = c_{glob} = 6.08\epsilon$, and given the activity factor, a, is approximately 0.10, then the total power dissipation in the signal interconnects for this system is 10.0W and the power density is 2.03 W/cm^2. The local interconnects dissipate 68.15% of the total power, and the global interconnects dissipate 31.9% of the total power. Even though the amount of area dedicated to the global tier and local tier is approximately the same, the global wires dissipate less power than the local wires because 1) the global and local distributed capacitance are identical and 2) the total length of the global interconnects is less than the total length of the local interconnects.

6.7 Optimal n-Tier Multilevel Interconnect Architectures

6.7.1 Introduction

Reverse-scaled multilevel wiring networks are extensively used in current VLSI systems to mitigate the impact of wiring on chip size and system performance [2], [9], [14]. A methodology to design the interconnect process parameters for current and future reverse-scaled multilevel interconnect networks is presented that uses the stochastic wiring distribution [15] and critical path model that is derived earlier in this chapter.

In addition, a repeater insertion methodology is also described that further enhances GSI system performance. Because repeater insertion reduces wire delay, repeaters can be an effective tool to reduce the stringent interconnect limits on future GSI systems [16]. A repeater insertion methodology is reviewed in this work that optimally inserts repeaters in a multilevel wiring network to decrease chip size, cycle time, number of metal levels or power dissipation.

6.7.2 n-tier Multilevel Architecture Design Methodology

A wiring tier is defined as a collection of metal levels that have approximated the same pitch and maximum timing constraints. In general a GSI multilevel metal stack can have n tiers. In this section, we will discuss the opportunities to find the optimal dimensions of each tier such that key system parameters, such as clock frequency, power dissipation, and die size, are optimized.

The key assumptions made in the n-tier design methodology are:

1. The interconnects in a system obey the stochastic interconnect wiring distribution in (6.47) and (6.48).

2. Shortest wires are routed on the tier (collection of levels with the same wiring pitch) with the smallest pitch and successively longer wires go on tiers with progressively larger pitches.

3. The tier pitch is chosen based on a given maximum performance constraints (for the example case study, the maximum permissible time delay is 25% of the clock period for the lowest tier and 90% of the clock period for all other tiers).

4. For the example case study, the aspect ratio is chosen to be unity. However, the n-tier methodology is independent of the aspect ratio chosen; a non-unity aspect ratio requires a slight modification to the interconnect delay equation.

5. The wiring efficiency factor (e_w) is assumed to be constant for all the levels (for the example case study, it is assumed to be 40%).

To reduce wiring layout problems, interconnects on adjacent pairs of metal levels are assumed to be routed orthogonally with the same wiring pitch. A collection of pairs having the same pitch is identified as a tier. In general, the range of interconnect lengths on the tth tier is calculated by equating the area available for wiring, A_{av}, to the area that is required for wiring, A_{req}, which is

$$A_{av} = n_t e_w A_m = \chi p_t \sqrt{\frac{A_m}{N_g}} \int_{L_{t-1}}^{L_t} l i(l) dl = A_{req} \tag{6.82}$$

The number of metal levels in the tth tier is n_t, e_w is the wiring efficiency factor [9], [12], χ converts point-to-point interconnect length to wiring net length [15], p_t and L_t are the wire pitch and longest interconnect length on the t_{th} tier in microns and gate pitches, respectively.

The wiring pitch of the local tier is equal to twice the minimum feature size ($p_t = 2F$); for all non-local tiers (i.e. $p_t > 2F$), the pitch is obtained by equating the RC time delay of the longest interconnect to an acceptable fraction of the cycle time [16], [17]:

$$\tau = \frac{\beta}{f_c} \approx 4 \frac{1.1 \rho \varepsilon_r \varepsilon_o 6.2}{p_t^2} \frac{A_m}{N_g} L_t^2 \tag{6.83}$$

Therefore,

$$p_t = 2\sqrt{\frac{1.1\rho\varepsilon_r\varepsilon_o 6.2 f_c}{\beta}}\sqrt{\frac{A_m}{N_g}}L_t \tag{6.84}$$

where τ is the interconnect time delay, β is the interconnect time delay expressed as a fraction of the cycle time and ρ is resistivity of metal. This formulation assumes that all the interconnect aspect ratios are unity i.e. $W = H_\rho = S = H_\epsilon = p_t/2$, where W, H_ρ, S and H_ϵ are the metal width, metal thickness, spacing between the interconnects and height of the inter-level dielectric, respectively.

For a given value of A_m and f_c, the n-tier design methodology starts with the lowest tier and moves upwards, filling each tier with interconnects. To maximize wire density, it tries to employ minimum width interconnects on the local tier; solving (6.82) gives the longest interconnect length on the local tier. For non-local tiers, (6.82) and (6.84) are solved simultaneously, while scaling the pitch of each tier so that the longest interconnect on that tier satisfies the timing constraint (delay $\tau = \beta/f_c$). Thus the wire density is maximized resulting in a minimum number of metal levels. Once the twice-die-edge-long interconnect has been accommodated on a tier, the algorithm stops, and counts the total number of levels. This procedure is repeated for different values of the macrocell area and clock frequency, to determine the various optimizations described in the following section. A flowchart showing the complete n-tier design methodology is shown in Figure 6.20

Case Study Example. The performance enhancement achieved by the optimized n-tier architecture is demonstrated for an ASIC logic macrocell using 100 nm technology projections from the International Technology Roadmap for Semiconductors (ITRS) [18]. In this case study, the macrocell has 11.3M logic gates (assuming the use of 3-input 6-transistor NAND gates, this corresponds to approximately 68M transistors), a low permittivity dielectric ($\epsilon_r = 2$) and copper interconnects (for the system to meet the ITRS projections of a 2GHz clock frequency and a maximum of 8 metal levels for the 100 nm technology generation, the maximum size of the macrocell is 11.3M logic gates). The following values were assumed for the other parameters in (6.47), (6.48), and (6.82)-(6.84): Rent's exponent $p = 0.6$, Rent's coefficient $k = 4$ [10], $\alpha = 0.75$ (α = fanout/(fanout+1); for 3-input NAND gates, fanout = 3), $\chi = 0.667$ [15] and $e_w = 0.4$ [12]. Shorter interconnects (first tier) are more likely to constitute critical paths and hence are assigned a smaller β (= 0.25), so that the critical path gates can have a larger (remaining) time delay. A larger β (= 0.9) is assigned to longer interconnects because they would most likely be used for cross-chip communication only (requiring more delay). Three optimizations are defined in this section that minimize the number of metal levels, the macrocell area, or the cycle time. To demonstrate the advantages of an

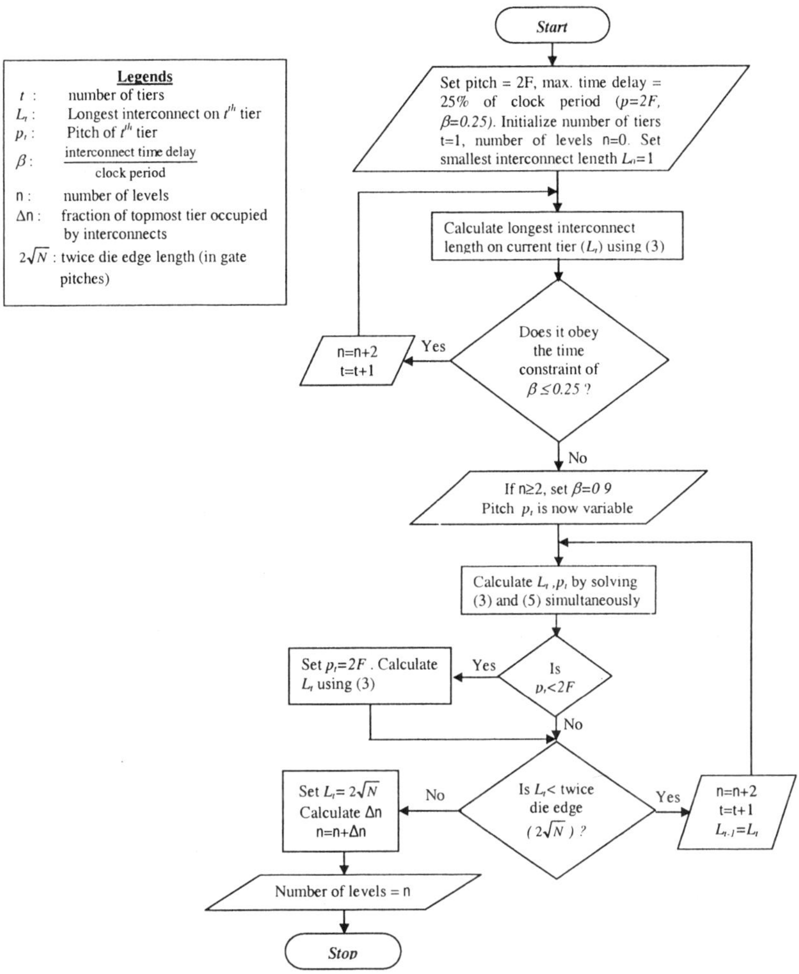

Figure 6.20. Flowchart for n-tier design without repeaters

n-tier architecture, a comparison is made with conventional designs where the wiring pitch is arbitrarily scaled from one tier to the next.

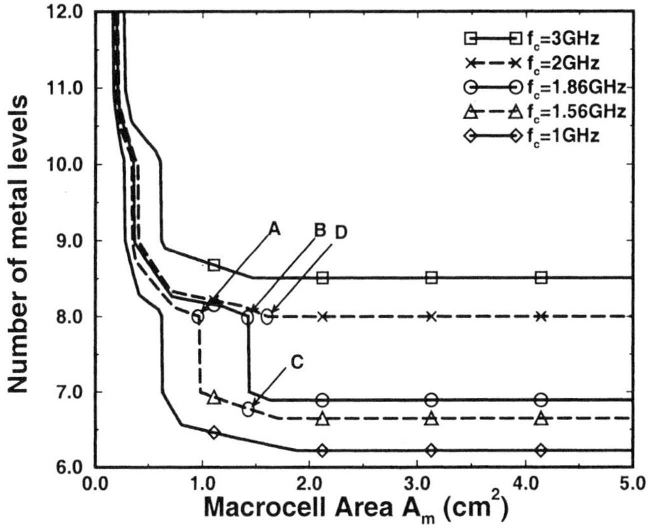

Figure 6.21. Number of metal levels vs. macrocell area for clock freqencies of 3Ghz, 2Ghz, 1.86Ghz, 1.56Ghz, and 1Ghz.

The n-tier multilevel architecture is designed with the wire pitch of each orthogonal pair of levels dependent on the time-delay of the longest interconnect routed in that pair. This ensures that the time delay constraints are exactly satisfied and also the wiring density is maximum. Figure 6.21 plots the number of metal levels versus macrocell area of the n-tier design for clock frequencies of 1, 1.56, 1.86, 2 and 3 GHz (the optimized n-tier designs have clock frequencies of 1.56, 1.86 and 2GHz). The curves in Figure 6.21 saturate when the minimum feature size width and spacing can no longer be used for any tier (i.e. $p_t > 2F$ for all tiers). This is because, when (6.84) is substituted in (6.82), the longest interconnect length L_t of the lowest tier also becomes independent of macrocell area A_m. Therefore, the number of metal levels remains constant because the interconnect pitch and length increase in the same ratio as the macrocell area is increased. The improvement in system design using the n-tier methodology is quantified in the following three optimizations and the results are summarized in the right half of Table 6.4. Detailed comparison between the conventional baseline design and the optimized n-tier design is shown in Table 6.5.

Minimum Area Optimization. For the same number of metal levels and clock frequency, the optimized n-tier architecture reduces the macrocell area by decreasing the wiring pitch resulting in greater packing density for the interconnects. The minimum macrocell area optimization is shown by point A on the curve in Figure 6.21 which plots the number of metal levels against macrocell area for the n-tier design at the baseline clock frequency of 1.56GHz. For a maximum of n=8 metal levels, the conventional baseline design requires a macrocell area of $1.45cm^2$ whereas the n-tier design requires a macrocell area of only $0.98cm^2$, which is a 32% reduction in cell size.

Maximum Clock Frequency Optimization. For the same number of metal levels and macrocell area, the optimized n-tier architecture improves the clock frequency by increasing the wiring pitch, so that the area wasted by the conventional baseline design is utilized. The maximum clock frequency optimization using the n-tier design is shown by point B on the curve in Figure 6.21 which plots the number of levels versus macrocell area for the n-tier design at a clock frequency of 1.86GHz. This corresponds to a 16% reduction in the cycle time over the baseline clock frequency of 1.56GHz for the conventional baseline design. This is the maximum clock frequency achievable using the n-tier designs for n=8 metal levels and $A_m = 1.45cm^2$. To achieve higher clock frequencies, the curves will shift upward and to the right, as seen in Figure 6.21, necessitating a greater number of metal levels or larger area.

Minimum Number of Levels Optimization. The optimized n-tier architecture reduces the number of metal levels, for the same area and clock frequency, by decreasing the wiring pitch resulting in greater packing density for the interconnects. The reduction in the number of metal levels using the n-tier design is shown by point C on the curve in Figure 6.21 which plots the number of levels versus macrocell area for the n-tier design at the baseline clock frequency of 1.56GHz. Since the levels are grouped in x-y orthogonal pairs, the number of metal levels should be rounded off to the next higher even integer. For the baseline macrocell area of $A_m = 1.45cm^2$, the n-tier design requires 6.77 metal levels (which when rounded off to 8 metal levels, is the same as for the conventional baseline design). Thus, for this example, although there is no actual reduction in the number of metal levels, 62% of the wiring tracks on the topmost tier have been freed and can be used to accommodate additional power, ground and clock wiring resources.

Thus, the n-tier design methodology can be used to determine the interconnect pitches for different tiers so that optimal performance is extracted from the system for the available resources. Additional improvements in performance can be achieved through the insertion of repeaters. The following sections investigate an optimum repeater insertion methodology in the n-tier design process and quantify its impact.

Table 6.4. Comparison of conventional designs with new optimized n-tier designs.

	Conventional Designs	Optimized n-tier design: Min. area optimization	Max. frequency optimization	Min. # levels optimization
	2- Tier (2F,2F,8F,8F)			
$A_m =$	0.82cm^2	0.37cm^2	0.82cm^2	0.82cm^2
$f_c =$	217MHz	217MHz	1.47GHz	217MHz
$n =$	8 levels	8 levels	8 levels	4.98 $\approx$ 6 levels
	3- Tier(2F,2F,4F,8F)			
$A_m =$	0.68cm^2	0.47cm^2	0.68cm^2	0.68cm^2
$f_c =$	672MHz	672MHz	1.16GHz	672MHz
$n =$	8 levels	8 levels	8 levels	6.57 $\approx$ 8 levels
	3- Tier(2F,4F,4F,8F)			
$A_m =$	1.01cm^2	0.48cm^2	1.01cm^2	1.01cm^2
$f_c =$	710MHz	710MHz	1.6GHz	710MHz
$n =$	8 levels	8 levels	8 levels	6.24 $\approx$ 8 levels
	4- Tier(2F,4F,8F,16F)			
$A_m =$	1.45cm^2	0.98cm^2	1.45cm^2	1.45cm^2
$f_c =$	1.56GHz	1.56GHz	1.86GHz	1.56GHz
$n =$	8 levels	8 levels	8 levels	6.77 $\approx$ 8 levels

6.7.3 Repeater Insertion Models and Methodology

Repeaters have been previously shown to improve the dependency of time delay on interconnect length from a square-law to a linear relationship [9], [19]. As mentioned, repeater insertion provides a viable option to relieve the demanding interconnect restrictions placed on future GSI systems. Repeaters are increasingly being used by chip designers to improve the performance of microprocessors [20]. The potential of repeaters to improve the n-tier architecture design is demonstrated in this section.

Repeater Models. Bakoglu [10], [14] derived an expression for the time-delay, τ, of an interconnect when the number of equi-spaced repeaters is "optimal", which minimizes the cumulative delay of repeaters and interconnect segments as

$$\tau = \frac{\beta}{f_c} = 2.46\frac{2}{p_t}\sqrt{6.2\rho\varepsilon_r\varepsilon_o R_o C_o}\sqrt{\frac{A_m}{N_g}}L_t. \tag{6.85}$$

where R_o and C_o are the output resistance and input capacitance of a minimum size inverter, respectively. However, if the number of repeaters is a factor ζ times the "optimal" number of repeaters, $0 < \zeta \leq 1$, then the time delay can be expressed as

Table 6.5. Interconnect parameters for the various design points

Tier # (n)	number of levels	L_{n-1} (Gate pitches)	L_{n-1} (cms)	L_n (Gate pitches)	L_n (cms)	p_n (μm)
Conventional baseline architecture : A_m=1.45cm^2, f$_c$=1.56GHz and n=8 levels.						
Tier 4	2	1884	0.67	6723	2.41	1.6
Tier 3	2	818	0.29	1884	0.67	0.8
Tier 2	2	174	0.06	818	0.29	0.4
Tier 1	2	1	0.0004	174	0.06	0.2
(n-tier) Minimum area : A_m=0.98cm^2, f$_c$=1.56GHz and n=8 levels.						
Tier 4	2	1760	0.52	6723	1.98	1.17
Tier 3	2	735	0.22	1760	0.52	0.61
Tier 2	2	93	0.03	735	0.22	0.26
Tier 1	2	1	0.0003	93	0.03	0.2
(n-tier) Maximum frequency : A_m=1.45cm^2, f$_c$=1.86GHz and n=8 levels.						
Tier 4	2	1847	0.66	6723	2.41	1.56
Tier 3	2	846	0.30	1847	0.66	0.85
Tier 2	2	174	0.06	846	0.30	0.39
Tier 1	2	1	0.0004	174	0.06	0.2
(n-tier) Minimum # of levels : A_m=1.45cm^2, f$_c$=1.56GHz and n=6.77 $\approx$ 8 levels.						
Tier 4	0.77$\approx$2	2000	0.72	6723	2.41	1.42
Tier 3	2	893	0.32	2000	0.72	0.85
Tier 2	2	174	0.06	893	0.32	0.38
Tier 1	2	1	0.0004	174	0.06	0.2

$$\tau = \frac{\beta}{f_c} = \left(1.4 + 0.53\zeta + \frac{0.53}{\zeta}\right)\frac{2}{p_t}\sqrt{6.2\rho\varepsilon_r\varepsilon_o R_o C_o}\sqrt{\frac{A_m}{N_g}}L_t. \tag{6.86}$$

This sub-optimal design in (6.86) utilizes a smaller number of repeaters resulting in a larger delay. The trade-off between the performance and the number of repeaters is shown in Figure 6.22, which plots the ratio of the sub-optimal delay in (6.86) to the optimal delay in (6.85) versus ζ. A 50% reduction in the number of repeaters from the optimal number imposes a performance penalty of only 10%. The considerable savings in silicon area, wiring complexity and power dissipation encourages this worthwhile trade-off. Assuming the area of the chip is wire-limited, there is un-utilized silicon area that is available for repeaters and is given by

$$A_{rep} = e_{rep}A_{free} = e_{rep}\left(A_m - A_{logic\_gates}\right), \tag{6.87}$$

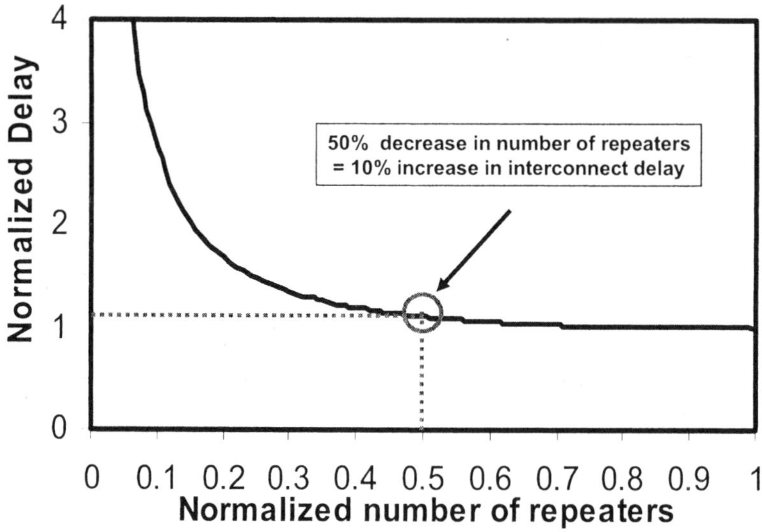

Figure 6.22. Time delay versus sub-optimal number of repeaters normalized to optimal time delay and optimal number of repeaters, respectively.

where A_{rep} is area occupied by repeaters, e_{rep} is the repeater insertion efficiency, A_{free} is free area and $A_{logicgates}$ is area occupied by logic gates. Only a fraction ($e_{rep} = 0.6$) of the free area is assumed available for repeater insertion to account for practical routing and placement constraints, and additional silicon area needed for on-chip decoupling capacitors. The area occupied by logic gates and repeaters is estimated using the following gate area models.

Gate Area Models. The area of a gate (either a logic gate or a repeater), A_g, is calculated as [21] :

$$A_g = k_I \left(1 + \frac{4\sqrt{G_{ar}}\,(f_i - 1)}{\sqrt{k_I}}\right)\left(1 + \frac{(1+\beta_g)\,(w_k - 1)}{\sqrt{k_I G_{ar}}}\right) F^2 \qquad (6.88)$$

where k_I is the area of a minimum sized inverter with respect to F^2, G_{ar} is the gate aspect ratio, f_i is the number of inputs, β_g is the ratio of pFET to nFET width and w_k is the NFET width to feature size ratio. The PFET width is constrained to satisfy equal worst case rise and fall times, and w_k is calculated by equating the critical path delay to the cycle time ($= 1/f_c$) [22],

$$\frac{1}{f_c} = \frac{n_{cp} T_{PDn} f_{ineff}}{b}, \qquad (6.89)$$

where T_{PDn} is the NFET propagation delay including the transition time effect that is derived from the physical alpha-power law model [22], f_{ineff} is the effective fan-in factor for series connected MOSFETs [23], n_{cp} is number of gates in a critical path and b is the clock skew factor (=0.9). Using (6.86), (6.87) and (6.88), the total logic gate area is estimated, which defines the limit on the maximum amount of area available for repeaters.

Repeater Insertion Methodology. For a given macrocell area A_m and clock frequency f_c, the n-tier methodology described in Section 6.7.2 is first used to design its multilevel interconnect architecture. The models described in (6.86)-(6.88) are then used to determine the maximum number and size of repeaters that can be inserted in a multilevel architecture. Since thicker global interconnects benefit the most by using repeaters, a top-down repeater insertion methodology is adopted. Repeaters are first inserted in the uppermost tier. Repeater insertion then continues downward to the lower tiers depending on the amount of free silicon area available. This algorithm tries to insert 50% of the optimal number of repeaters (ζ=0.5) in all the wires on a given tier. If the area required for repeaters is not available, ζ is systematically decreased until either the repeater area constraint is satisfied or the number of repeaters reaches zero. If the interconnect width decreases to minimum width, then repeater insertion is discontinued. The resulting architecture is the new multilevel interconnect architecture with repeaters (if any have been added). The complete repeater insertion methodology for the n-tier architecture is shown in Figure 6.23.

Case Study Example. The maximum clock frequency achievable by the optimized n-tier architecture (without repeaters), for the macrocell case study described in Section 6.7.2, using 8 metal levels (and area $A_m = 1.62cm^2$) is 2GHz, as shown by point D in Figure 6.21. This design is henceforth called the n-tier baseline design. Repeaters are used to minimize the area, cycle time, number of metal levels or power dissipation of the macrocell case study. These optimizations are compared against the n-tier baseline design (without repeaters) and the results are summarized in Table 6.6.

Minimum Macrocell Area. The area of a macrocell with 8 metal levels and f_c=2GHz is minimized using repeaters as shown in Figure 6.24. The wire-limited and transistor-limited areas are the minimum areas required to have a maximum of 8 metal wiring levels, and for accommodating logic gates and repeaters, respectively. The power limited area is the minimum area required to keep the power dissipation density within the specified upper bound of 50 W/cm^2, which is the assumed maximum heat removal capacity. This is calculated by dividing the total power dissipation of the macrocell by 50 W/cm^2. The macrocell area is the maximum of these three areas. Initially, the macro-

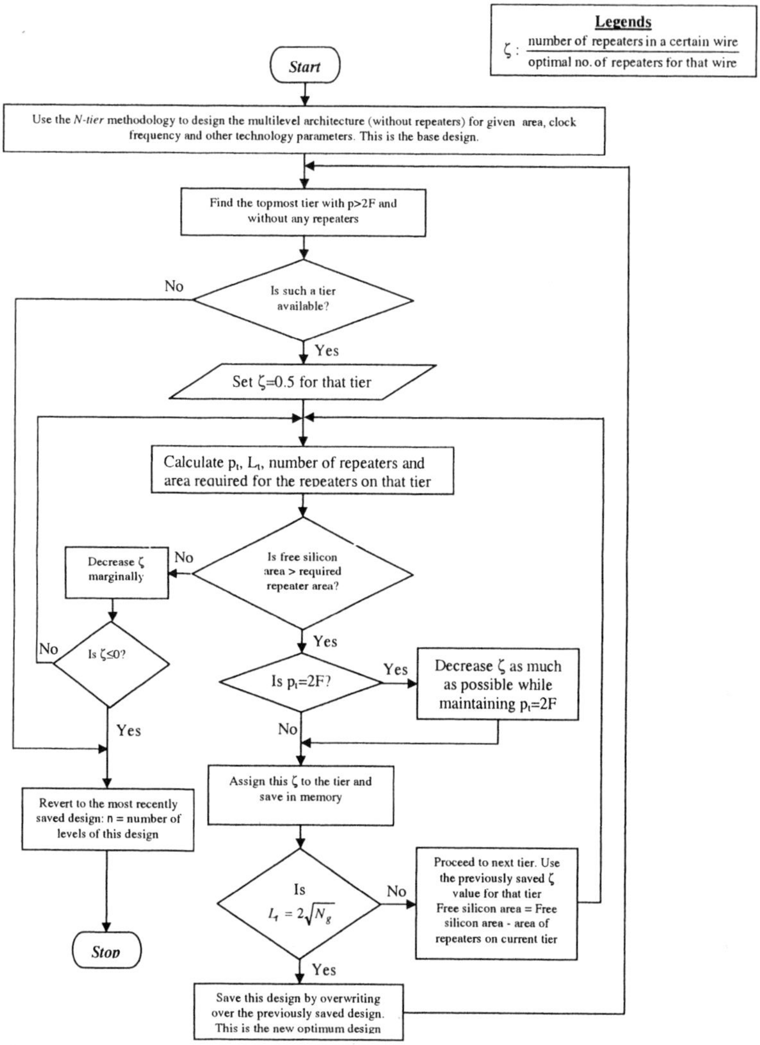

Figure 6.23. Flowchart for n-tier design with repeaters

cell area is wire-limited as illustrated in Figure 6.24. By inserting repeaters, the interconnects become narrower which decreases the area required for wiring and reduces the wiring capacitance, C_w. As C_w decreases, the logic gates can be made smaller thereby decreasing the transistor limited area. Increasing the number of power-dissipating repeaters increases the power-limited area. As the number of repeaters increases, the wire-limited macrocell area decreases until it equals the power-limited area. If more repeaters are inserted, then the macrocell area becomes power-limited and begins to increase. Thus, for this example, the macrocell area is minimized when the wire and power-limited areas become equal. As seen from Figure 6.24, optimal repeater insertion decreases the macrocell area from $1.62cm^2$ to $0.48cm^2$, almost a 70% reduction in the cell size. If more advanced heat removal mechanisms are used, such as liquid or two phase cooling, then the power limited area curve would shift lower. Then the wire-limited area may decrease until it equals the transistor-limited area. From here the macrocell area would start to increase due to an increase in the number of repeaters and the size of logic gates required to achieve the desired clock frequency for an increased wiring capacitance.

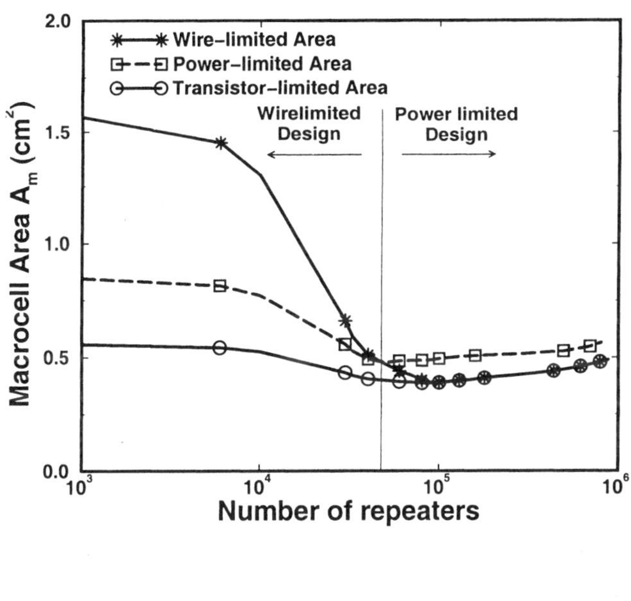

Figure 6.24. Macrocell area as a function of the number of repeaters with n-tier design

Minimum Power Dissipation. The total power dissipation of the macrocell with 8 metal levels and operating at f_c=2GHz can be significantly reduced using repeaters. Figure 6.25 plots the power dissipation versus the number of repeaters for this design. Comparing Figure 6.24 and Figure 6.25, as the macrocell area decreases with an increase in the number of repeaters, the power dissipation in logic gates and interconnects also decreases. Reducing macrocell area decreases the average interconnect length and hence reduces the average wiring capacitance. Therefore, the size of the logic gates can be reduced, which decreases the gate capacitance, reducing the total logic gate power. In Figure 6.25, the repeater power dissipation increases monotonically because the effect of decreasing repeater size is overshadowed by the increase in the number of repeaters. The total interconnect power is also decreased because of a reduction in the macrocell area. Because the repeater power is small compared to the logic gate and interconnect power, the total power dissipation decreases and reaches a minimum when the macrocell area is minimized. Beyond this point, the increase in the macrocell area and number of repeaters causes the power dissipation in the logic gates, interconnects, and repeaters to increase and results in an increase in the total power dissipation. Moreover the design becomes power-limited because the power dissipation density is at the maximum permissible limit of 50 W/cm^2. From Figure 6.25, repeater insertion decreases the total power dissipation from 43W to 24W, a reduction of 44%, which corresponds to the minimum area design point of A_m=0.48cm^2.

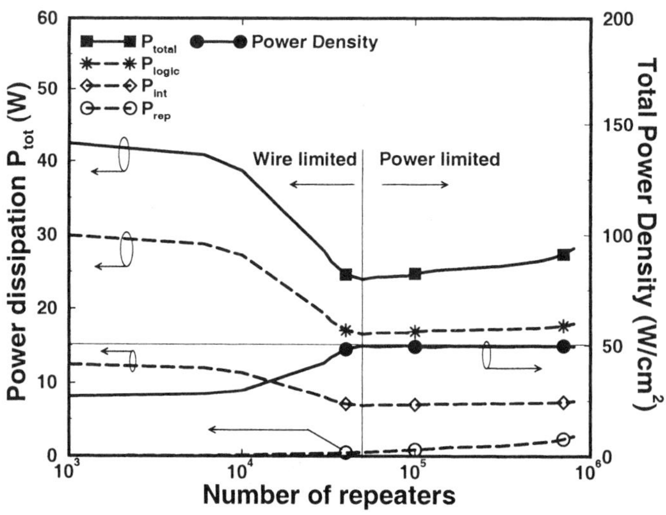

Figure 6.25. Minimum power optimization using repeaters for $n \leq 8$ levels and f_c=2GHz.

Maximum Frequency Optimization. Figure 6.26 shows a plot of clock frequency versus number of repeaters to maximize the clock frequency through repeater insertion for a macrocell with A_m=1.62cm^2 and 8 metal levels. The wire-limited clock frequency is the maximum frequency for which all the transistors can be wired within 8 metal levels for the specified macrocell area. The wire-limited clock frequency increases with an increase in the number of repeaters because repeaters decrease wire delay for a constant pitch. The transistor-limited clock frequency is the maximum frequency at which the logic critical path transistors can operate if they are enlarged to occupy all remaining macrocell area after accounting for repeaters. Transistor-limited clock frequency decreases with an increase in the number of repeaters due to a decrease in the area available for logic gates. Initially, when the macrocell area is wire-limited, the clock frequency is also wire-limited. As the number of repeaters increases, the wire and transistor limited-frequencies eventually converge when the macrocell area becomes transistor-limited. More repeaters can be added only by shrinking the logic transistors which reduces the transistor-limited frequency. Hence, the clock frequency peaks at the point where the wire and transistor limited frequency curves intersect. Therefore, repeater insertion increases the maximum clock frequency from 2GHz to 2.44GHz, which is a 22% improvement.

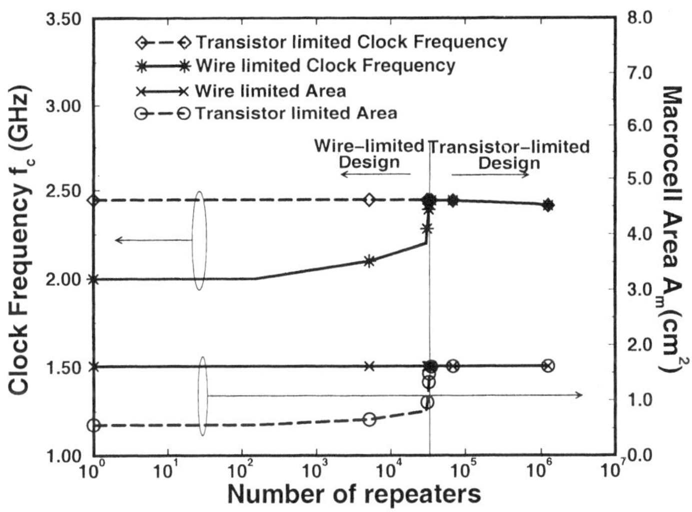

Figure 6.26. Maximum clock frequency optimization using repeaters for $A_m = 1.62cm^2$ and $n \leq 8$ levels.

Minimum Number of Metal Levels. The number of metal levels required for a macrocell with A_m=1.62cm^2 and f_c=2GHz is minimized using repeaters. Fig.6.27 plots the number of metal levels versus the number of repeaters for this design. Increasing the number of repeaters decreases the number of metal levels and increases the transistor-limited area. Once the wire-limited and transistor-limited areas become equal, continued repeater insertion will increase macrocell area. However, in Figure 6.27, the minimum number of levels optimization occurs before the convergence of wire and transistor-limited areas when all tiers become satiated with repeaters (point A), such that either they have the quasi-optimum number of repeaters (ζ=0.5) or have reached minimum wire pitch. Since the levels are grouped in $x - y$ orthogonal pairs, the number of levels should be rounded to the next higher even integral value. Therefore, in this example, optimal repeater insertion decreases the number of levels from 8 to 6, which is a reduction of 2 metal levels. From a designer's perspective, it would be prudent to choose point B, in Figure 6.27, as the design point since it has the same number of levels (six levels) but requires a far fewer number of repeaters than point A. The specifications for point B are tabulated in Table 6.6.

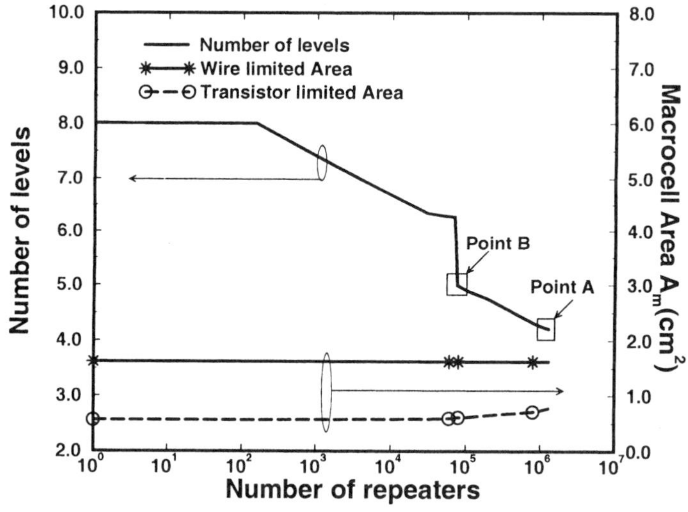

Figure 6.27. Minimum number of metal levels

6.7.4 Discussion of Assumptions

One of the key assumptions used in the n-tier methodology is that the wiring efficiency factor is constant for all metal layers. However the wiring efficiency factor can be expressed as a product of three factors:

$$e_w = e_{rout} \cdot e_{p/c} \cdot e_{via} \tag{6.90}$$

Table 6.6. Interconnect parameters for the various design points

Tier # (n)	No. of levels	L_n gate pitches	L_n (cms)	p_n (μm)	# of rep.	ζ	Dynamic Power
Without repeaters : *n-tier* baseline design							
A_m=1.62cm^2, f$_c$=2GHz and $n_{n_{levels}}$=8 metal levels.							
Tier 4	2	6723	2.55	1.71	0	0	P_{total}=43W
Tier 3	2	1871	0.71	0.95	0	0	P_{logic}=71%
Tier 2	2	885	0.34	0.45	0	0	P_{int}=29%
Tier 1	2	208	0.08	0.20	0	0	P_{rep}=0%
With repeaters : Minimum area and power							
A_m=0.48cm^2, f$_c$=2GHz and $n_{n_{levels}}$=6.97≈8 metal levels.							
Tier 4	0.97≈2	6723	1.39	0.73	49840	0.216	P_{total}=24W
Tier 3	2	1003	0.21	0.28	0	0	P_{logic}=69%
Tier 2	2	283	0.06	0.20	0	0	P_{int}=29%
Tier 1	2	34	0.007	0.20	0	0	P_{rep}=2%
With repeaters : Maximum frequency							
A_m=1.62cm^2, f$_c$=2.44GHz and $n_{n_{levels}}$=6.49≈8 metal levels.							
Tier 4	0.49≈2	6723	2.55	1.36	34855	0.308	P_{total}=150W
Tier 3	2	1684	0.64	0.94	0	0	P_{logic}=89%
Tier 2	2	810	0.31	0.45	0	0	P_{int}=10%
Tier 1	2	193	0.07	0.21	0	0	P_{rep}=1%
With repeaters : Minimum no. of levels							
A_m=1.62cm^2, f$_c$=2GHz and $n_{n_{levels}}$=4.99≈6 metal levels.							
-	-	-	-	-	-	-	P_{total}=44.4W
Tier 3	0.99≈2	6723	2.57	1.21	75390	0.26	P_{logic}=69%
Tier 2	2	885	0.34	0.45	0	0	P_{int}=28%
Tier 1	2	208	0.08	0.20	0	0	P_{rep}=3%

where e_{rout} is the router efficiency, $e_{p/c}$ is the power/ground/ clock efficiency (= 1 - power/ground/clock blockage) and e_{via} is the via efficiency (= 1 - via blockage). The router efficiency is assumed to be a constant for all metal layers. Since power/ground/clock lines are primarily routed on global layers, the blockage caused because of these lines is greater on the upper layers and lesser on the lower layers. Chen et al. [24] have described a new via blockage model that improves Sai-Halasz's empirical model in [12], according to [25]. Using an n-tier case study from [26], similar to the ones described in Section 6.7.2, it is shown in [24] that the via blockage is significant for the lowest two layers and decreases rapidly for the upper metal layers. Because via blockage dominates at lower levels and power/ground/ clock blockage dominates at higher levels, assuming a constant wiring efficiency factor for all levels is a reasonable first order approximation. An enhancement to the n-tier methodology would entail the use of experimentally observed values for power/ground/clock blockage and the use of models in [24] to calculate the wiring efficiency factor for each metal level.

Another assumption is that the maximum permissible time delay is 90% of the clock period for interconnects in all the non-local tiers. However, for chips with large areas and high clock frequencies, this constraint might be impossible to achieve for full chip wires. One of the solutions would be to pipeline the interconnects by inserting latches as has been indicated in [20]. This would make the global interconnects much thinner but communication latency would be increased.

6.8 Summary

Based upon Rent's Rule, a new complete stochastic wiring distribution is rigorously derived that determines wire length frequency for a homogeneous array of random logic gates, and it enables a priori estimation of the wiring requirements for future GSI systems. Unlike previous distributions that describe only local interconnect requirements [10, 11], the new distribution provides a complete description of local, semi-global, and global wiring requirements. The new wire length distribution is verified through comparisons to actual data from real systems, and a methodology to calculate the wire length distribution for future GSI products is proposed. This interconnect density function is incorporated into critical path and power dissipation models for advanced models of GSI systems.

A new n-tier multilevel interconnect optimization technique has been described in this paper. The methodologies presented in this chapter demonstrate an a priori (i.e. before physical design and layout) optimization of a multilevel interconnect architecture. A general conclusion of the analysis in this chapter is that detailed system level optimization, instead of conventional design heuristics, is necessary to extract maximum system performance for future GSI designs. The methodology has been demonstrated to reduce macrocell area by 32%, cycle time by 16% or number of wiring tracks required on the topmost tier by 62% (although there is no reduction in the actual number of wiring levels), when compared to a conventional baseline design where wire pitches are doubled for every pair of levels.

Also, a top-down repeater insertion methodology has been developed that uses the free silicon area for inserting repeaters in the interconnects. This methodology has been demonstrated to reduce the macrocell area by 70%, cycle time by 18%, number of metal levels by 25% or power dissipation by 44%, when compared to an n-tier baseline design without repeaters. These results illustrate the main advantages of extensively utilizing repeaters in a n-tier multilevel interconnect architecture to alleviate the restrictive wiring demands of future GSI systems.

Acknowledgments

The authors wish to express their appreciation to DARPA, contract F33615-97-C-1132, MARCO, contract MDA 972-99-1-002, and the SRC, contract 448:048, NSF 0009240 and 0098227 for their generous support.

References

[1] Semiconductor Research Association, San Jose, CA, *National Technology Roadmap for Semiconductors*, 1997.

[2] J. Meindl, "Low power microelectronics: Retrospect and prospect," *Proceedings of IEEE*, vol. 83, pp. 619–635, Apr. 1995.

[3] R. Keyes, "Physical limits in semiconductor electronics," *Science*, vol. 195, pp. 1231–1235, Mar. 1977.

[4] A. A. E. Gamal, "Two-dimensional stochastic model for interconnections in master slice integrated circuits," *IEEE Transactions on Circuits and Systems*, vol. CAS-28, pp. 127–138, Feb. 1981.

[5] H. Bakoglu and J. Meindl, "A system level circuit model for multi- and single-chip cpu's," *Digest of Technical Papers of the International Solid-State Circuit Conference*, pp. 308–309, 1988.

[6] B. Landmann and R. Russo, "On a pin versus block relationship for partitions of logic paths," *IEEE Transactioins on Computers*, vol. C-20, pp. 1469–1479, Dec. 1971.

[7] P. Christie, "A fractal analysis of interconnection complexity," *Proceedings of the IEEE*, vol. 81, pp. 1492–1499, Oct. 1993.

[8] W. Donath, "Placement and average interconnect lengths of computer logic," *IEEE Transactions on Circuits and Systems*, vol. CAS-26, pp. 272–277, Apr. 1979.

[9] H. Bakoglu, *Circuits, Interconnections, and Packaging for VLSI*. Reading, Mass.: Addison-Wesley Company, 1990.

[10] W. Donath, "Wire-length distribution for placement of computer logic," *IBM Journal of Research and Development*, vol. 25, pp. 152–155, May 1981.

[11] J. Brews, *Sub-micron Integrated Circuits*, ch. Electrical Modeling of Interconnections. New York: John Wiley and Sons, 1989.

[12] G. Sai-Halasz, "Performance trends in high-end processors," *Proceedings of the IEEE*, vol. 83, pp. 20–36, Jan. 1995.

[13] J. Chern, J. Jurang, L. Arledge, P. Li, and P. Yang, "Multilevel metal capacitance models for cad design synthesis systems," *IEEE Electron Device Letters*, vol. 13, pp. 32–34, Jan. 1992.

[14] J.A.Davis *et al.*, "Interconnect limits on gigascale integration (gsi) in the 21st century," *Proc. IEEE*, vol. 89, pp. 305–324, Mar. 2001.

[15] J.A.Davis, V.K.De, and J.D.Meindl, "A stochastic wire-length distribution for gigascale integration (gsi) - parts i and ii," *Trans. Electron Devices*, vol. 45, pp. 580–597, Mar. 1998.

[16] J.A.Davis, *A hierarchy of interconnect limits for gigascale integration.* PhD thesis, Georgia Institute of Technology, July 1999.

[17] T. Sakurai, "Closed-form expression for interconnection delay, coupling, and crosstalk in vlsi's," *IEEE Transactions on Electron Devices*, vol. 40, pp. 118–124, Jan. 1993.

[18] Semiconductor Research Association, San Jose, CA, *International Technology Roadmap for Semiconductors*, 1999.

[19] H. Bakoglu and J. Meindl, "Optimal interconnection circuit for vlsi," *IEEE Transactions on Electron Devices*, vol. ED-32, pp. 903–909, May 1985.

[20] R.McInerney *et al.*, "Methodology for repeater insertion management in the rtl, layout, floorplan and full chip timing databases of the itanium microprocessor," in *Proceedings of ISPD*, pp. 99–104, 2000.

[21] J.C.Eble, *A generic system simulator with novel on-chip cache and throughput models for gigascale integration.* PhD thesis, Georgia Institute of Technology, Nov. 1998.

[22] K.A.Bowman, B.L.Austin, X.Tang, J.C.Eble, and J.D.Meindl, "A physical alpha-power law mosfet model," *IEEE J. Solid-State Circuits*, vol. 34, pp. 410–414, Oct. 1999.

[23] T. Sakurai and A.R.Newton, "Delay analysis for series -connected mosfet circuits," *IEEE J. Solid State Circuits*, vol. 26, pp. 122–131, Feb. 1991.

[24] Q.Chen, J.A.Davis, P.Zarkesh-Ha, and J.D.Meindl, "A compact physical via-blockage model," *IEEE Trans. VLSI Systems*, vol. 8, pp. 689–692, Dec. 2000.

[25] A.B.Kahng, S.Mantik, and D.Stroobandt, "Requirements for models of achievable routing," in *Proceedings of ISPD*, pp. 4–11, 2000.

[26] R.Venkatesan, J.A.Davis, and J.D.Meindl, "Performance enhancement through optimal n-tier multilevel interconnect architectures," in *Proceedings of ASIC*, pp. 19–23, 1999.

Chapter 7

INTERCONNECT-CENTRIC COMPUTER ARCHITECTURES

S.M. Chai and D. Scott Wills
Georgia Institute of Technology

7.1 Introduction and Motivation

Portable multimedia systems are becoming increasingly important as applications that handle image, video and audio dominate processor workloads. Low-cost portable communication devices such as cellular phones are already commonplace, and the extension to more capable devices that handle electronic video will be inevitable. The proliferation of these multimedia systems is on the verge of a growth rate that will match the desktop computer industry's success. To lower cost for the mass consumer market, these systems require efficient system performance because of stringent resource limitations in portable environments. Processor architectures must provide high computation throughput while minimizing chip area and power.

To provide efficient performance in future VLSI technology, new architectures must have the following characteristics:

1. Overcome limits of interconnect technology
2. Exploit physical locality of data
3. Avoid the need for low latency, long distance (global) communication
4. Limit circuit complexity with domain specific instruction set architectures
5. Exploit inherent data parallelism in multimedia applications for high computational throughput.

7.1.1 Interconnect Delay

In past designs, gate delay is the dominating performance factor. However, technology scaling is making wires, rather than transistors, the limiting factor [1]. Poor wire technology imposes high communication costs and makes contemporary architectures with extremely wide data-paths, centralized structures, and complex control schemes unsuitable. New architectural innovations must be "interconnect-aware" to continue historically established performance trends. These architectures must localize computation and reduce demand for low latency, global communication [2]. They must exploit physical locality by processing data where it is stored [3] or where it enters the chip. Communication distances are reduced with proper data partitioning and mapping.

7.1.2 Architecture Complexity

Current microprocessors are less efficient because of their generality. Application specific architectures specially designed for multimedia applications make better use of silicon area and energy for high performance, highly efficient computation [4]. These architectures have modest programming flexibility for their target applications. The instruction set architecture (ISA) is specialized for multimedia applications to maintain high code density and low-power consumption. Limiting complexity and generality provide an added benefit of reducing processor size and wire demands, making scaling for the architecture easier in future technologies.

7.1.3 Increasing Computational Throughput through Data Parallelism

To maintain high computation performance, portable multimedia systems must exploit the explicit data parallelism in multimedia applications. Sequential processor performances are approaching a point of diminishing returns [5], and designers are extending their efforts in obtaining performance from instruction-level or thread-level parallelism. However, multimedia applications contain large amounts of inherent data parallelism [6] and the processors should remain small (fine-grain) to integrate more processing elements in a single chip. Large amounts of chip area and power spent on cache hierarchies, complex issue logic, and branch prediction circuitry should be used to increase the number of processing elements.

7.1.4 Existing Approaches

Fine-grain processors such as MGAP [7] and ABACUS [8] address efficiency and portability issues. However, these systems suffer from a bottleneck of poor I/O throughput. Furthermore, since they are reconfigured for specific applications, they suffer from configuration latency as well as potentially low resource utilization.

Other architectures specialized for media processing are digital signal processors (DSPs) such as TMS320C6201B [9] and media processors like MPACT [10]. In addition, conventional processor manufacturers have added SIMD-like instructions to handle packed-data. Examples of these multimedia extensions include Intel's MMX, Hewlett-Packard's MAX2 for the PA-RISC architecture, Sun Microsystems' VIS for the SPARC, MIPS's MDMX, the Alpha's MVI, and Motorola's ALTIVEC for the PowerPC architecture. Both DSP and microprocessors only capture sub-word level parallelism [6], and lack the ability to capture full data parallelism in multimedia applications. Their designs are not well suited for the stream nature in video related applications.

Application specific VLSI systems [11,12] represent the opposite end of the design spectrum. These systems sacrifice flexibility for a more compact, efficient implementation. Application specific systems lack programmability required in the evolving multimedia application standards.

7.2 Interconnect-Aware Architectures

Designing "interconnect-aware" architectures requires a study of algorithm, architecture, and technology to localize computation and communication. The multimedia application domain provides new opportunities to define an efficient computation substrate for embedded systems that scale well with poor wires. These architectures have simpler, distributed processing elements for fine-grain processing. This chapter presents system models that capture interconnect demand and uses these models to explore the potential of different architectures and technology.

Needed architectural research employs these approaches to localize computation and communication in an efficient computing substrate. Projected technology capabilities from the National Technology Roadmap for Semiconductors(NTRS) [13] are used to make realistic design choices and to offer a view of performance in future designs. Key performance metrics such as chip area, cycle-time, and power dissipation are used for comparison.

The main tasks to study the relationship between architecture and wire demands are the definition and implementation of system models that ties architecture description to wiring models. The validity of the models is verified with comparisons against actual chip data. Architectural explorations can then be performed to provide insights into the performance of different designs.

Current research on better material and fabrication processes to improve wire performance does not alter wire demands. A study of processor architectures and their use of interconnects is needed to find system solutions that scale well with poor wires. This study is important since modern processor architectures are developed under the assumption of ideal on-chip interconnects that scale accordingly with decreasing feature sizes. Local and global wire demands must be estimated to construct a view of key performance metrics such as chip area, cycle time, and power dissipation.

7.3 Interconnect Demand Models

The study of interconnect modeling, scaling and prediction has gained new resurgence in the advent of Gigascale Integration (GSI) operating at gigahertz clock frequencies. With decreasing feature sizes, interconnect channels for signaling, clock distribution, and power distribution impose a greater cost in chip area, signal latency, and power dissipation, than transistors. As wire characteristics do not scale well with feature sizes [1], interconnect delay becomes the larger percentage of overall signal delay, and the chip distance reachable in a single clock cycle shrinks with greater wire demands at higher operating frequencies [2]. While many studies have given attention to better material and fabrication processes to improve wire performance, these techniques only decelerate the progress towards the fundamental electromagnetic limit of wires [18], and they do not alter wire demands. A broader understanding of processor architectures and their use of interconnects is needed to find system solutions that scale well with poor wires.

Local and global wire consumptions must be estimated to construct a view of key performance metrics such as chip area, cycle time, and power dissipation. This study is important since modern processor architectures are developed under the assumption of ideal on-chip interconnects that scale accordingly with decreasing feature sizes. The relationships between architecture and wire demands must be investigated to find architectural innovations that maintain good locality and limit the need for long-distance, fast interactions.

This chapter introduces a set of heterogeneous architecture models, designed to unite architecture descriptions with interconnect parameters. The architecture models provide important parameters to Rent's Rule wiring models, such as number of logic gates, Rent's exponent and coefficient, number of signal terminals, and fan-out distribution, for various functional units in a modern processor. Empirical descriptions from cell structure, placement, and routing, are extracted to provide realistic wire demands. These architecture models correlate estimated wire demands with the appropriate architecture rather than a simplistic random logic block. Different system designs are created from smaller cell models using the Heterogeneous Rent 's Rule [19], allowing a comparison of wire demands with varying architectural changes.

These architecture models build upon the Generic System Simulator, called GENESYS, a hierarchical tool for exploring future processor architecture and technology [14,15]. GENESYS incorporates a complete description of a synchronous ASIC chip with models for a hierarchy of physical and practical limits [18]. The heterogeneous architecture models engage all levels of the

GENESYS hierarchy for an in depth analysis of the architecture-technology relationships.

The verification process against extracted data from chip layout shows highly accurate wire-demand estimations for local wires, thereby enabling realistic area and cycle time predictions. In addition, accurate global wire demand estimations for intra-unit interconnects enable predictions for across-chip clock frequencies and total chip area. High correlation among other design styles and implementation technologies demonstrates the effectiveness of this simulation methodology for architectural explorations to uncover limiting factors in prospective designs.

Two candidate architectures, superscalar processor and parallel SIMD (Single Instruction Multiple Data) array, are modeled and simulated to find the relationship between architecture and wire demands. Results show that across-chip frequencies dominate over gate delays in current superscalar processor, requiring architectural techniques to alleviate poor wire performance. Parallel arrays offer performance with computational locality, maintaining short global wire lengths with smaller feature sizes. Increasing clock frequencies without long global wires allows projected parallel array performance from 0.3 to 1.5 Tops/s by the year 2012.

7.4 Related Work

Many models of wire length distribution are based on Rent's Rule [20, 21, 22, 23, 24], a simple relationship between the number of signal terminals, T, and the number of logic gates, N. This correlation is given as, $T = kN^p$, where the parameters k and p are empirical constants. This power-law relationship is shown in [23] by partitioning existing logic circuits for scientific computing. A similar power-law relationship exists to describe external I/O of gate array chips, random logic circuits, memory, and microprocessors [20]. By separating logic circuits into hierarchical divisions, average wire length can be estimated via Rent's Rule by calculating the number of connections and distance between partitions [22].

A more accurate wire length distribution has been derived for a uniform homogeneous logic block [24]. For a heterogeneous system, global wire length can be derived from stochastic models for net-list, placement, and routing information [25,26,27]. Together with a recently derived Heterogeneous Rent's Rule that provides equivalent pin versus gate relationship of a heterogeneous system [19], these wire models have been incorporated into GENESYS for use with the heterogeneous architecture models.

GENESYS extends the capabilities of previous system-level performance simulators [20,28,29], allowing a more thorough exploration of the entire hierarchy of physical and practical limits. While other simulators focus exclusively on interconnect technology [30], GENESYS describes the complete chip design for a more realistic investigation. Furthermore, GENESYS incorporates superior Rent's Rule wiring models to accurately predict wire demands of various computer architecture designs. Other simulators [31] that consider only random logic blocks with the theoretical upper bound on wiring [22] may not produce accurate results for different system designs.

Recent architectural research has begun to explore architecture-technology relationships. Processor complexity and cycle time relationships are reported in [32,33]. [34] considers single chip multiprocessors in 250 nm technology, while [35] proposes optimal processor configurations in 350 nm technology. This study extends the collection of research by investigating wire demands and technology capabilities using the heterogeneous architecture models and GENESYS.

7.5 GENESYS Organization and Models

GENESYS embodies five model levels to represent the key limits of the following hierarchy: fundamental, material, device, circuit, and system. The first three model levels capture physical effects of material properties and switching device behaviors. Technology parameters such as interconnect metal, dielectric materials, current drain, and submicron effects, serve as inputs to the simulator. Circuit models incorporate gate level characteristics such as signal propagation delay and power dissipation. The system module consists of architecture, interconnection and packaging models to complete the description of a single ASIC chip. More information on the GENESYS organization is available in [14,15]. As this study is focused on interconnect issue, the remainder of this section will summarize the interconnection models that describes the complete on-chip wiring demand.

A Rent's Rule based stochastic wirelength distribution model [24] is used to estimate interconnection requirements for a homogeneous logic circuit. Since signal propagation delay at the gate level is composed of device switching delay and interconnect delay (RC delay, time-of-flight, and rise-time delay of input signals), the estimated wire-lengths accurately represent physical circuit characteristics and enable optimal partitioning of multilevel interconnect levels. Effects from interlevel blockage, routing efficiency, wiring structure, peak crosstalk noise, and clock skew are included in the interconnection models. Long-distance interconnect driving schemes [20] and clock distribution networks are included to determine cycle time, chip area and power consumption.

Global wire lengths for a heterogeneous system can be derived from stochastic net-list distributions, random placement of terminals, and routing information [27]. The net-list information is estimated using the Heterogeneous Rent's Rule [19] to define the connectivity between cells. Heterogeneous Rent's rule's

Rule is useful to describe equivalent Rent's parameters in heterogeneous systems where logic blocks are not uniform. Placement information gives the wire bounding-area while routing information translates bounding-area dimensions to wire lengths.

7.6 Heterogeneous Architecture Models

As described earlier, the heterogeneous architecture models enable a futuristic probe to uncover limiting factors in a system design. This section describes the modeling efforts to bind architecture description to wiring demands. Comparisons against actual data are also provided to show the high accuracy of the models in GENESYS and to confirm that important architectural characteristics have been captured. Local and global wire demands are predicted to construct a view of key performance metrics such as chip area and cycle time.

The architecture models consider a chip design as a collection of cells, as shown in Figure 7.1. For a processor array, the cells consist of single processors joined together with an interconnection network. A single processor consists of a collection of random logic cells (functional units) or local memory (cache). An internal data bus and control-signal wires connect the functional units and cache together in the single processor model.

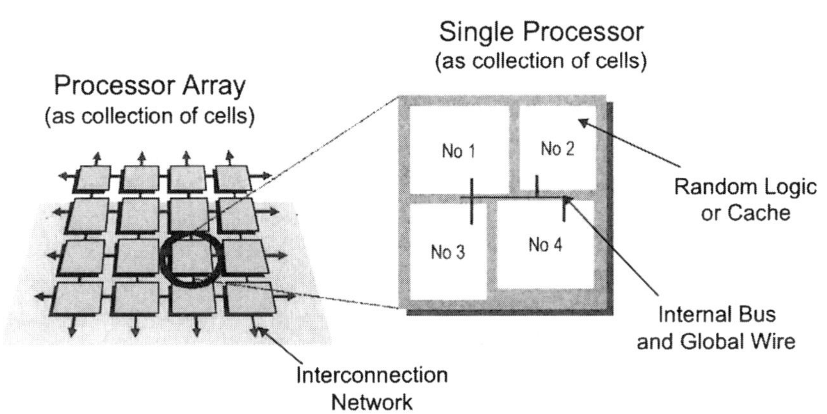

Figure 7.1. The architecture models incorporate a hierarchical collection of cells. For a processor array, parallel processors are tiled with an interconnection network. For a single processor, random logic or cache cells are connected with a data bus and global control signals.

The number of logic gates (N), Rent's exponent (P) and coefficient (K), number of signal pins (T), and number of gates in the critical path (N_{cp}) are modeled for different functional units in a modern processor. The fan-out distributions (FO), describing the data bus or interconnection network between cells, are also provided. These values are used by GENESYS wiring models, and they are extracted from actual designs or modeled as functions of architectural parameters shown in Table 7.1. Equivalent Rent's parameters are calculated for the single processor using the Heterogeneous Rent's Rule [19] for GENESYS to partition wire demands into multi-tier interconnection layers and to find the longest global wire. Clock frequency and chip area for the single processor are then estimated after wire partitioning. For a processor array, wiring demands for the interconnection network are also included. The heterogeneous architecture models extend GENESYS's single homogeneous cell model with a generic specification for a heterogeneous system-on-a-chip, providing greater architectural details and better performance predictions. In addition, the architecture models allow designers to vary contents and connectivity of cells with more flexibility and accuracy in system description.

Table 7.1. Parameters for heterogeneous architecture models

Symbol	Description
W_s	Datapath width in bits
R_p	Register file read ports
W_p	Register file write ports
N_r	Number of registers
C_w	Number of channels in the communication ports
N_w	Number of words in local memory
N_n	Number of processing elements in a tiled array
N_b	Number of bits for instruction word
Op	Number of bits in instruction opcode

7.6.1 Model Derivations

Elements in the heterogeneous architecture models are extracted from gate level schematics [36,37,38] and chip layout [39] to capture details on cell organization, circuit placement, and internal routing within a functional unit. The number of logic gates (N) and the number of pins (T) are tabulated in closed-form equations as functions of architectural parameters shown in Table 7.1. The number of gates in the critical path (N_{cp}) is also determined from the gate level schematics to find functional unit latency. Rent's coefficient (K) is calculated as the average I/O pins per gate in the gate level schematic. Rent's exponent (P) is extracted in a process similar to [22], by partitioning the logic circuits into hierarchical divisions and finding the value that matches the power-law relationship between pins and gates. These models are reported in Table 7.2 for a set of functional units in modern superscalar processors and parallel SIMD processors [36,39]. To reduce gate delays, execution units are pipelined [36,38] by adding the number of gates in pipeline latches to total gate count (N) and by reducing N_{cp} accordingly. While other functional units exist, each with different design styles and structures, this modeling approach captures the important architectural characteristics and enables detailed system description with the selection of functional units.

The data bus and control signal wires can be described using the stochastic net-list distribution described earlier [27]. However, if the connectivity between the functional units is known, the net-list distribution can be specified as the number of wiring nets per fan-out. For example, control signals in a single SIMD processor [39] are modeled as follows: $FO(2) = 4C_w + N_b + 1$. Since control signal connectivity is point-to-point between the instruction decoder and functional units, only fan-out distribution of two is tabulated. The data bus connects all functional units (excluding the instruction decoder) with the register file and is modeled as follows: $FO(7) = 3W_s$. The number of wires for the data bus is three times the data path width because there are two source channels and one result channel in the processor implementation. Placement and routing information for the SIMD processor are utilized in GENESYS to convert the net-list distribution to wire lengths.

Wiring demand for the interconnection network in a processor array can also be included. For a mesh interconnection network, inter-processor signals exist only between near-neighbors and is modeled as $FO(2) = 4C_w + N_b + 6$. This demand is scaled to the size of a single processor and then added to the internal processor wiring to obtain the total on-chip interconnect demand. Different interconnection networks and internal processor wiring can be modeled using different net-list distributions.

7.6.2 Verification

The rest of this subsection describes the verification process for the heterogeneous architecture models. Comparisons against actual data from a chip prototype [39] show the high accuracy from using the models in GENESYS because important architectural characteristics have been captured. The chip prototype has been developed independently to the modeling effort. Figure 7.2 shows area comparisons and Figure 7.3 shows gate-delay comparisons for various functional units. Actual chip areas are extracted from layout, while gate delays are simulated using a switch level simulator with RC loads.

Functional units are modeled in GENESYS as homogeneous units to find silicon area and gate delay without global wiring effects. Estimated average gate-delay is multiplied with the number of gates in the critical path (N_{cp}) to obtain functional unit latency. Minor functional units such as the activity/sleep unit, communication unit, serial shifter, and instruction decoder, are modeled

Table 7.2. Heterogeneous architecture models of single processor functional units

Functional Units	**K**	**P**	**No. of signal pins** $T^{*}*$	**Gates in critical path** N_{cp}	**Number of Gates** N
Register File	2.25	0.63	$2 + (R_p + W_p)(log_2 N_r) + W_s$	$log_2 N_r + 4i$	$(R_p + W_p)(3 - N_r + (3 + N_r)(log_2 N_r)) + (N_r(6W_p + W_s(2 + R_p + W_p) + W_p + 3 + 2(R_p + 2))) + (2(W_s(W_p + R_p) + 4 + W_p))$
*ALU	2.90	0.58	$3W_s + 6i$	$2log_2 W_s$	$8W_s + 14(W_s - 1)/3 + 15W_s + 8$
*MAC***	3.37	0.63	$3W_s + 5$	$5W_s + 8$	$7W_s^2 + 22W_s$
* Barrel Shifter	4.90	0.68	$2W_s + log_2 W_s + 4$	$3log_2 W_s + 3$	$(9W_s + 10)(log_2 W_s) + 16W_s - 13$
Local Memory	5.20	0.43	$log_2 N_w + 2W_s + 4$	$[log_2(log_2 N_w)] + 5$	$(N_w/2)(log_2 N_w) - 1) + W_s(N_w + 5) + 3$
Sleep/Activity	2.92	0.63	$W_s + 12$	$log_2 W_s + 9$	$W_s + 101$
Communication Unit	3.25	0.63	$4Cw + 2log_2(W_s/C_w) + W_s + 5i$	$2log_2(W_s/C_w) + 2$	$W_s(12 + log_2(W_s/C_w)) + 23C_w + 3(W_s/C_w) + 2log_2(W_s/C_w) - 1$
Serial Shifter	3.00	0.63	$2W_s + log2Nn + 6$	$[log_2(log_2 N_n - 1)] + 8$	$13log_2 N_n + 6W_s + 26$
Instruction Decoder	2.30	0.63	$[log_2 N_b] + Op$	$[log_2(log_2 N_b)] + 2$	$3log_2 N_b + 2N_b - 1$

* Non- pipelined execution units
* Equations are given as functions of architectural parameters (Table 7.1).
**Multiple-Accumulator

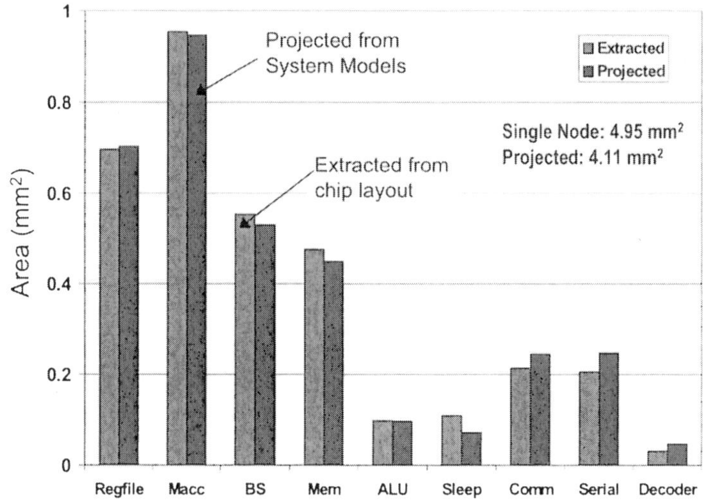

Figure 7.2. Functional unit area comparisons against extracted data

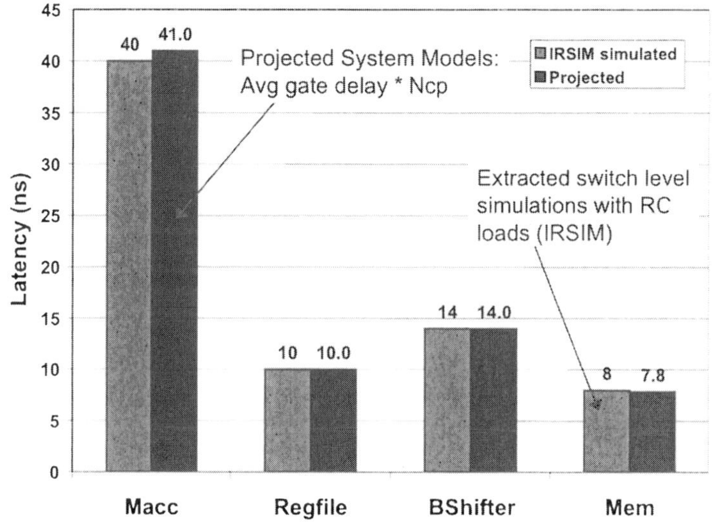

Figure 7.3. Functional unit latency comparisons against extracted data.

with a gate-pitch proportional to the barrel shifter because these units are pitch-matched to the data bus width and are less compact in the chip layout.

Figure 7.4 presents the wire demand comparisons for internal and global wiring. To find actual wiring demand, node capacitances were extracted from chip layout after removing non-wiring layers such as diffusions and wells. An average unit-capacitance per length was then determined to calculate total wiring length. Each functional unit was extracted separately to find individual wiring demand. Global wiring demand was found from extracting the chip layout with functional units removed. Internal wiring demand predicted by GENESYS models correlates well with extracted data.

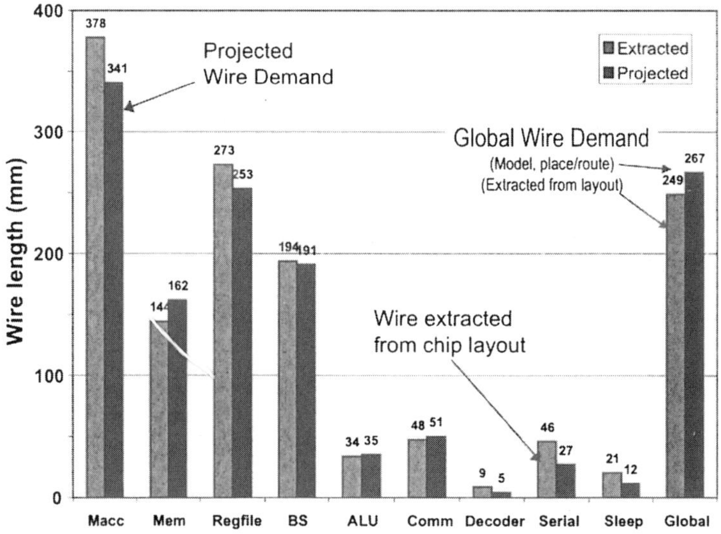

Figure 7.4. Wire demand comparisons against extracted data from chip layout.

The architecture models are verified against other design styles and structures. Figure 7.5 shows register file area comparisons against other reported designs [39, 40, 41, 42]. In addition to the different implementation technologies, the chosen register files have varying architectural structure (physical registers, word size and read/write ports). Predicted silicon areas relate well to actual data, with an average deviation of only 17%. For the Hwang register file, a memory cell Rent's exponent ($P = 0.43$) is used because the design style is similar to a memory cell. For each register file simulation, the proper technology parameters are used to accurately estimate chip area. These verification results in chip area, wiring demands, and latency for different functional units, varying in technologies and design styles, show the effectiveness of the architecture and wiring models as a predictive tool for architecture explorations.

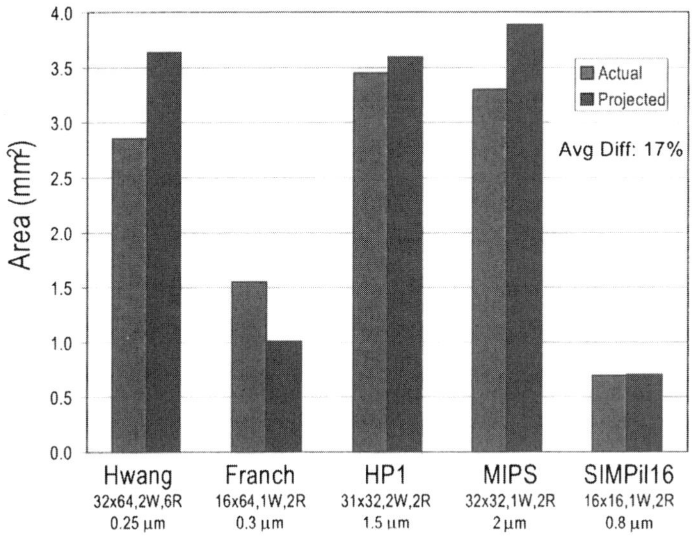

Figure 7.5. Area comparisons against actual data [39, 40, 41, 42]. Register files are chosen for their different structures and implementation technologies to show the effectiveness of the architecture models.

7.6.3 Model Applicability

This section offers a discussion on the scope of applicability for the heterogeneous architecture models. A review of the modeling approach and the effects of different design styles are provided. Asymptotic behaviors of four different adder architectures are provided to illustrate the bounds on the models.

The heterogeneous architecture models are built from gate-level schematics found in the following published texts [36, 37, 38]. Details from these schematics such as cell organization, circuit placement, and internal routing between circuit blocks are obtained to derive the models. As such, the heterogeneous architecture models are based upon this set of design style and CMOS logic family. This approach allows a designer to predict performance for systems with similar designs in future gigascale technology.

While the chosen design style is widely used, there are other design styles and logic families that offer different characteristics and performances. Different design styles affect the models in the following manner:

1. Variation on Rent's Rule parameters. The average gate fan-out sets the value for Rent's constant, k. Therefore, variance on k is small for regular designs. Rent's exponent, p, is derived from the interconnectivity of the cells, and for the set of functional units described in this chapter, the p exponents follow published empirical data [22]. Because future designers will use custom libraries of cells rather than fully customized cells, functional block designs should be regular, and Rent's parameters (k and p values) should not have significant deviations. Different design styles and logic families will require a new set of Rent's parameters to accurately characterize the cells. The verification task shown in Figure 7.5 illustrates this process.

2. Number of signal terminals T. Because the number of signal pins T is set for individual functional block to describe the external connectivity, the value should not deviate for different internal designs. The number of signal terminals will change only for designs with different connectivity between functional blocks.

3. Number of gates, N, and gates in the critical path, N_{cp}. These two aspects of the model are perhaps the most variant for different design styles and logic families. New models of N and N_{cp} should be derived to properly characterize new cells.

The following paragraphs review four different adder architectures to offer a view on the bounds of the models with respect to size and latency. Adder size and latency are directly related to N and N_{cp}, which were previously described as the most variant aspects of the heterogeneous architecture models.

The four types of adders considered in this section are Ripple-Carry, Carry-Look-Ahead (CLA), Carry-Skip, and Carry-Select. Details on these adders are available from [36, 37]. Table 7.3 illustrates the asymptotic size and latency. The choice for a type of adder is dependent on technology and application requirements. In general, Ripple-Carry adders are the most compact, but they are relatively slow. Carry-Select is attractive in technologies when a signal can drive many multiplexors. The Carry-Skip adder works well when signal rippling is fast. The CLA adder trades off chip area for speed. They are regular, fast, and particularly capable for word widths of 16 to 32 bits [37].

The asymptotic behavior shown in Table 7.3 shows that there is a clear choice for type of adders to build (speed or area) for large word width sizes ($n \gg 32$, where n is the word width). For image-processing domains (with required precision < 32 [43]) considered here, the Manchester adder is the most effective [37]. This adder type is modeled in the heterogeneous architecture models and has an asymptotic behavior like the Ripple-Carry, O(n) for latency and chip area.

Table 7.3. Asymptotic latency and area requirements [36]

Adder Type	Latency	Chip Area
Ripple-Carry	$O(n)$	$O(n)$
Carry-Look-Ahead	$O(log_2 n)$	$O(nlog_2 n)$
Carry-Skip	$O(n0.5)$	$O(n)$
Carry-Select	$O(n0.5)$	$O(n)$

The above discussion offers insights on the utility of the heterogeneous architecture models, and describes the variation of model outputs. Results in the following sections predict trends and performances for systems that are similar in design style. For different design styles and logic families, the models should be extended to accurately describe each functional unit.

7.7 System Design Analysis

This section describes the exploration of two candidate architectures using the architecture models in GENESYS for different GSI technologies. Superscalar processor designs are investigated to find processor complexity versus cycle time relationships. In addition, parallel SIMD processor arrays are modeled to find optimum data path width and system size in future technologies.

7.7.1 Superscalar processor design

Current industry trend has been to design microprocessors with multiple instruction issue width to extract instruction-level-parallelism. These designs are pipelined, employing hardware bypassing to eliminate data hazards, to avoid interlock stalls from dependent instructions, and to provide results to execution units in the following cycle [36]. While logically simple, hardware bypassing can be costly with deeply pipelined, wide instruction issue processors [44]. Bypass delays have been estimated to dominate over other pipeline stages by 180 nm technologies [32].

The data bypass hardware is among the most wire demanding units, requiring $(2 \times IW^2 \times S)$ bypass paths [44], where IW is the instruction issue width, and S is the number of pipeline stages after the first result-producing stage. The formula assumes a 2-input functional unit, and is scaled with data path width (W_s) to obtain total wire demand. Because the number of bypass paths grows quadratically with issue width (IW), a study is required to uncover limits in processors with wide degree of instruction issue.

Architecture and wiring models have been engaged in GENESYS to estimate clock frequencies for processors with varying issue width. Wire demand

Table 7.4. Simulation parameters for wide issue superscalar processors

Instruction Issue Width	**4**	**6**	**8**
Number of ALU units	2	3	5
Number of shifters/branch	1	1	2
Number of Load/Store	2	3	4
Number of Integer Multipliers	1	1	1
Register file (Size, Read, Write ports)	80 x 64, 8, 4	101 x 64, 12, 6	120 x 64, 16, 8

for hardware bypass is modeled as bus channels, connecting the functional units shown in Table 7.4. The number of functional units is increased according to issue width, following existing simulation parameters in [32]. The Load/Store unit consists of an adder for computing memory addresses, while other functional units have been described previously. Register file parameters are extracted from previous studies in optimal configurations [45]. Technology parameters are obtained or extracted from the National Technology Roadmap for Semiconductors (NTRS) [13].

Figure 7.6 presents simulation results for various superscalar processor designs. The number of critical path gates (N_{cp}) is extrapolated from [3] to represent pipeline delay from logic circuits. The family of solid-line curves represents local clock frequency for processors with different issue widths, while the dashed line represents cycle time for across-chip global wires. A processor design with a single synchronous clock must operate at a frequency that is equal to the smaller of the two clock frequencies (local and across-chip). In Figure 7.6, these two values are shown separately to illustrate the divergence of these two operating frequencies. The simulated values were estimated by GENESYS with optimized partitioning of the multi-tier interconnection network.

NTRS projected on-chip local and across-chip clock frequencies form two operating regions: Region-I represents a region of operation below the NTRS projected across-chip clock frequencies; Region-II represents a region of operation above across-chip clock frequencies but below local clock frequencies. Processor designs with issue widths smaller than four are able to track projected local clock frequencies, operating in Region II by the year 2003. Wider issue processors are more complex, requiring larger chip area and longer wires. Simulation results show that by 2006, across-chip clock frequencies (dashed line) may not be able to meet NTRS projections, as wire lengths and demands for large, dense chips exceed any performance gains from wire technology. In 250 nm technology (1997), a four-issue processor design is not yet limited by across-chip clock frequencies. However, across-chip delays begin to dominate

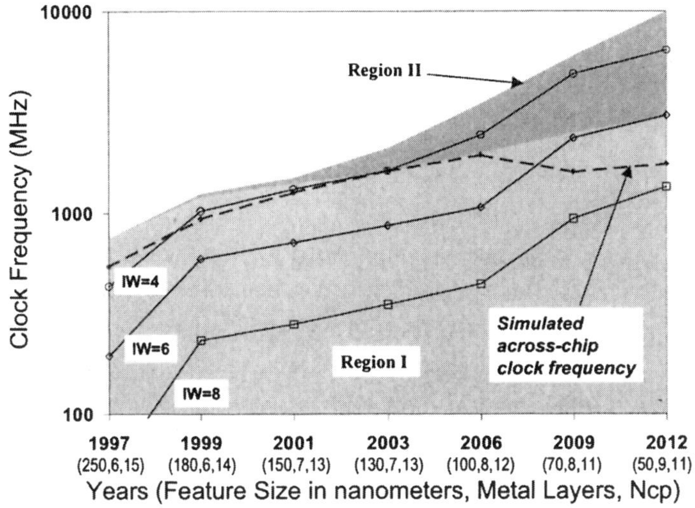

Figure 7.6. Clock frequency projections for superscalar processor designs with varying instruction window (IW) sizes. NTRS projected on-chip local and across-chip clock frequencies form the two operating regions.

local gate delays by today's technologies (180 nm, 1999) and by 2006, global wire performance can no longer track a four-issue processor. By the year 2012, synchronous processor designs with issue widths of four and six will be dominated by long global wire delays.

These results clearly illustrate the need for architectural techniques to reduce poor wire performance. In order to maintain aggressive operating clock frequencies, future processor designs must be locally synchronized with multicycle access to regions across the chip. Clustered organizations with smaller computation regions, similar to schemes used in the Alpha 21264 [46], are required to maintain short wire lengths. These clusters should be limited to an issue width less than six to maintain aggressive clock frequency as projected by NTRS. Other techniques such as incomplete bypass structures [44] and single-chip multiprocessor [34] alleviate the need for fast global wires.

7.7.2 Parallel Array

This section describes architectural exploration of a SIMD parallel array to find optimal data path width (W_s) in different technologies. Larger data path width (W_s) improves computation precision but requires larger chip area and reduces the number of processing elements (PEs) that fits on a single chip.

SIMD arrays are well suited for multimedia applications with large amounts of explicit parallelism because the same operation is performed over the entire data set [43]. Programming is thereby simplified and the array becomes easily scalable for large data sets. As future processor workloads become increasingly dominated by multimedia applications, understanding performance characteristics of parallel SIMD arrays becomes vital for successful deployment of such systems.

The SIMD architecture consists of a tiled array of PEs with a mesh interconnection network, following the design in [39]. Each PE is composed of the functional units listed in Table 7.2. In order to determine the effects of long wires, the functional units are pipelined to shift from performance reliance on gate delay to inter-processor wire delay. Wiring demand has been described previously in this chapter. Cycle times with varying data path widths are estimated using architecture and wiring models in GENESYS.

Figure 7.7 presents simulation results for the parallel SIMD array. The number of gates in the critical path is set to eight to account for unit selection, execution and latch delays. The family of solid-line curves represents the estimated gate latencies for different data path widths, while dashed lines represent inter-processor wire delays. The results clearly show the large difference between transistor and wire improvements for the projected technologies. By 2012, gate delay reductions are two to three times more than long wire performance improvements for 128-bit SIMD processors. Because denser transistors allow for smaller, closer units, total gate delays can reduce even for the same data path width and critical path gates. Wire performance for inter-processor communication, however, does not scale as well.

Figure 7.8 shows the optimal data width size for the parallel SIMD array. The optimal value is determined as the data width size where total gate delay of a PE is equal to inter-processor wire delay. Since SIMD arrays have a synchronous global clock, designs with data widths larger than the optimal value will have clock frequencies that are dominated by long wire delays. A peak value of 55 is found as wire performance after 2003 reduces optimal data path width.

While general applications probably require about 53 bits of precision [36], multimedia applications only need 8 or 16 bits of computation precision [43]. This information, along with simulation results, motivates two design configurations for parallel SIMD arrays: 16-bits for multimedia processing, and 64-bits for high performance scientific computation (where precision is critical).

Figure 7.9 shows the GENESYS predicted chip area. For a 16-bit PE, a tiled array of 14,600 SIMD PEs fits in a single chip by 2012. With a NTRS projected power budget of 3.2 W for portable devices, the 16-bit multimedia SIMD array provides 372 Gops/s at 50 MHz clock frequency. Today's DSP performance is in the range of 1 to 2 Gops/s. For high-performance scientific

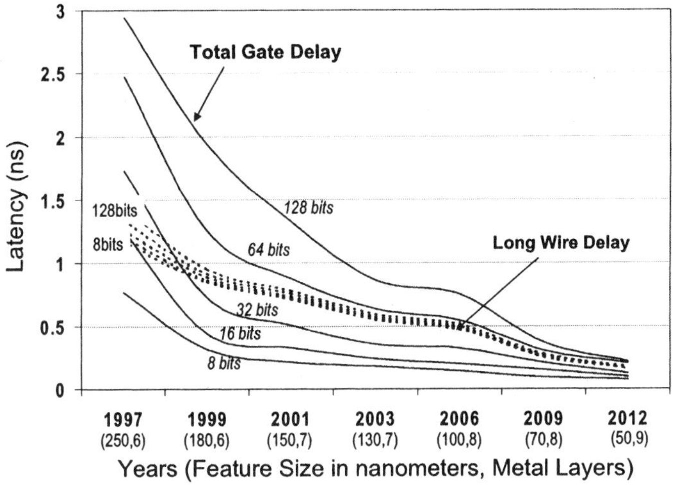

Figure 7.7. Projected wire and gate delays for a pipelined SIMD processor with varying data width size (8 to 128 bits).

applications, 2500 tiled 64-bit SIMD processors offer 1.5 Tops/s at 1.2 GHz clock frequency.

The computational throughput for the parallel array is computed as follows:

$$\text{Throughput} = IPC \cdot \nu \cdot U \cdot N_{PE} \cdot f_c \text{ [ops/sec]} \quad (7.1)$$

where IPC is the average issue width per cycle, ν is the fraction of instructions that can execute in parallel, U is the average array utilization, N_{PE} is the number of processing elements, and f_c is the clock frequency. A realistic multimedia application suite is used to determine the fraction of parallel instructions ($\nu = 0.6895$) and array utilization ($U = 0.725$) [47]. IPC is set to one for the single-issue processor design. Clock frequency is set to operate below power density limits in 2012 [39]. The number of PEs (N_{PE}) is the maximum allowed as pad areas and other interface units are not considered in the architecture model.

While this analysis provides performance throughputs for a SIMD array, the models have also been applied to systolic arrays [39]. Simulation data for systolic array assumes a larger memory size, making projected systolic array sizes smaller. The difference is used to show that performance can be obtained from dense parallel arrays to capture data parallelism, for both SIMD and systolic computing models. The different computation models affect the fraction of parallel instruction, v, and the average array utilization, U. Results for both

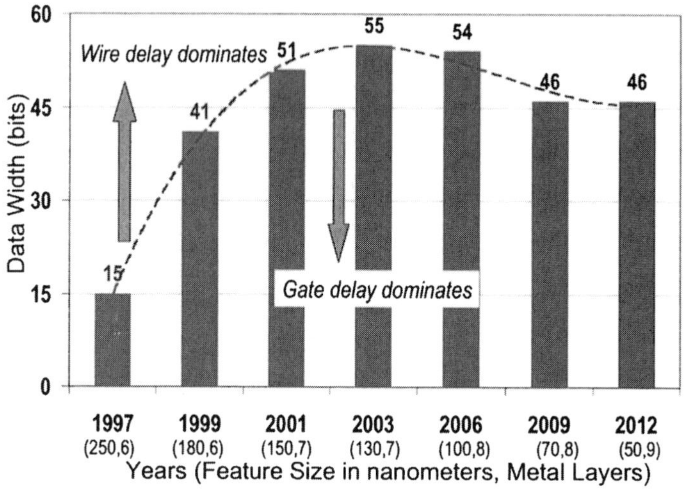

Figure 7.8. Optimal data width size determined by the threshold between wire and gate delay

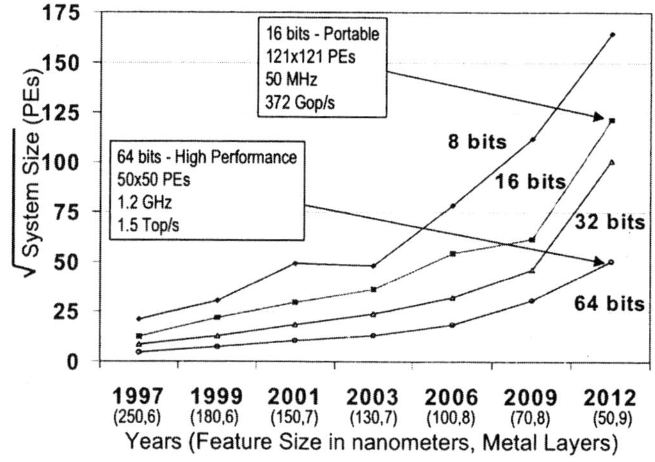

Figure 7.9. Projected array size and performance for different data width size

SIMD and systolic parallel arrays show that their computation throughputs are sufficient for the real-time workload in high-end digital cameras.

High computational performance of a parallel array comes from the large number of processing elements, shifting the reliance from clock frequency to explicit data parallelism. Silicon area taken by complex bypass structures, branch prediction logic, and large on-chip caches is better utilized by increasing the number of PEs per chip, creating a dense computational system. With small local memory and specialized data paths for multimedia applications, each PE forms a localized computational region that limits the need for fast, long global wires.

7.8 Wire Demands and their Relation to Architecture

A-priori predictions of interconnect parameters are performed before actual layout to determine total wire demand and length early in the design process. They can be used to aid floor-planning, placement, and routing tools, and also to determine critical bottlenecks in a design. However, the effectiveness of a-priori wire estimations depends heavily on the model's capability to capture important architectural characteristics. This section offers insights on the relationships among Rent's Rule, wiring models and architecture to better characterize the heterogeneous architecture models. Local wire demands are then reviewed for functional units. Global wire demand from internal bus and interconnection networks are then discussed.

7.8.1 Rent's Rule and wiring models

In [22], Donath suggested that the Rent's exponent p can be interpreted as the degree of parallelism of complex logic. Although this may be true for the circuits he considered, designers must be careful in using p literally as an index for architecture parallelism. Figure 7.10 provides an example to illustrate this caution. Consider a single bit-slice of a functional unit with Rent's constant k, exponent p, and N gates. Increasing word size to four bits increases word level parallelism in the circuitry, but k and p does not change. This example can be extended to architectures with multiple functional units. Just like the example in Figure 7.10, these functional units operate in parallel but their Rent's parameters do not change. As dictated by the Heterogeneous Rent rule [19], only the number of gates changes.

Rent's Rule wiring models [24] can generate larger wire demand because they do not have architectural information to provide accurate results. In the example, wiring models can provide a larger wire demand for the four-bit circuit because the models assumed a circuit with more gates would have similar connectivity among all bit-slices. In fact, the wire connectivity internal to each bit-slice is not the same as the wire connectivity between the bit slices. This

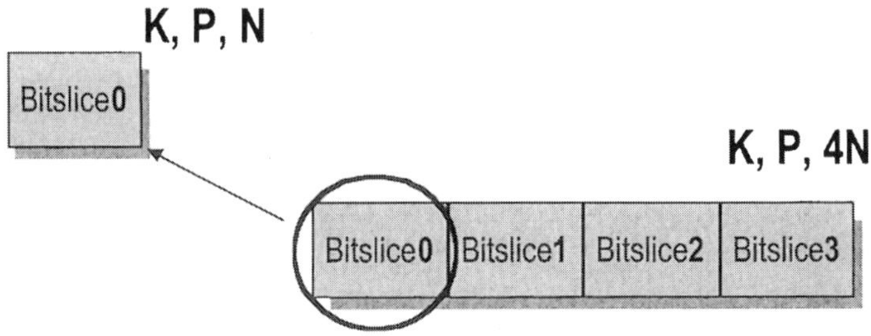

Figure 7.10. Rent's constant and exponent do not vary with increasing data path width (single bit and four-bit circuit).

trend is extended to parallel execution cores with multiple functional units. In the heterogeneous architecture models, the connectivity between cascaded cells are considered to properly use the homogeneous wiring models. For larger cells, the heterogeneous wiring models are used.

Rent's Rule based models rely heavily on previously established empirical data, making predictions for new architectures difficult. The heterogeneous architecture models can provide architecture details as a substitution for empirically based models to increase prediction accuracy. Figure 7.11 shows the predicted global wire demands against an extracted value [39]. The dashed line represents the stochastic fan-out model for a pipelined processor [27]. The solid line represents the fan-out model provided by the heterogeneous architecture model for a non-pipeline system. While stochastic models can predict wire demands in the absence of design specifications, the figure shows high accuracy when architectural data is present in the model. For a large data selection (data widths ranging from 8 to 1024 bits), an average difference of 60% is found between the two global wire demands. The stochastic fan-out models were based on pipelined processors and can provide accurate predictions for similar processor designs [27]. Different fan-out distribution must be provided (in a process similar to those described earlier in this chapter) if the design changes.

7.8.2 Local Wire Demand

Wire demand can be categorized into local, intermediate, and global wires. Local wire demands are wires internal to the functional units. Intermediate wires are associated with wires that connect the functional units together in a single processor such as signal and data bus wires. Global wires are the longest wires in the architecture, which includes the interconnection network between processors (and caches).

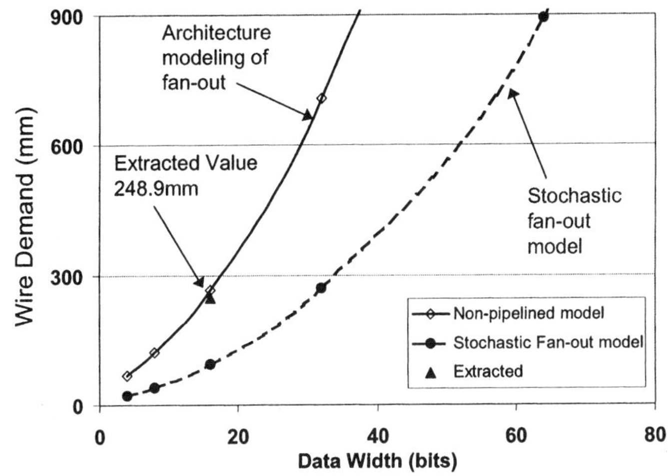

Figure 7.11. Heterogeneous architecture models provide more accurate results for global wire demands.

Figure 7.12 shows the local wiring demand of different non-pipelined functional units with varying data widths. The multiply-accumulator (MACC) unit requires the most wires because its complexity increases quadratically with data width. In comparison, the decoder unit does not vary much with data width because it is only dependent on the instruction set architecture. While a large data width ($W_s \gg 128$) processor may never be built, the heterogeneous architecture models can predict the wire demands at different data width values to determine scalability. The silicon area for each unit follows the wire consumption because an interconnect-limited area model is used.

7.8.3 Intermediate and Global Wire Demand

Figure 7.13 shows projected global wire length for superscalar processor and parallel arrays. This section offers a discussion of wire length rather than wire demand because intermediate and global wires are much longer than local wires and they become the dominant performance factor. The dashed line represents the chip width extracted from NTRS projected maximum die size. The solid lines (without triangles) represent the long wires for superscalar processor. The line labeled "across-chip global wire" represents the length of wires that connect the processing core and caches. This length scales proportion-

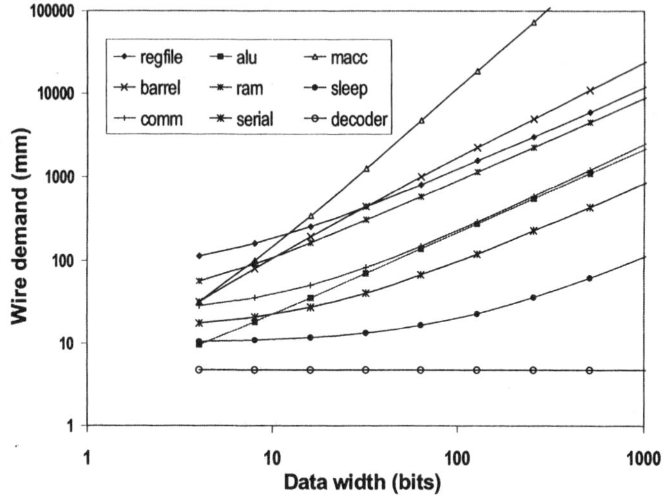

Figure 7.12. Local wire demand for non-pipelined functional units

ally to the chip width because long distance (global) communication may be required to fetch data from caches.

Intermediate wire lengths for the data bypass units are shown for different issue width (IW). Unlike the across-chip global wire, intermediate wire lengths reduce with smaller feature size and more metal layers. This result further justifies the need for future superscalar processor designs to synchronize locally with multi-cycle access on the across-chip global wire. Intermediate wire lengths for 8-issue processors ($IW = 8$) can exceed the chip width and across-chip global wire length because the architecture requires full connectivity between all functional units (as described earlier).

In contrast, the near-neighbor-mesh interconnection networks for parallel arrays have wire lengths that remain relatively constant with technology. Clock frequencies for parallel arrays are not dependent on the across-chip global wire because data communications are localized for the architecture either by compilers or by programmers. For interconnection networks that do not have near neighbor connections, wire lengths will increase and scale with the across-chip global wire. Examples of these networks include shared-memory busses and multi-stage interconnection networks (MINs) [48].

A review of total wire demand reveals three distinct levels of wire lengths and their relation to architecture. Local wire-demand comes mostly from functional units, while intermediate level wires come from internal buses and con-

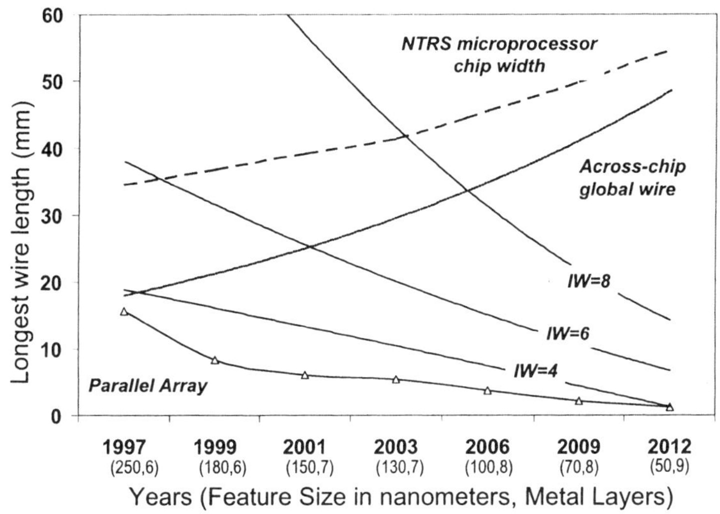

Figure 7.13. Projected longest wire length in superscalar processor and parallel array

trol signals. These two categories scale with feature size and more metal wires. However, global wire length for across chip communication does not scale down with projected technology. Interconnection networks such as shared-memory busses and multi-stage interconnection networks should be avoided because they exhibit long-distance communications that are difficult to scale in future technologies. Designers should consider near-neighbor mesh-like networks to minimize across-chip communications. While circuit techniques such as wave-pipelining [49] and high-speed serial lines [50] can improve data throughput to the processor core, architectural changes in the interconnection network reduce the physical communication distances and provide a distributed computation model that scale well with poor wires.

7.9 Conclusion

Heterogeneous architecture models capture the relationship between wire demand and architecture. These models provide the following key parameters for use with Rent's Rule wiring models:

1. Rent's constants and exponents

2. Number of external signal pins

3. Number of gates in the critical path

4. Total number of gates in functional unit

5. Fan-out distributions for non-pipeline internal-bus, control signals and mesh interconnection network.

The models can be used to aid floor-planning, placement, and routing tools. In addition, they offer a futuristic probe to uncover critical bottlenecks in a design. The architecture models enable realistic estimations of interconnect demand by providing a flexible specification for a heterogeneous system-on-a-chip. Comparisons against actual data show the high accuracy from using the models for architectural explorations.

The architecture models can capture interconnect requirements at an average of 60% more accurate than existing stochastic models because of better architectural details describing functional unit connectivity. Simulation results show that across-chip frequencies in conventional processors may not be able to track NTRS projections, increasing the gap between local and across-chip clock frequencies by the year 2006. Across-chip wire delays already dominate superscalar processor with issue width greater than four in today's technology.

Parallel arrays offer performance with computational locality, maintaining short global wire lengths with smaller feature sizes. Optimal data width size for the parallel array is found to peak at 55 bits as wire performance overwhelms execution delay after the year 2003. A review of wire demands shows the need for architectural techniques to reduce physical distances between processors. Interconnection networks that require across-chip connectivity will not scale in future technology. The parallel array offers such computation platform with an array of fine-grained processors. Communication is localized with a mesh-interconnection network.

References

[1] M. Bohr, "Interconnect Scaling - The Real Limiter to High Performance ULSI," *Proc. Intl' Electron Devices Meeting* , IEEE Press, New York, 1995, pp. 241-244

[2] D. Matzke, "Will Physical Scalability Sabotage Performance Gains?" *IEEE Computer*, vol.30, no.9, Sept.1997, pp. 37-39

[3] W. J. Dally and S. Lacy, "VLSI Architecture: Past, Present, and Future,"*20th Anniversary Conference on Advanced Research in VLSI*, pp. 232-241, IEEE Computer Society, March 21-24, Atlanta, Georgia, 1999

[4] A. Abnous, J. Rabaey, "Ultra-Low-Power Domain Specific Multimedia Processors," *VLSI Signal Processing IX*, edited by W.Burleson, IEEE Press, November 1996, pp. 459-468

[5] W. Dally, "Directions in Concurrent Computing," *IEEE International Conference on Computer Design*, 1986, pp. 102-106

[6] R. B. Lee, M. D. Smith, "Media Processing: A New Design Target," *IEEE Micro*, vol. 16, no. 4, August 1996, pp. 6-9

[7] M. J. Irwin, R. M. Owens, "A Two-Dimensional, Distributed Logic Processor," *IEEE Transactions on Computers*, v. 40, n. 10, 1991, pp.1094-1101

[8] M. Bolotski, R. Armithrajah, W. Chen, "ABACUS: A High Performance Architecture for Vision," *Proceedings of the International Conference on Pattern Recognition*, 1994

[9] Texas Instruments Incorporated, www.ti.com/sc/docs/products/dsp/overview.htm

[10] P. Kalapathy. "Hardware-Software Interactions on MPACT," *IEEE Micro*, 1997, pp. 20-26

[11] C. Mead, *Analog VLSI and Neural Systems*, Addison-Wesley, Reading Massachusetts, 1989

[12] C. B. Kuznia, A. A. Sawchuk, and L. Cheng, "FET-SEED Smart Pixels for Free-Space Digital Optics Systems," *Optical Computing*, 1995 Technical Digest Series, vol. 10, 1995, pp. 108-110

[13] Semiconductor Industry Association, *The National Technology Roadmap for Semiconductors*, 181 Metro Drive, Suite 450, San Jose, California 95110, 1997

[14] J. C. Eble, V. K. De, D. S. Wills, J. D. Meindl, "A Generic System Simulator (GENESYS) for ASIC Technology and Architecture Beyond 2001," *Proceedings of the Ninth Annual IEEE International ASIC Conference*, Rochester, NY, September 1996, pp. 193-196

[15] J.C.Eble III, *A Generic System Simulator with Novel On-Chip Cache and Throughput Models for Gigascale Integration*, Ph.D. Dissertation, Georgia Institute of Technology, Atlanta, Georgia, 1998

[16] H.T. Kung, "Why systolic architectures?" *Computer*, vol. 15, no.1, pp.37-46, January 1982

[17] H.T. Kung and C. Leiserson, "Systolic arrays for VLSI," $Sparse Matrix Proceedings$, Society of Industrial and Applied Mathematicians, Philadelphia, pp. 245-282, 1978

[18] J. D. Meindl, "Low Power Microelectronics: Retrospect and Prospect," *Proceedings of the IEEE*, vol. 83, no. 4, 1995, pp. 619-635

[19] P. Zarkesh-Ha, J. A. Davis, W. Loh, and J. D. Meindl, "On a Pin versus Gate Relationship for heterogeneous Systems: Heterogeneous Rent's Rule," *Proceedings of the IEEE 1998 Custom Integrated Circuits Conference*, Santa Clara, CA, May 1998, pp. 93-96

[20] H. B. Bakoglu, *Circuits, Interconnections, and Packaging for VLSI*, Addison-Wesley Company, Reading, MA, 1990

[21] P. Christie, "A Fractal Analysis of Interconnection Complexity," *Proceeding of the IEEE*, vol. 81, no. 10, Oct 1993, pp.1492-1499

[22] W. E. Donath, "Placement and average interconnect lengths of computer logic," *IEEE Transactions on Circuit and System*, vol. CAS-26, April 1979, pp. 272-277

[23] B. S. Landman, R. L. Russo, "On a pin versus block relationship for partitions of logic paths," *IEEE Transactions on Computers*, vol. C-20, Dec. 1971, pp. 1469-1479

[24] J. A. Davis, V. K. De and J. D. Meindl, "A stochastic wire-length distribution for gigascale integration (GSI): Part I: Derivation and validation," *IEEE Trans. Electron Devices* , vol. 45, no. 3, March 1998, pp. 580-589

[25] R. M. Karp, F. T. Leighton, R. L. Rivest, C. D. Thompson, U. Vazirani, and V. Vazirani, "Global wire routing in two-dimensional arrays," *24th Annual Symposium on Foundations of Computer Science*, Tucson, AZ, Nov. 1983, pp. 453-459

[26] G. B. Sorkin, "Asymptotically Global Routing: A Stochastic Analysis," *IEEE Transaction Computer Aided Design Integrated Circuits and Systems*, vol. CAD-6, no. 5, Sept. 1987, pp. 820-827

[27] P. Zarkesh-Ha and J. D. Meindl, "Stochastic Net Length Distribution for Global Interconnects in a Heterogeneous System-on-a-chip," *1998 IEEE Symposium on VLSI Technology Digest of Technical Papers*, Honolulu, HI, June 1998, pp. 44-45

[28] A. Masaki, "Possibilities of deep-submicrometer CMOS for very high speed computer logic," *Proceedings of the IEEE*, vol. 81, Sept 1993, pp. 1311-1324

[29] G. A. Sai-Halasz, "Performance trends in high-end processors", *Proceedings of the IEEE*, vol. 83, Jan. 1995, pp.20-36

[30] R. Mangaser and K. Rose, "Estimating interconnect performance for a new National Technology Roadmap for Semiconductors," *Proceedings of the IEEE 1998 International Interconnect Technology Conference*, San Francisco, CA, June 1998, pp.253-255

[31] D. Sylvester and K. Keutzer, "Getting to the Bottom of Deep Submicron," *Proceedings of International Conference on Computer-Aided Design*, Nov. 1998, pp.203-211

[32] S. Palacharla, N. P. Jouppi, J. E. Smith, "Complexity-effective superscalar processors," *Computer Architecture News*, 24th Annual International Symposium on Computer Architecture, vol.25, no.2, May 1997, pp. 206-218

[33] T. Hara, H. Ando, C. Nakanishi, M. Nakaya, "Performance comparison of ILP machines with cycle time evaluations," *Computer Architecture News*, 23rd Annual International Conference on Computer Architecture, vol.24, no.2, 1996, pp.213-224

[34] K. Olukotun, B. A. Nayfeh, L. Hammond, K. Wilson, K. Chung, "The case for a single-chip multiprocessor," *SIGPLAN Notices*, 7th International Conference on Architectural Support for Programming Languages and Operating Systems Cambridge, MA, Oct. 1996, vol. 31, no. 9, pp. 2-11

[35] S. Fu and M. Flynn, "Performance and Area Analysis of Processor Configurations with Scaling of Technology," Stanford Technical Report: CSL-TR-94-605, March 94

[36] J. L. Hennesy and D. A. Patterson, *Computer Architecture: A Quantitative Approach* , Morgan Kaufmann Publishers, San Mateo, California, 1990

[37] N. H. E. Weste and K. Eshraghian, *Principles of CMOS VLSI System Design: A System Perspective*, Addison-Wesley, Reading, Massachusetts, 1993

[38] I. H. Unwala and E. E. Swartzlander Jr., "Superpipelined Adder Designs," *IEEE International Symposium on Circuits and Systems*, vol. 3, May 1993, pp. 1841-1844

[39] S. M Chai, A. Gentile, D. S. Wills, "Impact of power density limitation in Gigascale Integration for the SIMD Pixel Processor", *20th Anniversary Conference on Advanced Research in VLSI* , pp. 57-71, IEEE Computer Society, March 21-24, Atlanta, Georgia, 1999

[40] R. L. Franch, J. Ji, C. L. Chen, "A 640-ps, $0.25 - \mu m$ CMOS, 16x64-b Three-Port Register File," *IEEE Journal Solid-State Circuits*, vol. 32, no. 8, August 1997, pp. 1288-1292

[41] W. Hwang, R. V. Joshi, W. H. Henkels, "A 500 MHz, 32-Word x 64-Bit, Eight-Port Self-Resetting CMOS Register File," *IEEE Journal of Solid-State Circuits*, vol. 34, no. 1, January 1999, pp. 56-67

[42] J. M. Mulder, N. T. Quach, M.J. Flynn, "An area-utility model for on-chip memories and its applications," Stanford Technical Report: CSL-TR-90-413, Feb 1990

[43] K. Diefendorff and R. Dubey. "How multimedia workloads will change processor design," *IEEE Computer*, vol. 30, no. 9, September 1997, pp. 43-45.

[44] P. Ahuja, D. Clark, A. Rogers, "Performance impact of incomplete bypassing in processor pipeline," *IEEE Proceedings of the 28th An-*

nual International Symposium on Microarchitecture, Ann Arbor, MI, Nov.1995, pp. 36-45

[45] K. I. Farkas, N. P. Jouppi, P. Chow, "Register file design considerations in dynamically scheduled processors," *IEEE Symposium on High-Performance Computer Architecture*, San Jose, CA, February 1996, pp. 40-51.

[46] R. E. Kessler, "The Alpha 21264 microprocessor," *IEEE Micro*, vol. 19, no 2, 1999, March 1999, pp. 24-36

[47] A. Gentile, J. L. Cruz-Rivera, D. S. Wills, L. Bustelo, J. J. Figueroa, J. E. Fonseca-Camacho, W. E. Lugo-Beauchamp, R. Olivieri, M. Quinones-Cerpa, A. H. Rivera-Rios, I. Vargas-Gonzles, M. Viera-Vera, "Real-Time Image Processing on a Focal Plane SIMD Array", *Parallel and Distributed Processing - Lecture Notes in Computer Science* (Eds. Jose' Rolim et al.), vol. 1586, Springer Verlag, New York, 1999, pp. 400-405.

[48] J. Duato, S. Yalamanchili, L. Ni, *Interconnection Networks: An Engineering Approach*, IEEE Computer Society Press, 1997

[49] K.J. Nowka, M.J. Flynn, "Wave Pipelining of High Performance CMOS Static RAM," Stanford Technical Report: CSL-TR-94-615, January 1994

[50] W. Dally and J. Poulton, "Transmitter Equalization for 4 Gb/s signaling," *IEEE Micro*, Jan/Feb 1997, pp. 48-56.

Chapter 8

CHIP-TO-MODULE INTERCONNECTIONS

P. Kohl
Georgia Institute of Technology
paul.kohl@che.gatech.edu

8.1 Introduction

The architectural construction of an electronic system from individual integrated circuits (ICs) and other components is dependent upon effective interconnection substrates and means of connecting to them (chip-to-module connections). Traditionally, general packaging, assembly, and chip-to-module connections were available and applied to devices. A close interaction between IC design and substrate technology was not required. Early ICs used transistor-transistor logic which had long internal delays (e.g. 15 ns) making the interconnect and module delays irrelevant. The number of input and output (I/O) connections was small, the DC power was minimal, and packaging density of silicon on boards was sparse. This made cost, reliability, and yield the focus of the chip-to-module connection technology.

The advancement in CMOS transistor technology, principally through the shrinking of device dimensions, has created a significant challenge for connecting the integrated circuit's I/O to the module. It has also created a pressing need for substrates (or interconnection boards) to provide the IC with the services it requires. Today, the chip-to-module connection methodology and substrate technology are a critical part of the IC design and partitioning. There is an ever-greater risk that inadequate I/O connection and inferior substrate technologies will be the limiting factor in system performance and reduce the overall utility of the system.

The role of the chip-to-module connection technology is to provide vital services and communications to the IC and inflict a minimum degradation in performance and minimum yield loss at the lowest possible cost. The methods of chip-to-module interconnection for gigascale integration (GSI) is largely evolutionary nature because they depend on complimentary advances in substrates (i.e. printed circuit board) and ICs. In this chapter, the goals and requirements for an effective chip-to-module connection technology will be presented. This will be developed by first providing a brief historical review of substrate and connection methodologies which have served the industry. The review will provide a background as to the current infrastructure. Then, the approaches to compliant chip-to-module interconnections will be presented along with future challenges.

8.1.1 GSI Requirements for Chip-to-Module Interconnection

Access to the substrate (electronic package) is provided by the chip-to-module interconnections. The functions provided by the supporting substrate and package can be divided into four classes: electrical, novel communication mechanisms (e.g. optical and RF), mechanical & chemical, and thermal. The International Technology Roadmap for Semiconductors (ITRS) provides guidance for GSI needs. Select values from the ITRS are presented in Table 8.1 [1].

Table 8.1. ITRS(2000 update) goals for high performance ICs.

Generation (node), nm Year	130 2001	90 2004	60 2008	40 2011	30 2014
DC Power, W	140	170	171	177	186
DC Voltage, V	1.5	1.2	0.9	0.6	0.6
DC Current, A	93	142	190	295	310
Package Pin Count	2248	3158	4437	6234	8758
Size, mm^2	325	372	427	489	561
Speed, MHz (across chip clock speed and chip-to-board speed)	1,600	2,155	2,655	3,190	3,825

Electrical Functions. The electrical I/O provide signal, power and ground, and clock distribution for the IC. The critical parameter for signal and clock I/O is delay, as determined by the path length, and electrical characteristics (R, C, and L). For power and ground, the cross sectional area of the metal is critical because the current density cannot exceed about 1 MA/cm^2. Also, the voltage drop along power and ground lines as transistors switch are important because if the voltage is below a critical value then transistor switching may not occur and swings in voltage appear as noise.

Novel Communications. The benefits of intra-IC and inter-IC optical and RF communication have been discussed elsewhere in this text. The materials, processing, and spatial requirements are significant and come as added requirements to the pre-existing electrical, mechanical, and thermal requirements. In the worst-case scenario, optical requirements add considerable constraints. Issues such as planarity of the topography, surface roughness and temperature gradients may cause little or no concern for electrical interconnections, but can significantly effect the confinement of an optical signal in a waveguide.

Mechanical and Chemical. The mechanical requirements include the physical support for the packaged die, and protection from environmental agents. The suitability of the die to failure due to mechanical shock, atmospheric or residue induce corrosion, and failures due to thermal cycling (in-use temperature excursions) must be considered in designing the chip-to-module interconnections. For example, the mechanical impact of a device hitting a hard object after being dropped from several meters (dropping a component onto a hard floor) can produce deceleration values of 10,000 times that of gravity. Minimal mechanical and thermal shock protection needs to be built into the devices. As interconnect and module connections, layers of packaging will disappear bringing the IC and attachment metallization closer to atmospheric exposure. Hermetic packages using ceramic and metal enclosures are not viable for direct chip attachment to substrates.

8.1.2 Thermal Issues

Table 8.1 shows the heat transfer (cooling) needs for high performance GSI devices. Since CMOS devices consume electrical energy only when a change in state occurs (transition from on to off, or off to on), as the clock speed increases so does the power consumed. CMOS devices are also irreversible devices resulting in the conversion of electrical energy into thermal energy, which making the cooling needs the same as the power needs. The heat transfer problem is exacerbated by close packing of silicon ICs. While the latency in metal interconnects can be addressed by keeping interconnects short, it comes with an ever higher heat transfer challenge.

Successful chip-to-module interconnection methods must address these device needs along with meeting system cost and form-factor constraints. Examination of evolution path which packaging, substrates, and interconnection have followed is constructive in understanding the infrastructure for GSI and cost/form-factor issues. The next section will present a brief historical perspective on substrate, packaging, and chip-to-module interconnections.

8.2 Packaging and Chip-to-Module Trends

The substrate and chip-to-module connection technologies (via the electronic package) are intimately tied together, and they evolve according to the needs of the IC technology. The historical evolution of electronic packages and chip-to-module interconnections will be presented first followed by substrate trends.

8.2.1 Perimeter Attachment Packages

The dual in-line packages (DIP), and quad flat package (QFP) found decades of service. The low I/O count ICs had wire bond pads distributed around the perimeter of the die. The package size was up to 100 times the size of the IC. Typically, the die was attached to a copper-based lead frame and the I/O pads were wire bonded to the lead frame. The die was molded onto the lead frame and the metal leads trimmed so as to be inserted into a printed wiring board (through-hole technology) or attached to the surface of the board (surface mount). Figure 8.1 shows a drawing from U.S. Patent 5,455,462 where the IC (part 101) is wire bonded to the platform (part 102) and the package leads were fanned out on all sides. The DIP or QFP package, and wire bond provided several important functions. First, the circumference of the die had sufficient space to provide adequate I/O for the IC. Second, the wire bond connections provided a low stress electrical connection for power & ground, and signal I/O. Wire bond automation reached amazing speed and reliability. Third, the copper lead frame or organic platform was mechanically matched to the printed wiring board. That is, coefficient of thermal expansion (CTE) of the board was adjusted to match that of copper by mixing different materials, such as low CTE filler in epoxy. Last, the pin-in-hole (DIP package) or surface mount attachment was robust against mechanical shock. The stress on the solder attachment was minimal. Figure 8.2 is a cross sectional sketch of a surface mount device from U.S. Patent 5,629,561 showing the low stress wire bond connections to the IC and surface mount connections to the board.

8.2.2 Area Array Packages

Numerous problems developed with the perimeter I/O and wire bonded packages. As the area of the IC increased, the number of I/O that could be fit on the perimeter was not adequate. This is evident in Figure 8.1 where only a limited number wires can be positioned around the IC and package. Artificially increasing the area of the die so as to increase the number of I/O is not an acceptable option, and there is a physical limit as to the density of wire-bonded pads. Second, the parasitic R, L, and C of the wire bonds and their length became an important issue because the package because a limiting fac-

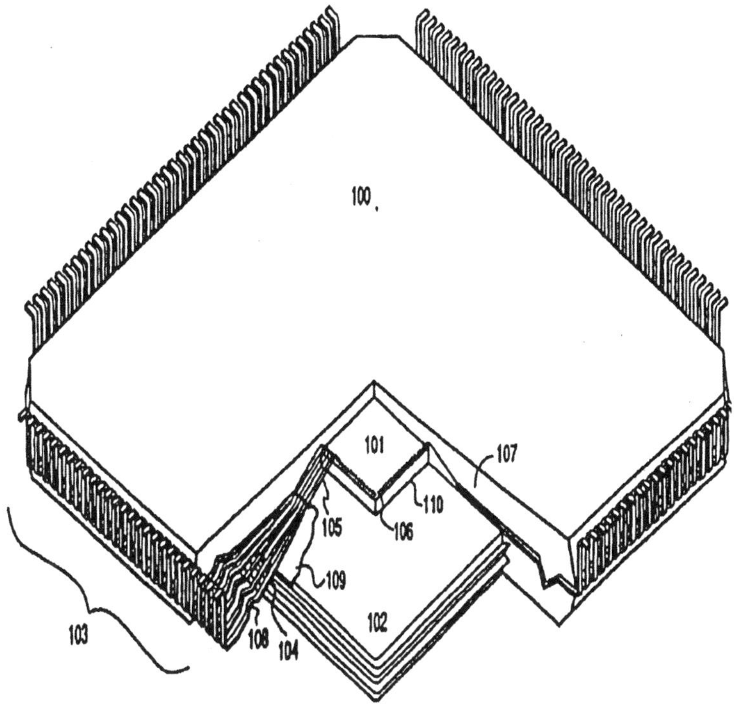

Figure 8.1. Perimeter attached QFP package.

tor in system performance. The third and fourth columns of Table 8.2 show the evolution to area array devices as a means of increasing the package to board density. The pin grid array (PGA) and ball grid array (BGA) provided a high number of connections to the board. Implementation took place with first wire bonding of the IC to the package. Later, area array flip-chip attachment of the IC to the package has lead to the very dense packaging. It is instructive to examine specific technologies at the transition point to high density, compliant area array. Figure 8.3 shows a sketch of Intel Corporations U.S. patent #5,894,410. The IC (part 18) is wiring bonded onto a multilayer printed circuit substrate (part 12). A detailed discussion of substrates is presented in the next section. An area array of solder balls (part 34) is used to attach the package to the substrate. The multilayered package substrate provided a high density of fan-out to match the printed circuit board. High-density printed circuit boards are a critical issue. The CTE mismatch between the silicon IC and the organic substrate (no solid copper lead frame is present like the simpler DIP) is ad-

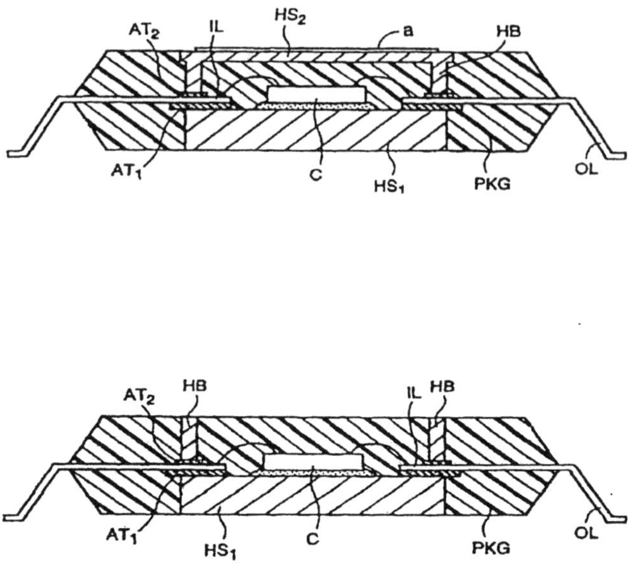

Figure 8.2. Sketch of surface mount package.

dressed in this package. The area array of solder balls has an inner and outer array because a high level of failures were observed in solder balls directly under the outer perimeter of the die. The package also pays attention to the need for high heat transfer. The inner array of solder balls is available heat transfer due to its close proximity to the printed wiring board. The top of the electronic package is the primary heat transfer pathway is such packages (see heat transfer section below). Further enhancements of the BGA have the IC flip-chip attached to the high-density package substrate, as shown in Figure 8.4. Shrinkage of the package platform allows the footprint of the packaged die to approach that of the IC. This provides ever shorter chip-to-module interconnects, however, the CTE mismatch issue becomes ever more pressing. Also, the reduced area of the package and increase in IC power, exacerbates the heat transfer issues, as will be discussed.

8.2.3 Chip Scale Packaging (CSP)

The package shown in Figure 8.5 allowed for a large number of chip I/O to be distributed in a array where the packaged served as an intermediate between the fine pitch on the IC and coarse pitch on the printed wiring board. As boards advanced to finer pitch (see next section), the package was allow to shrink. High density printed wiring boards allowed the package size to approach that of the IC. These chip-scale packages enabled a high silicon packing density,

Table 8.2. IC packaging historical trends.

Technology	DIP	QFP	BGA /CSP	DCA
Number I/O	64	500	1600	3600
Package area, cm^2	20	25	6.5	5.6
Lead pitch, μm	250	400	400	150
Heat transfer density, W/cm^2	0.025	0.3	4.5	20

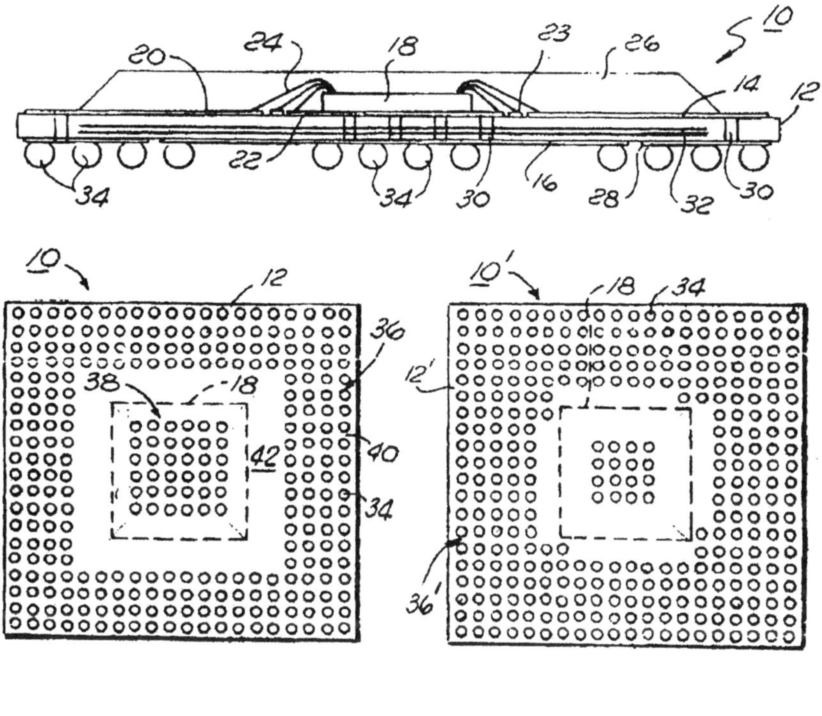

Figure 8.3. Sketch of BGA package, U.S. Patent 5,894,410.

however, CTE mismatch between the silicon and the board needed to be addressed [2, 3 and references therein]. Numerous approaches have been taken. Pre-fabricated film or tape-based interconnections have been used to connect the ICs to printed wiring boards. Amkor and Toshiba have used this style of interposer. The interposer (tape) is applied to the die followed by assembly. The IC can be wire bonded to the interposer followed by flip chip attachment of the interposer-plus-IC to the printed wiring board. A degree of compliance has been given to the interposer by use of 3-dimensional, spring-like structures. Tessera has made a compliant vertical link technology where the metal

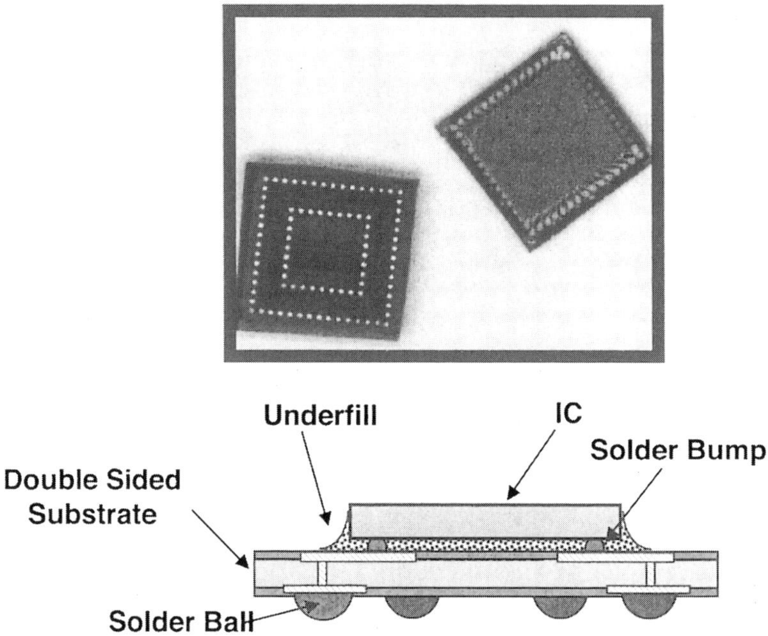

Figure 8.4. Ball grid array package used by Motorola.

interconnection in the interposer lifts off the flexible foil by the injection of an encapsulant between the interposer and the IC. The device is then attached to the PWB by use of solder balls. Figure 8.5 shows the Tessera μBGA for perimeter I/O and Figure 8.6 shows the WAVE package for area array devices. FormFactor has developed a micro-spring technology using nickel-plated gold wires. The micro-springs are formed on the wafer using a wirebonder, Figure 8.7. These two examples have shown that adequate chip-to-module compliance can be used to compensate for coefficient of thermal expansion (CTE) mismatches between polymer-based boards and silicon. They have further shown that higher density I/O, and additional redistribution capabilities are necessary in many cases.

Redistribution of the I/O can be performed in wafer-form by extension of the back-end-of-the-line processing. After normal IC fabrication, a final polymer-metal build-up process can be used to redistribute the I/O. Flipchip Technologies fabricates copper redistribution traces in wafer form. The devices are flip-chip attached using solder paste, which is especially cost-effective compared to other solder deposition techniques. Compact electrical devices, such as cellular telephones, are primary applications. Unitive Electronics and Apack provide

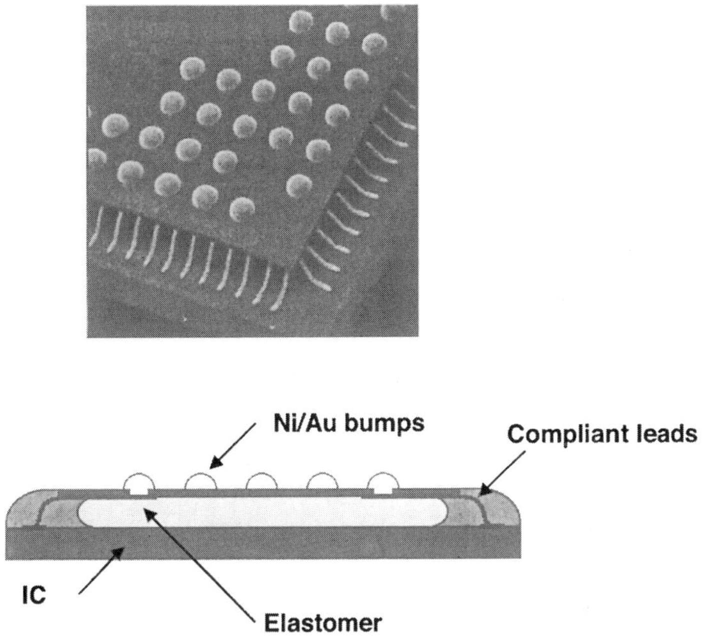

Figure 8.5. Tessera μBGA.

wafer redistribution and bumping based on aluminum metallization. The use of an underfill material between the packaged IC and the PWB is necessary because the compliance of the interconnections is not adequate to take-up the CTE mismatch between the polymer board and the silicon.

8.2.4 Wafer-Scale Packaging (WSP)

Wafer level batch packaging is an approach to forming chip-to-module interconnections that can address many of the current and future needs of ICs. Wafer level processing is where the traditional back-end-of-the-line IC interconnect processes (multi-layer metal build-up) are extended to include all packaging functions. This includes chip-to-module interconnection and full wafer-scale test and burn-in. Once the wafer level batch packaging is completed, the die are separated and ready for assembly onto a printed wiring board. The early uses have occurred primarily in lower pin-count devices and are driven by cost reductions. Since a traditional package is not required and the size of the printed wiring board can be reduced (smaller IC footprint), the packaging costs are lower. Several companies have tried to address the need

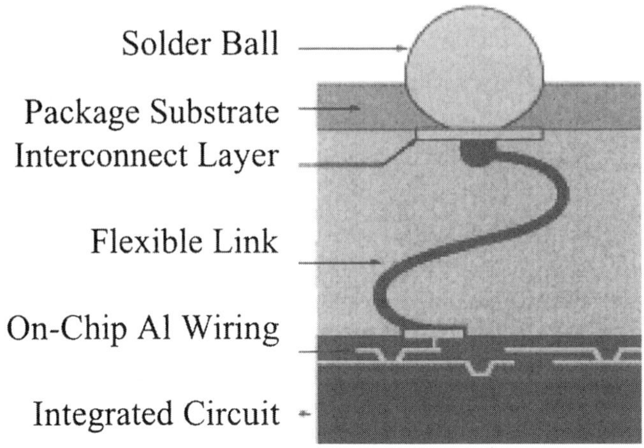

Figure 8.6. Tessera WAVE package.

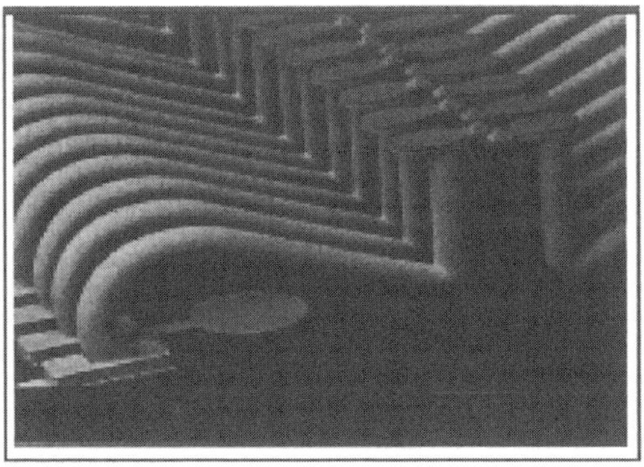

Figure 8.7. FormFactor MOST packaging technology.

for compliance between the IC and printed wiring board by fabricating metal posts in wafer form. A polymer then encapsulates the posts. The structure has a degree of compliance because the post can elastically deform and the polymer encapsulation can help distribute the stress. Additional compliance is provided by conventional solder bumps attached to ends of these posts. Fujitsu's SuperCSP is perhaps the best know wafer level ' copper-post' technology. Copper posts (350 μm diameter, 100 μm tall) are fabricated on top of an I/O redistribution layer. Special equipment has been developed to injection mold a polymer encapsulant around the posts followed by solder ball attachment. Oki and Casio have developed a similar structure for use in low pin count devices. Ibiden has developed a thin, flexible post in the second layer of polymer used in the redistribution build-up. These 'post' technologies show the importance of in-plane (x-y direction) compliance in producing a reliable, wafer-level packaged device.

Many other issues remain to be addressed for wafer level packaging. Among those are creation of an infrastructure for processing equipment, design tools for interconnection and I/O, wafer-scale test and burn-in equipment, and assembly tools that can handle different size die, especially as die-shrinks occur.

8.2.5 Heat Transfer Issues

The electrical energy consumed in microelectronic devices is converted into thermal energy as transistors change state in CMOS devices, and Joule heating occurs in metal wires and other materials. Cooling is required in microelectronic systems to transfer the heat from the transistors and wires to outside the system [3, 4]. In a desktop computer, forced convection (via a fan) is used to facilitate the transfer of heat from the IC to the outside air. In portable electronic devices, natural convention is the primary means of transferring heat to the outside air. In a small fraction of electronic systems, liquid cooling can be used, however, the complexity and cost of the system is significantly higher.

The heat transfer requirements for high performance ICs are shown in Table 8.1. The power consumed in high performance devices has risen to extremely high values (e.g. 100 to 200 W). CMOS devices consume electrical energy only when they change state. Thus, as the clock frequency increases, transistor switching occurs more frequently resulting in nearly proportional power increases.

The dramatic increase in heat transfer needs shown in Table 8.1 are exacerbated by three factors. Conductive heat transfer, Equation 8.1, will be used to illustrate the issues.

$$Q = (kA/L)(\Delta T) \tag{8.1}$$

where Q is the heat transfer rate in Watts or Joule/s, k is the thermal conductivity of the material in W/mK, A is the area, and L is the path length.

First, the CMOS junction temperature cannot rise over current levels and it would be desirable to have lower junction temperatures. Since the driving force for heat transfer is a temperature gradient, the ΔT available is at best constant with time, and will decrease in worst case situations. Second, area available for a single device to transfer heat is shrinking by a very large factor. Microprocessors and other high-power ICs could be spaced far apart on a printed wiring board with a large-area, expanded surface heat spreader. Thus, the area available for heat transfer can be made 100 times the surface area of the IC ('A' in Eq. (8.1). In order to keep interconnects from chip-to-chip short, IC will need to be placed close together eliminating much of the area available for heat sinks and heat spreaders. Last, changes in IC technology will contribute to increased difficulty for heat transfer. These changes include three-dimensional stacking of die (making heat from one die flow through another die), and replacement of some insulators (e.g. SiO_2) with low-k insulators (e.g. polymers or air) which have significantly lower thermal conductivity (decreasing 'k' in Eq. 8.1).

Figure 8.8 shows an example of the heat sink needed for the graphics processor G-Force 3 from nVidia. The large, aluminum heat sink has an integrated fan, a superficial area of 70 cm^2, and 1 cm high fins which expand the superficial area by about $10\times$. The processor is about 1.4 cm^2.

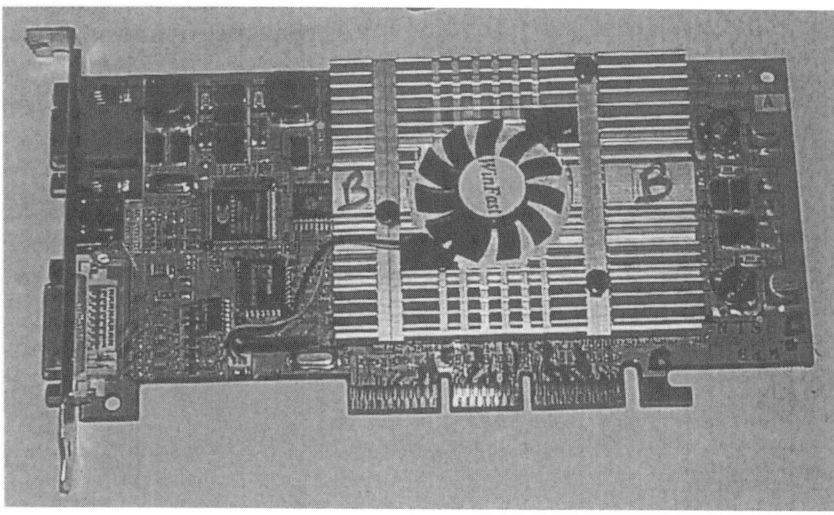

Figure 8.8. nVidia G-Force 3 board. The processor is cooled with the assistance of the large aluminum heat sink with integrated fan.

A thermal resistance model and approach can be taken to viewing the impedance to heat transfer. Figure 8.9 shows the two paths for transferring heat from the IC in a densely packed system. The heat sink on the IC is contained to superficial area of the die (worst case scenario for a two-dimensional IC). A three-dimensional IC would compound the problem because of the small surface area available. The cooling of 186 W from a 561 mm^2 IC would require a maximum thermal resistance of 0.35 K/W ((85-20K)/186W). There are two pathways for heat to flow, which occur in parallel, as shown in Figure 8.9. Generally, the thermal resistance for heat to flow through the Si, the interconnect buildup, solder bumps, and through the heat sink is small. For example, the resistance of 200 mm of silicon is 0.004 W/K ((2x10^{-4} m)/(84 W/mK)(5.61x10^{-4} m^2)). Table 8.3 shows thermal conductivities for a variety of materials used in microelectronics. The limiting factor in cooling is the removal of the heat from the back of the IC or back of the printed wiring board. For most systems, air cooling via convective heat transfer from the solid to the gas phase is used to transfer the heat. Equation 8.2 is the rate of heat transfer, Q, for convection.

$$Q = hA\Delta T \tag{8.2}$$

Where h is the convective heat transfer coefficient in W/m^2K. Table 8.4 gives

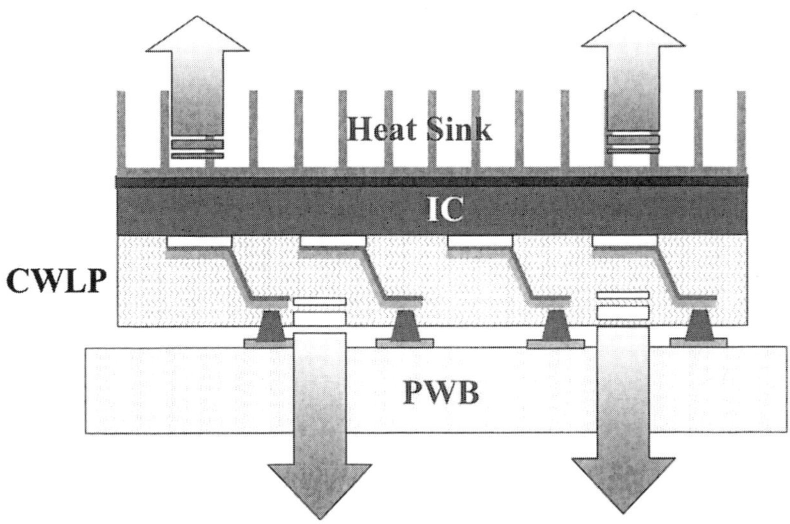

Figure 8.9. Thermal paths for densely packed ICs.

values of h for several convective heat transfer conditions. Given the boundary conditions of Q=186 W, A=561 mm^2, and ΔT=65K, one would need H>5100 W/m^2K (if heat were transferred from only one surface). If heat were equally discharged from the front and back of the IC, then H>2550 W/m^2K would be required. The convective heat transfer coefficient for solid-to-air heat transfer is too small to be used for cooling GSI devices. Even very high velocity air would not be sufficient. Using cold air (cooled via external means) would help, however, lowering the temperature below the ambient dew point presents significant challenges due to the collection of condensation and draining of condensed water. Traditional vapor-compression refrigeration systems using freon as the refrigerant can provide cooling capacities (186 W = 630 BTU/hr) and adequate heat transfer coefficients (h of a boiling liquid is sufficient); however, the infrastructure of miniature compressors and heat exchangers, and internal plumbing of freon would need to be developed.

Table 8.3. Thermal conductivity of electronic materials.

Material	Conductivity, W/mK
aluminum	216
alumina	29
copper	398
diamond	2000-2300
epoxy	0.23
GaAs	50
gold	297
lead	34
nickel	92
polyimide	0.33-0.4
quartz	3-5
sapphire	30
silicon	84
SiO2	10
silver	418
tantalum	50
titanium	22
tungsten	170
water	0.6
air	0.028

Table 8.4. Heat transfer coefficients, W/m^2.

Mode of heat transfer	h, W/m^2
Air: natural convection	5-9
Air: forced convection , velocity =1 to 3 m/s	7 to 47
Liquid water, velocity=3 m/s	9,000
Boiling water	>25,000

8.3 Microvia Printed Wiring Board Technologies

Traditional printed wiring boards are fabricated by laminating multiple layers of copper-clad insulator together to form layers of conductive traces [5-9]. The copper is patterned by contact printing and wet-etching. Layer-to-layer connections are made by mechanical drilling holes and metallization of the inside of the holes through electroplating. Figure 8.10 shows a cross section of a board with different size through-holes. Spatial resolution is limited by contact lithography, the wet etching of copper, and mechanical drilling of holes. Small holes and high aspect ratio holes are especially difficult and expensive to fabricate. Hole drilling is a sequential process whose cost rises proportionately with density of holes. The need for higher spatial density has been met by the development of microvia board technologies. Figure 8.10 shows a sketch of a traditional printed wiring board with finer line, sequential build up metallization on each side. The fabrication steps for making the build-up layers are different from the core printed wiring board. Figure 8.11 shows a blind-via made during build-up without the use of mechanical drilling. Contact to the printed wiring board can be made directly to the fine-line micro-via metallization, in addition to the drilled areas. A board suitable of flip-chip and direct chip attachment (without an intermediate package) would have a pitch approaching that of the top layer of metalliztion on the IC. Numerous approaches have been followed in the development of high-density boards. In this section, several high density microvia fabrication technologies will be surveyed [5-9].

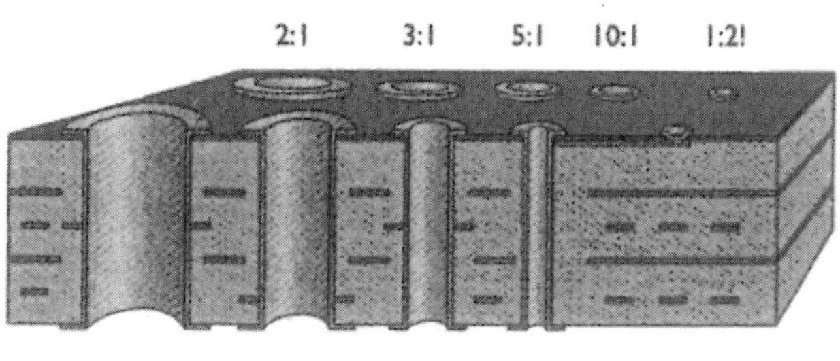

Figure 8.10. Generic printed wiring board with drilled and metalized through-holes.

Ibiden has developed a double sided, four layer build-up, board (8 total layers of build-up) using fully additive copper plating and photosensitive epoxy. The microvia layers are on top of a drilled, though-hole board. The technology is capable of producing 40 μm copper lines with 40 μm spaces. The vias can be as small as 80 μm diameter on 125 μm diameter pads. Figure 8.12 shows two layers of microvia build-up on a board. The substrate for the fine-line build-up is a BT resin, copper-clad board with drilled through-holes. A photosensitive epoxy is used as the build-up dielectric material. The fully-additive copper process provides higher resolution metal patterns than subtractive copper processing using wet etching. Fully additive processing uses electroless copper plating and does not require the presence of a seed-layer of copper for electroplating. Excellent adhesion is achieved between the epoxy and the copper by

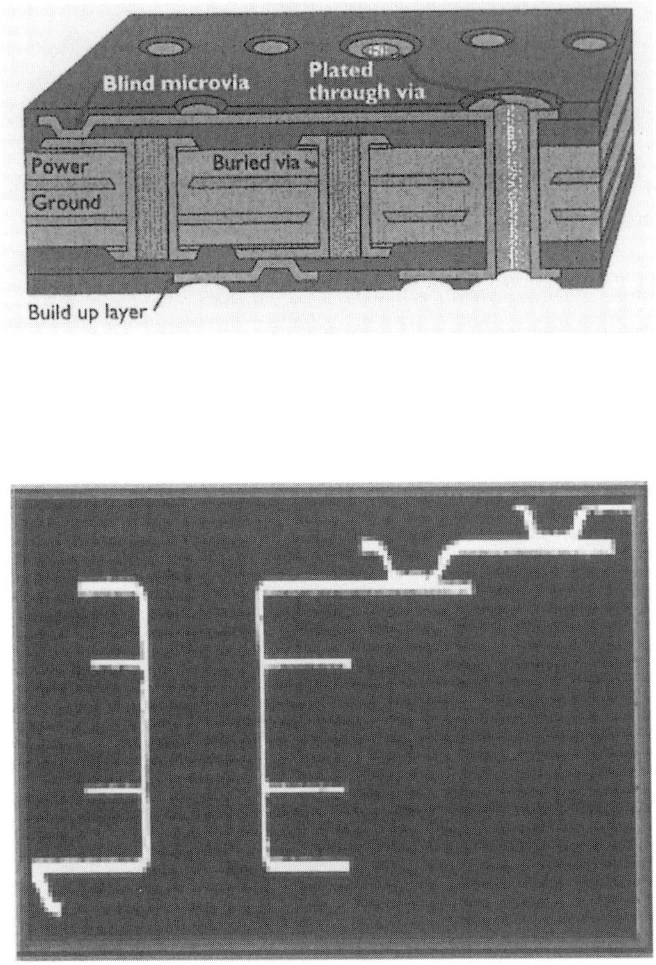

Figure 8.12. Cross section of two layer of microvia interconnect on a traditional microvia board. (Courtesy of Ibiden)

using a blend of two epoxies – one with an imidazole hardener and one with an amine hardener. The amine serves as a soluble filler in the epoxy creating a rough surface when chemically oxidized during the surface-roughening step. Excellent adhesion of the copper is achieved by mechanical anchoring in the small cavities of the roughened epoxy. Peel strengths of 1.2 kg/cm have been reported. The electroless seeding of the epoxy is followed by photoresist mask coating to form the shape of the metal lines. The photoresist mask is a permanent dielectric and left in place. The epoxy resin does not contain glass cloth. The glass cloth assists in providing crack-resistance of the dielectric layers in conventional boards. In place of the glass cloth, Ibiden uses an interpenetrated network structure of a linear, thermoplastic polymer interdispersed in an epoxy. The material composition was optimized for tensile strength (800 kg/cm^2) and elongation-to-break (8%). Ibiden has also used laser drilling to form the vias in the insulator.

NEC Toyama has developed high-density microvia (μV) printed wiring board for use in high-density, advanced microelectronic packaging applications. The μV technology is built on a conventional four-layer PWB with through-holes. Two layers of fine-line interconnection are fabricated on the board using CO_2 laser ablation for the micro-vias and additive copper plating for the conductors. A permanent photoresist is used defining the conductor patterns. High electroless copper plating rates were obtained (5 μm/hr) by automated control and micro filtering. The technology is capable of producing 50 μm copper lines with 50 μm spaces. The vias can be as small as 50 μm on 150 μm pads. NEC Toyama foresees that UV lasers will replace the CO_2 lasers when future miniaturization is needed.

The IBM surface laminar circuit (SLC) also uses liquid dispensed photo-imaged epoxy in the build-up layers on FR4 or BT substrates. Figure 8.13 shows a cross section of the design and SLC device. Four build-up layers on each side of the board can be produced. The negative-tone epoxy dielectric is deposited via curtain or slot coaters. The trade off between forming microvias by laser ablation or by use of a photosensitive polymer has been shifting in favor of laser ablation. This is due to the rapid progress in laser ablation facilities and difficulty in controlling, and the cost of photosensitive polymers.

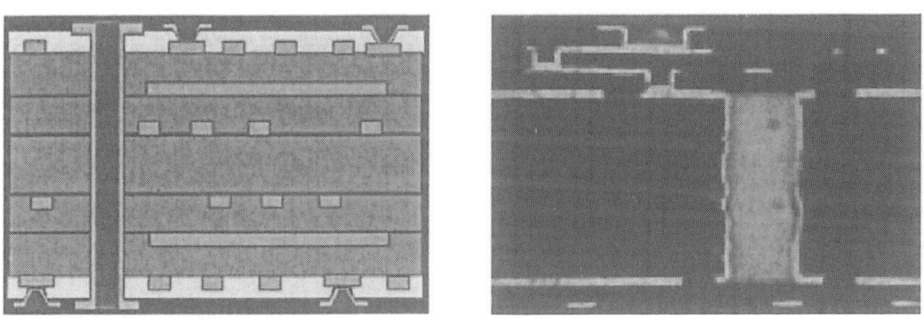

Figure 8.13. IBM surface laminar circuit (SLC) technology. (Courtesy of IBM Microelectronics)

A feature that distinguishes apart high-density boards is the structure of the vias. Figure 8.14 shows one of the problems with conformal, nonfilled vias. The vias cannot be stacked on two of each other because the nonplanarity causes problems with photolithography. In the filled via structure in Figure 8.14, the next layer of wiring can be directly routed on top of a via, or an-

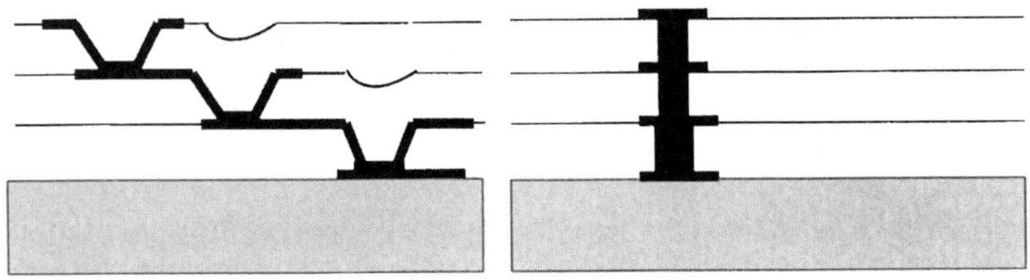

Figure 8.14. None-filled vias on left side must be off-set because of nonplanar surface, whereas the filled vias on right side can be stacked.

other via can be formed. The wiring-blockage from nonfilled vias reduces the function area of the board and causes design and layout problems.

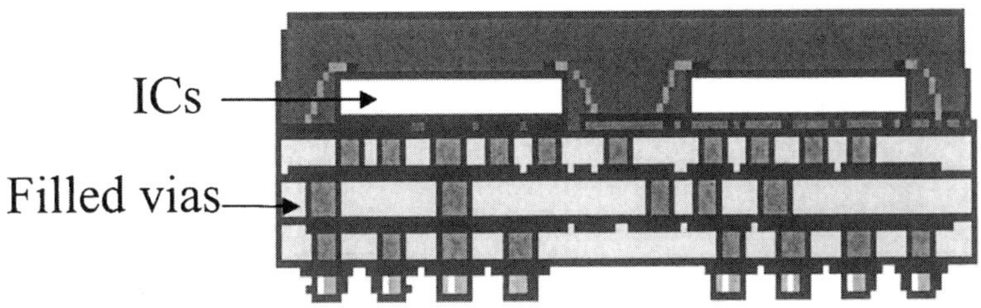

Figure 8.15. Two ICs mounted on an Alivh substrate.

Matsushita has developed unique filled-via board made by stacking substrates called "ALIVH" (any layer inner via hole structure). This technology has several advantages over conventional PWB manufacturing process. A sketch of two ICs mounted on an ALIVH substrate is shown in Figure 8.15. First, completely dry processes are used to form interconnections using laser ablation of an insulating film (currently CO_2 laser) and Cu paste for forming buried via-holes. After laser drilling holes in the insulator, a copper paste is squeezed into the hole forming what will become the layer-to-layer via-connection. Second, the conductor lines are formed hot-press lamination of copper foil onto the insulator (with vias). No plating processes are necessary to form conductive lines. Third, the structure is lightweight due to the use of epoxy impregnated aramid non-woven fiber. Fourth, there are no limitations on the layout of interconnections because each layer is planar. Lastly, the design time is shorter than conventional products. Matsushita has been successfully manufacturing ALIVH substrates with the following attribute: 30-50% substrate size reduction, 50% cost reduction, and 67% design time reduction compared to conventional PWB manufacturing method. Matsushita reported that its share in the Japanese cellular phone market has risen to 60% with the introduction of ALIVH substrates. The market share of ALIVH in high-density substrate market is 38%, in terms of production capability. Matsushita 's target for conductor line and space resolution is 50 μm/50 μm in 1999, and 25 μm/25 μm in 2000.

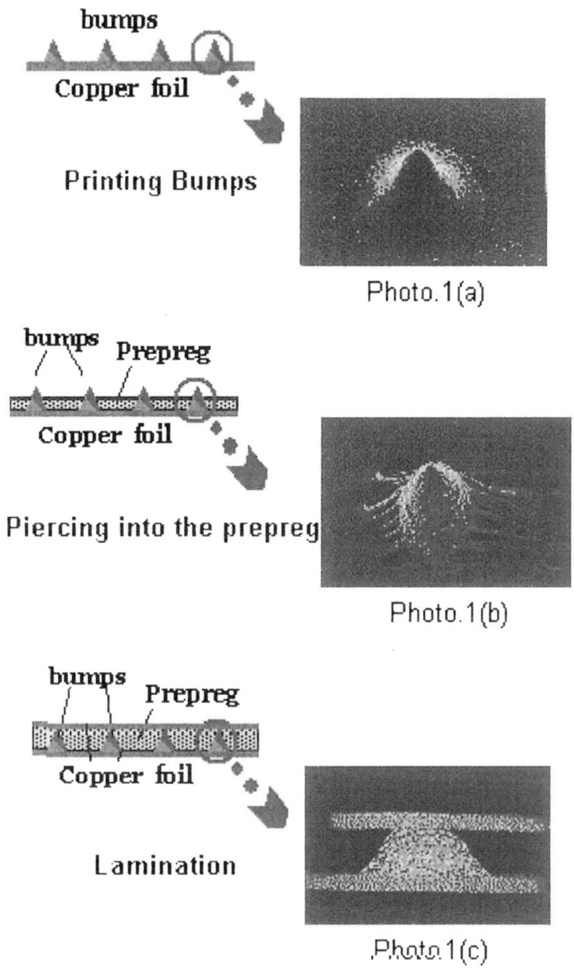

Figure 8.16. Process of one-layer of build-up for Toshiba B^{2}it. (Courtesy of Toshiba)

Toshiba has introduced a unique buried bump interconnection technology (B^{2}it). This technology is an adaptation of multi-layered ceramic processing.

Silver paste is used to form conductive bumps on copper foil, as shown in Figure 8.16. The bumps have a conical shape and pressed into B-staged prepreg. The bumps penetrate the insulator and form a conductive path through the prepreg. The B^2it can be used by itself to form the multilayered substrate. Alternatively, B^2it can be fabricated on a conventional PWB core or microvia layers can be fabricated on B^2it using laser ablation or photosensitive insulators. B^2it has several advantages over conventional PWB technologies. First, the silver paste printing method used to form interconnecting Ag bumps is simpler than drilling, using photovias or laser ablation vias. Second, no plating is necessary to form conductive lines in B^2it (only lamination using prepreg material). Third, there are no limitations on the placement of position of the interconnecting bumps. Fourth, there are few requirements on the interlayer dielectrics allowing new materials to be used. Lastly, the design time is shorter because there are no limitations on interconnection layout. Toshiba has been manufacturing 2 to 8 layer substrates with a 40% reduction in substrate size compared to those conventional boards. The via and land diameters are currently 200 μm and 400 μm, respectively. Toshiba is planning to reduce these dimensions to 100 μm/200 μm and 50 μm/75 μm in the future. The conductive lines and spaces, currently 75 μm width each, are being formed by subtractive processing using photoresist. Toshiba is pursuing 30 μm lines and 30 μm spaces by combining microvia build-up processing on B^2it substrates.

The microvia feature sizes range from 30 μm to 100 μm and continuing in a steady progression towards smaller values. The products are extremely cost sensitive and their evolution the needs of the IC product lines. The dimensions are in line with current needs as described by roadmap activities.

The question that remains is the ability of electronic packaging in meeting the needs of GSI, which will require the direct connection to the printed wiring board, or substrate. The task of providing high I/O count is nontrivial because of I/O pitch (center to center distance between two I/O pads) limitations of package interconnections and thermal mismatch issues between the IC and PWB. Wafer level packages use large solder bumps (150 μm diameter and in excess of 300 μm height) as their interconnection technology to withstand thermal and mechanical stress arising from thermal mismatch between the IC and the PWB. This limits the minimum pitch of these packages. The issues of heat transfer will be addressed in a separate section.

8.4 Chip-to-Module Interconnections for GSI

Future integrated circuit (IC) packaging requirements and the performance of chip-to-module interconnections are determined by the expected chip performance. For example, the International Technology Roadmap for Semiconductors [1] specifies that high performance chips at the 60 nm technology node will cover an area of 427 mm^2, have 4437 input/output (I/O) pins, operate with an on-chip local clock speed of 6 GHz, and have a chip-to-board speed of 2.6 GHz. These same chips will dissipate over 171 W of power and draw 190 A of DC current. At the 30 nm technology node, high performance chips will require interconnections that can support 8758 pins and operate at 16.9 GHz (on-chip) and 3.8 GHz (off-chip). They will dissipate 186 W of heat and consume 310 A DC current. The coefficient of thermal expansion(CTE) mismatch between silicon (CTE $\approx$ 2.6 $\times$ $10^{-6}/°C$) and typical printed wiring boards will only become more severe. The CTE of printed wiring boards, currently about 18 $\times$ $10^{-6}/°C$); may rise as the electrical performance becomes more critical. This translates into $\approx$ 20 μm of lateral mismatch at for a ΔT=45°C operation, at the 60 nm and 30 nm technology nodes. The chip-to-module interconnections must provide as much as 30 μm in vertical compliance (z-axis) to accommodate the non-planarity of the boards and to allow good electrical contact between the test fixture and packaged IC. Other IC categories, such as handheld devices (including wireless), will present similar challenges defined by the I/O density, operating speed, power density, and heat dissipation. Also, there will be special demands arising from unusual form factors (e.g. 3-D chips) and mixed technologies (digital, analog, RF, optical, MEMs). A final set of challenges (e.g. test fixture and assembly alignment and burn-in and attachment temperatures) will arise in the burn-in, testing, and assembly of these ICs.

8.4.1 Compliant Chip-to-Module Connections

The prospect of GSI and system-on-a-chip places enormous constraints on the package. The package must enable high DC current with minimal ground bounce, and operate at high frequency with minimal signal degradation. Global interconnections, which can span a chip's corner-to-corner distance, pose special problems especially for global clock distribution.

Sea of Leads (SoL). Sea of leads (SoL) technology is proposed as an enabling technology for future chip-to-module interconnections. SoL wafer level packaging technology provides an ultra-high I/O density of x-y-z compliant leads ($> 10^4$ per cm^2) and can enhance the performance of an system on a chip (SoC) by routing critical on-chip global interconnects off-chip to reduce signal delay and thus increasing global clock frequency [10].

Figure 8.17 is an SEM micrograph illustrating a portion of an SoL package with 12 $\times$ 10^3 x-y-z compliant leads per cm^2 [11-16]. In-plane and out-of-plane compliance were measured to be greater than 30 μm. While the leads are x y z axis compliant, they are short in length and thus exhibit minimal parasitics from DC to 45 GHz. The calculated resistance and inductance of

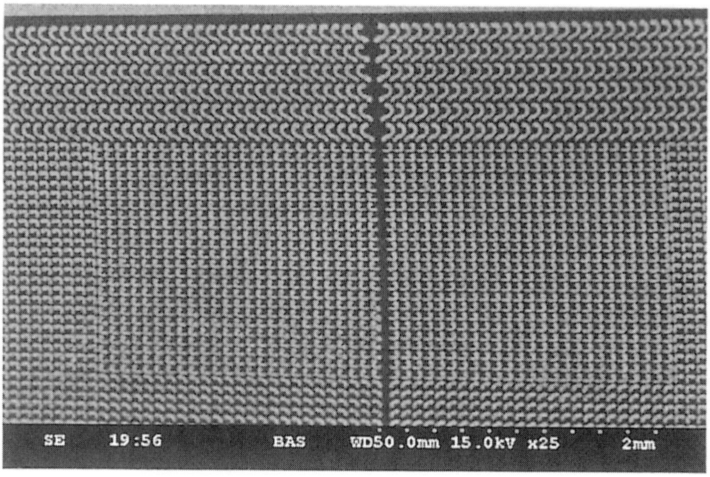

Figure 8.17. SEM micrograph showing portion of a SoL package with 12×10^3 x-y-z compliant chip I/O interconnects distributed across a cm^2.

the leads are less than 25 mΩ and 0.1 nH, respectively. Low electrical parasitics are desirable at both low and high frequencies for efficient conductive coupling of power, low power dissipation in the leads and thus low heat generation by the package. The microwave characteristics of SoL were measured at wafer-level using a two-port network analyzer with 150 μm coplanar ground-signal-ground (GSG) probes. To characterize the compliant interconnects, 15 thick Au leads were fabricated on a 15 μm thick polymer film. The return-loss and insertion-loss of the GSG lead interconnection were measured to be less than 20 dB and 0.2 dB, respectively, at 45 GHz. In comparison, for example, insertion losses before and after the addition of underfill within a flip-chip package mounted on an alumina substrate with 75 μm $\times$ 150 μm bumps interconnected by 600 μm long 50 Ω coplanar waveguides were found in [17] to be 0.6 dB and 1.8 dB, respectively, at 40 GHz. The normalized impedance of the load as seen by the microwave probes was derived. At low frequencies (up to a few GHz), the leads appear as shorts (i.e. good metal interconnects). Another microwave measurement was made on a fully processed package where the leads were connected in pairs by 100 μm and 1 μm thick copper interconnects routed along the silicon wafer surface. The worst case return-loss and insertion-loss for a pair of interconnected leads were measured to be less than 12 dB and 1.2 dB, respectively. These measurements include the losses due to the Cu interconnects. Crosstalk between adjacent parallel leads was measured to be less than 30 dB at 45 GHz, and crosstalk between orthogonal leads was approximately 40 dB, (i.e. 10 dB lower than the parallel leads). All interconnect structures were approximately 20 μm wide. All the compliant leads measured were approximately 110 μm long.

Similar microwave measurements for other packaging technologies exist in the literature. It appears that the SoL measurements reported above compare very well to flip-chip packages, which are widely used in microwave applications. However, the ultra-high I/O density of SoL can provide an exceptionally high I/O bandwidth that would be difficult to match using alternate packaging technologies. For example, if 8,000 leads are assigned as signal I/Os and operated at only 5 GHz, the SoL package shown in Figure 8.17 can yield an aggregate electrical I/O bandwidth of 40 Tb/(cm^{2}s). Moreover, an SoL package avoids microwave performance degradation caused by underfill, as no underfill is required during assembly. This is in contrast to assembly requirements of flip-chip and BGA packages, where the presence of underfill increases insertion-loss and shifts the return-loss frequency response to lower frequencies [17,18].

An important aspect of SoL is the set of performance enhancements it offers a mixed-signal SoC [1]. SoL can enhance the DC power distribution as well as satisfy 3D structure I/O requirements. Because SoL processing requires low temperature ($< 250°$C), it is expected that wafer-level package processing will not damage previously fabricated structures during front-end-of-line and back-end-of-line processing [11-13].

The addition of embedded air gaps into SoL adds vertical compliance (z-axis) needed for wafer level testing and mating to non-planar boards. Air-gaps also serve to lower the dielectric constant of the interconnect. Figure 8.18 shows a cross-section of process flow for an x-y-z compliant lead [11-16]. The use of a low modulus polymer encapsulating the air-gap forms a structure capable of elastically deforming in all three dimensions. An IC with exposed bond pads is shown in Figure 8.18a. A sacrificial material (e.g. UnityTM, Goodrich Corporation) is deposited and patterned onto the wafer, Figure 8.18b. The sacrificial material serves as a placeholder and will later be removed to form an air-cavity [14-16]. The sacrificial material is overcoated by a flexible material. and vias are opened to the bond pad in Figure 8.18c. When the overcoat elastomer is curing in Figure 8.18d, the sacrificial material simultaneously decomposes leaving a buried air-cavity. The exact compliance of the final structure depends on the size and shape of the air-cavity, and the elastic properties of the overcoat material. All these properties are within the control of the designer. Finally, the metal leads are fabricated in Figure 8.18e, and attachment solder added in Figure 8.18f. Fine-line structures can be processed by use of photolithography, photosensitive polymers, and reactive ion etching. Coarser features, such as for low I/O count ICs, can be processed by screen-printing and other similar processing techniques. It is important to note that the metal

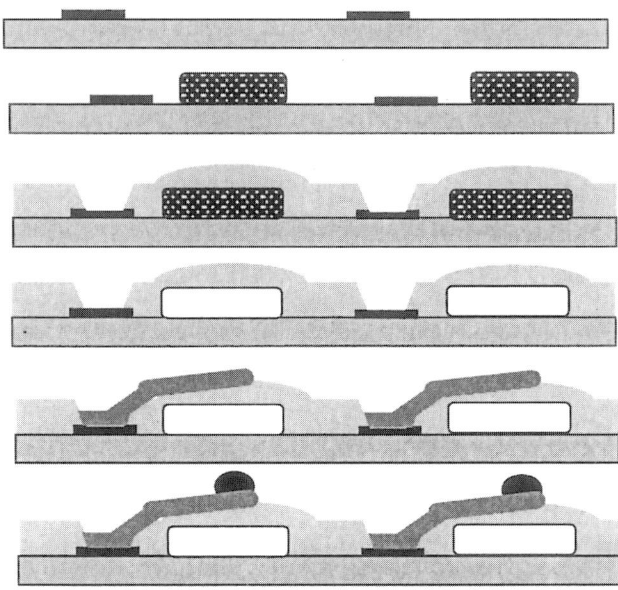

Figure 8.18. Build-up process for compliant wafer level interconnections.

traces are made not to adhere to the surface of the overcoat polymer in Figure 8.18, which can be accomplished in several ways, such as by using a polymer with poor adhesion. Figure 8.19 shows metal leads fabricated on buried air-cavities (the attachment solder has not yet been put in place). This type of structure eliminates the need for chip underfill during package assembly of the chip to the PWB because of the x-y-z compliance of the leads.

Wafer level batch packaging presents special opportunities and challenges to the burn-in and testing portion of the IC manufacturing process. On the one hand, wafer level packaging allows a complete wafer containing fully packaged die to be taken through each stage of the burn-in and testing process. This is in contrast to burn-in and testing of singulated die by inserting them into temporary test carriers and then removing them for final die-to-package integration and subsequent packaged part testing. Wafer level testing sharply reduces part handling and die-to-test fixture alignment. It also eliminates the need of inserting a die into a temporary carrier for testing purposes. A major challenge to taking full advantage of wafer level testing is to establish and maintain all the I/O connections during the burn-in and test process. The typical force applied to a single I/O by a test fixture can be from 5 - 10 mN, this translates into about 5 to 10 kN total force applied to a 300 mm wafer popu-

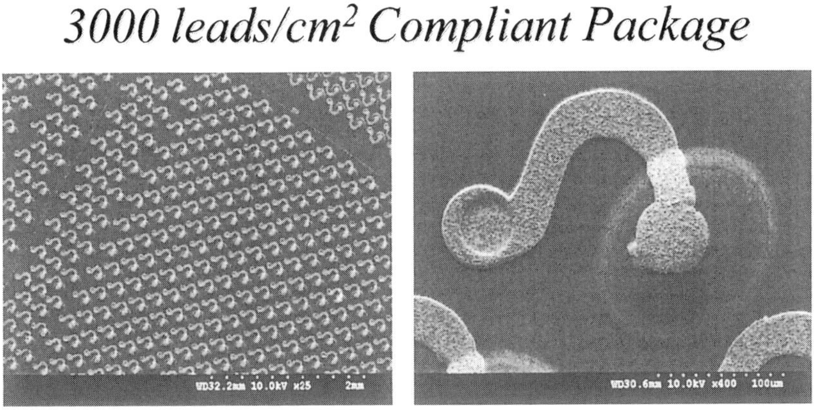

Figure 8.19. Sea of leads interconnect with compliant leads at 3000 leads/cm^2: (a) wide view and (b) close-up of one lead built over air-cavity.

lated by high performance chips at the ITRS 35 nm technology node. Fully powered, this same wafer would require 38 kA and dissipate 23 kW of heat. The test fixture and wafer must also accommodate the CTE expansion of the die during the burn-in and testing procedure (at 150°C an I/O at the edge of a 300 mm silicon wafer will be displaced by $\approx$ 50μm from its room temperature position). Testing algorithms need to be developed that speed up testing and allow the test equipment to efficiently examine a full wafer rather than be swamped by the bandwidth demands of distributing test vectors to and detecting responses from many die rather than one at a time. Testing a mixed signal SoC will present many challenges. Integrating different cores such as DSP, RAM, microprocessor, RF, and analog on a single chip requires careful testing considerations early during the chip design cycle. If 1,240 leads are allocated for only AC/DC testing a mixed-signal SoC, direct access to the various cores maybe possible and great insight can be acquired into failure mechanisms. In addition, this high number of leads will provide very high bandwidth to perform faster testing [11-16].

Next-generation SoL technology will incorporate optical waveguide interconnection to permit global optical clock distribution within high-performance chips. Optical waveguide interconnection allows for planar packaging of a hybrid electrical/optical system in a manner conducive to future heat removal and power supply requirements. In addition, polymer waveguide technologies offer immediate and low-cost compatibility with wafer-level fabrication processes. Minimal redesign of the global distribution would be required with increases in clock frequency, as design of the cross-sectional dimensions of an optical waveguide is independent of the clock frequency. Global propagation of the local clock frequency would be possible, thereby eliminating the need for global clock repeaters and allowing for the removal of cycle-to-cycle jitter from the global clock signal in its delivery to optoelectronic receivers for optical-to-electrical conversion. Finally, optical waveguide clock distribution would allow for enhanced predictability with respect to the arrival time of clock pulses at different optoelectronic receivers, thereby eliminating *unexpected* [19] components of global distribution skew.

A key feature of the proposed technology is the use of embedded air-gap regions as the upper cladding for optical waveguides. Embedded air-gap cladding regions allow for a maximization in refractive index contrast, Δn, between waveguide core and cladding, and hence permit smaller bending radii and higher waveguide densities due to tighter confinement within the waveguide core. By placing optical waveguides within the package, via blockage concerns are eliminated with respect to waveguide routing, leaving only electrical I/O interconnection to obstruct intra-chip waveguide routing.

The portion of the fabrication process associated with optical waveguide integration begins following the definition of bond pad regions for compliant interconnection. In recent experiments, optical-quality SiO_2 serves as the lower cladding region of optical waveguides (although this is not a requirement). Following deposition and curing of the waveguide material, which consists of an alkoxy-siloxane epoxy [20], channel waveguide regions are defined. To create the embedded air-gap cladding regions, a sacrificial photosensitive polymer composite (UnityTM 200) is applied and patterned. An overcoat polymer is then applied to embed the air-gap and waveguide regions, where, upon thermal cure, the remaining polymer regions decompose to leave an embedded air-gap. A micrograph of a single waveguide from an SoL package with 1000 electrical I/Os per cm^2 and 32 optical /indexSoL!optical waveguides waveguides is shown in Figure 8.20 where (i) individual compliant leads, (ii) an optical waveguide, and (iii) a buried air-gap region are shown. By incorporating a buried air-gap, a Δn = 0.52 is achieved between core and upper-cladding regions. Beam propagation methods reveal that, by incorporating an air-gap cladding region, the minimum allowed edge-to-edge spacing between adjacent waveguides is reduced by over three fold [21].

Compliant Contacts. Cantilever spring interconnects have been developed for use in probing and IC uses. Cantilever springs made from metal bilayers can be fabricated so that an internal stress gradient in the through plane direction causes a deflection of the beam. Figure 8.21 shows a row of springs made from a stressed Mo/Cr bilayer. The curvature of the springs can be adjusted by control thickness or internal stress gradient in the metal bilayer [22, 23]. The sharp tips on the springs can be used to probe ICs or can used as a contact by holding the spring in place with an adhesive. The cantilever can be fabricated

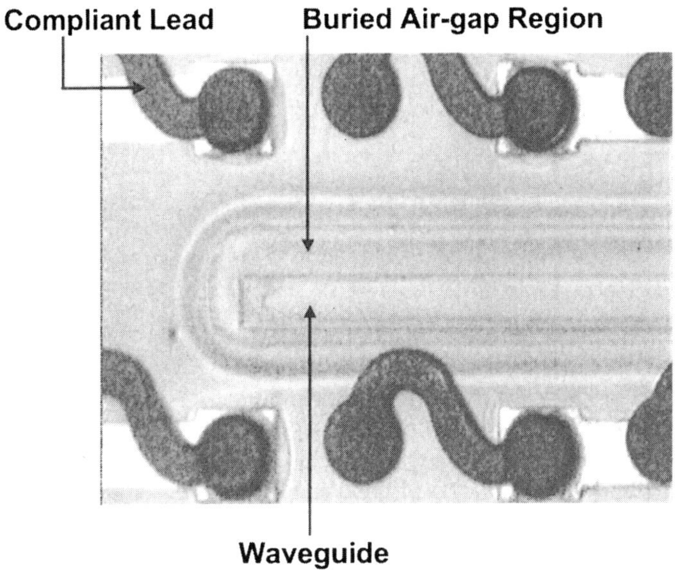

Figure 8.20. Embedded optical waveguide with air-gap within a SoL.

by use of a release layer under the deposited (and patterned) metals. Pure gold metal cantilever beams can also be fabricated by producing a stress gradient in the gold by altering the deposition conditions.

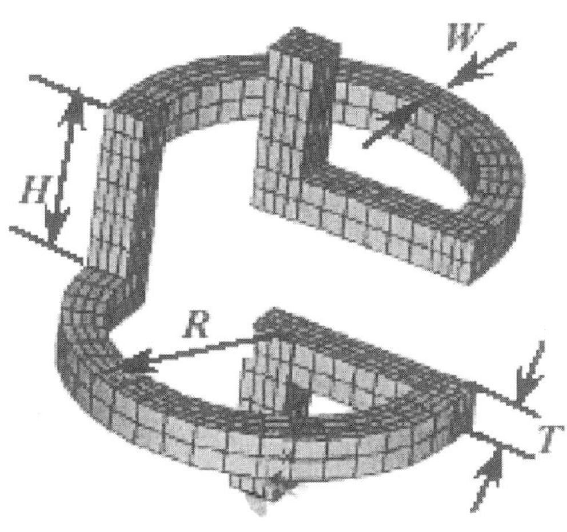

Figure 8.21. Design parameters of one turn helix structure.

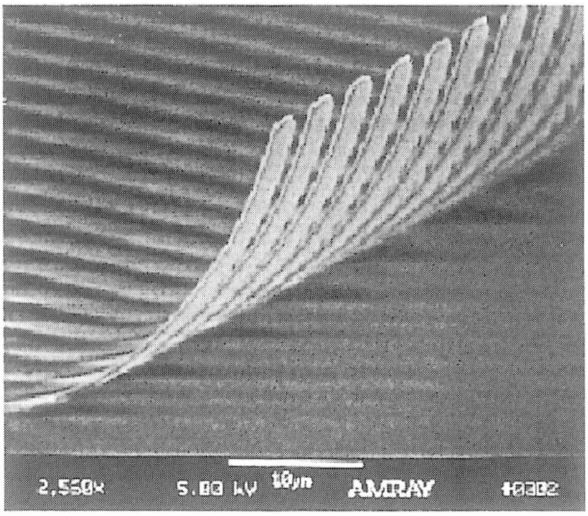

Figure 8.22. Micrograph of microspring contacts.

More complex, micromachined structures can be fabricated for custom designed x-y-z compliance of interconnects. Figure 8.22 shows a drawing for a one-turn helix structure that can serve and the x-y-z compliant element of a chip-to-module interconnection [24,25]. The helix is composed of five elements (three posts and two cantilevers), and can be fabricated with five lithography steps. The degree of compliance in the x, y, z directions are unequal, but can be adjusted in range of 2 to 70 mm/N. The mutual and self-inductance is typically in the range of 0.01 to 0.07 nH. There is a trade-off between high compliance and high electrical resistance.

In summary, wafer level packaging can be used to meet the performance and chip-to-module interconnection needs of future ICs. A variety of compliant structures can be fabricated in wafer form to reduce the need for underfill and provide improved electrical performance in a cost effective manner.

References

[1] International Technology Roadmap for Semiconductors (ITRS), SIA/Sematech International (ITRS, http://public.itrs.net/).

[2] P. Garrou, "Wafer Level Chip Scale Packaging", *Semi Chip Scale International '99*, page D-1 (1999).

[3] "Fundamentals of Microsystems Packaging", R. Tummala Ed., McGraw Hill, 2001.

[4] H. I. Rosten, and R. Viswanath, "Thermal Modeling of the Pentium Processor Package", in Proc. of ECTC, pp. 421, 1994.

[5] Printed Circuit Board Materials Handbook, Martin W. Jawitz, McGraw-Hill (1997).

[6] Y. Sato, K. Sasaoka, K. Shibayama, H. Hamano, and Y. Fukuoka, "A New PCB Utilizing Buried Interconnection Technology (B2it)", in *Proc. 7th Printed Circuit World Convention*, pp. 19-1 to 6, May 1998.

[7] H. Iwaki, Y. Taguchi, T. Shiraishi, Y. Bessho, and K. Eda, "High Frequency Electrical Characterization of a High Wiring Density Organic Substrate "ALIVH" and a Stud Bump Bonding SBB", in *Proc. IEEE International Symposium on Microelectronics*, pp. 341, 1998.

[8] R. Maniwa, "Finer Micro-via PWB by Laser and Additive Process", in *Proc. IEEE International Symposium on Microelectronics*, pp. 413, 1998.

[9] R. Enomoto, M. Asai, and N. Hirose, "High Density MLB using Additive and Build- up Process", in *Proc. IEEE International Symposium on Microelectronics*, pp. 399, 1998.

[10] A. Naeemi, P. Zarkesh-Ha, C. S. Patel, and J. D. Meindl, "Performance Improvement Using On-Board Wires for On-Chip Interconnects", in *Proc. IEEE 9th Topical meeting on Electrical Performance of Electronic Packaging*, pp. 325-328, Oct. 2000.

[11] M. S. Bakir, H. A. Reed, P. A. Kohl, K. P. Martin, and J. D. Meindl, "Sea of Leads ultra high-density compliant wafer level packaging technology," in *Proc. 52nd Electronics and Components Technology Conf.*, (San Diego, CA), May 2002.

[12] C. S. Patel, K. Martin, J. D. Meindl, P. A. Kohl, C. Powers, and M. Realff, "Low cost high density compliant wafer level package," in *Proc. International Conf. on High-Density Interconnect and Systems Packaging*, (Denver, CO), pp. 261-268, April 26-28, 2000.

[13] M. S. Bakir, H. A. Reed, A. V. Mule, P. A. Kohl, K. P. Martin, and J. D. Meindl, "Sea of Leads characterization and design for compatibility for board level optical waveguide interconnection," in *Proc. IEEE Custom Integrated Circuits Conference*, (Orlando, FL), May 2002.

[14] Kohl, P. A., Bhusari, D. M., Wedlake, M., Case, C., Lee, B. C., Gutmann, R. J., and Shick, R., "Air Gaps in 0.3 μm Electrical Interconnections", *Electron Device Letters*, vol. 21, pp. 557-560, 2000.

[15] H.A.Reed, C.E.White, V.Rao, S.A.Allen, C.L.Henderson, and P.A.Kohl, "Fabrication of Microchannels Using Polycarbonates as Sacrificial Materials", *Journal of Micromechanics and Microengineering*, vol. 11, pp. 733-737, 2001.

[16] D.Bhusari, H.Reed, M.Wedlake, A.Padovani, S.A.Bidstrup-Allen, P.A.Kohl, "Fabrication of Air-Channel Structures for Microfluidic,

Microelectro-mechanical and Microelectronic Applications", *Journal of Microelectromechanical Systems*, vol. 10, pp. 400-409, 2001.

[17] Z. Feng, W. Zhang, S. Bingzhi, K. C. Gupta, and Y. C. Lee, "RF and mechanical characterization of flip-chip interconnects in CPW circuits with underfill," *IEEE Trans. On Microwave Theory and Techniques*, Vol. 46, pp. 2269-2275, 1998.

[18] H. Kusamitsu, Y. Morishta, M. Ito, and K. Ohata, "The flip-chip bump interconnection for millimeter-wave GaAs MMIC," *IEEE Trans. Electron. Packaging Manufact.*, vol. 22, pp. 23-28, Jan. 1999.

[19] P. J. Restle, T. G. Mcnamara, D. A. Webber, P. J. Camporese, K. F. Eng, K. A. Jenkins, D. H. Allen, M. J. Rohn, M. P. Quaranta, D. W. Boerstler, C. J. Alpert, C. A. Carter, R. N. Bailey, J. G. Petrovick, B. L. Krauter, and B. D. McCredie, "A clock distribution network for microprocessors," *IEEE J. Solid State Circuits*, vol. 36, pp. 792- 799, May 2001.

[20] A. Jain, S. Rogojevic, S. Ponoth, N. Agarwal, I. Matthew, W. N. Gill, P. Persans, M. Tomozawa, J. L. Plawsky, and E. Simonyi, "Porous silica materials as low-k dielectrics for electronic and optical interconnects," *Thin Solid Films*, vol. 398-399, pp. 513-522, 2001.

[21] A. V. Mule, S. Schultz, E. N. Glytsis, T. K. Gaylord, and J. D. Meindl, "Input coupling and guided-wave distribution scheme for board-level intra-chip optical clock distribution network using volume grating coupler technology," in *Proc. IEEE International Interconnect Technology Conference*, (San Francisco, CA), pp. 128-130, June 2001.

[22] L. Ma, Q. Zhu, T. Hantschel, D. K. Fork, and S. K. Sitaraman, "J-Springs-Innovative Compliant Interconnects for Next-Generation Packaging", in *Proceedings ECTC*, 2002.

[23] L. Ma, Q. Zhu, W. K. Sitaraman, C. Chua, and D. K. Fork, "Compliant Cantilevered Spring Interconnects for Flip-chip Packaging", in *Procedings ECTC*, 2002.

[24] Q. Zhu, L. Ma, and S. K. Sitaraman, "design Optimization of One Turn Helix", in *Proc. Itherm*, 2002.

[25] Q. Zhu, L. Ma, and S. K. Sitaraman, "Mechanical and Preliminary Electrical Design of a Novel Complinat One-turn Helix Interconnect", in Proc. InterPACK-01, 2001.

Chapter 9

3-D ICS DSM INTERCONNECT PERFORMANCE MODELING AND ANALYSIS

S. J. Souri, T. Chiang, P. Kapur, K. Banerjee, and K. C. Saraswat
Center for Integrated Systems, Stanford University, Stanford, CA, 94305.

9.1 Introduction

The unprecedented growth of the computer and the information technology industry is demanding ULSI circuits with increasing functionality and performance at minimum cost and power dissipation. ULSI circuits are being aggressively scaled to meet this demand. This in turn has introduced some very serious problems for the semiconductor industry. Continuous scaling of ULSI circuits is reducing gate delays but rapidly increasing interconnect (RC) delays. Semiconductor Industry Association (SIA) roadmap predicts that, beyond the 130 nm technology node, performance improvement of advanced ULSI is likely to begin to saturate unless a paradigm shift from present IC architecture is introduced. Also, increasing interconnect loading significantly increases the power consumption in high-performance chips. In fact, around 40-70% of the total chip power consumption can be due to the wiring network used for clock distribution, which is usually realized using long global wires. Additionally, interconnect scaling has significant implications for traditional computer-aided-design (CAD) methodologies and tools which are causing the design cycles to increase, thus increasing the time-to-market and the cost per chip function. Furthermore, increasing drive for the integration of disparate signals and technologies is introducing various system-on-a-chip (SoC) design concepts, for which existing planar (2-D) IC design may not be suitable. This chapter analyzes the limitations of the existing interconnect technologies and design methodologies and presents a novel 3-dimensional (3-D) chip design strategy that exploits the vertical dimension to alleviate the interconnect related problems and to facilitate SoC applications. A comprehensive analytical treatment of these 3-D ICs has been presented and it has been shown that by simply dividing a planar chip into separate blocks, each occupying a separate physical level interconnected by short and vertical inter-layer interconnects (VILICs), significant improvement in performance and reduction in wire-limited chip area can be achieved, without using any other circuit or design innovations. A scheme to optimize the interconnect distribution among different interconnect tiers is presented and the effect of transferring the repeaters to upper Si layers has been quantified in this analysis for a two-layer 3-D chip. Furthermore, one of the major concerns in 3-D ICs arising due to increased power density has been analyzed and an analytical model has been presented to estimate the temperatures of the different active layers. It is demonstrated that advancement in heat sinking technology will be necessary in order to extract maximum performance from these chips. Implications of 3-D device architecture on several circuit designs and CAD methodologies and tools have also been discussed with especial attention to SoC design strategies. Finally, some of the promising technologies for manufacturing 3-D ICs have been outlined.

9.2 Motivation for 3-D ICs

9.2.1 Interconnect Limited VLSI Performance

In single Si layer (2-D) ICs, chip size is continually increasing despite reductions in feature size made possible by advances in IC technology such as lithography, etching etc., and reduction in defect density [1]. This is due to the ever-growing demand for functionality and higher performance, which causes increased complexity of chip design, requiring more and more transistors to be closely packed and connected [2]. Smaller feature sizes have dramatically improved device performance [3-5]. The impact of this miniaturization on the performance of interconnect wires, however, has been less positive [6-10]. Smaller wire cross-sections, smaller wire pitch and longer lines to traverse larger chips have increased the resistance and the capacitance of these lines resulting in a significant increase in signal propagation (RC) delay. As interconnect scaling continues, RC delay is increasingly becoming the dominant factor determining the performance of advanced ICs [1], [6-10]. Figure 9.1 illustrates this problem by plotting the gate and the interconnect (RC) delays as functions of various technology nodes based on the International Technology Roadmap for Semiconductors, 1999, (ITRS'99) [1]. It can be observed that at the 50 nm technology node the interconnect delay is a couple of orders of magnitude higher than the gate delay. Therefore, as feature sizes are further reduced and more devices are integrated on a chip, the chip performance will degrade, reversing the trend that has been observed in the semiconductor industry thus far.

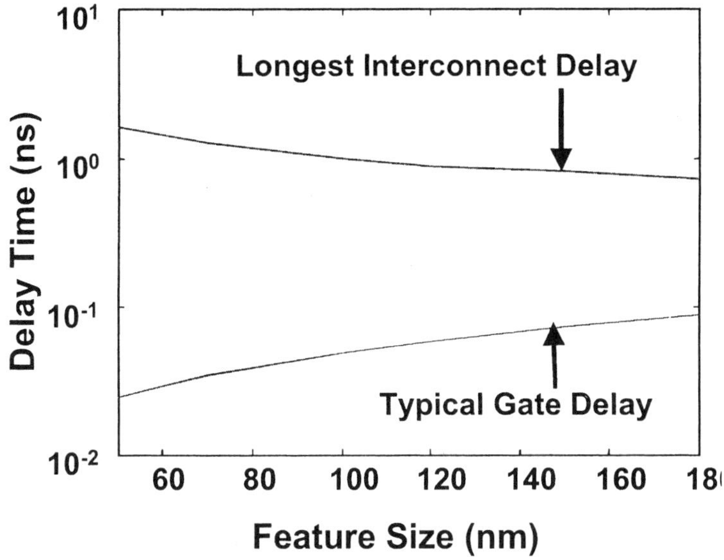

Figure 9.1. Typical Gate and Interconnect (RC) delays as a function of technology nodes. The RC delay includes the delay due to the repeaters.

9.2.2 Physical Limitations of Cu Technology

At 250 nm technology node, Cu with low-k dielectric was introduced to alleviate the adverse effect of increasing interconnect delay [11-15]. However, as shown in Figure 9.1, below 130 nm technology node, substantial interconnect delays will result in spite of introducing these new materials, which in turn will severely limit the chip performance [1]. Further appreciable reduction in interconnect delay cannot be achieved by introducing any new materials. This problem is especially acute for global interconnects, which typically comprise about 10% of total wiring, for current architectures. Therefore it is apparent that material limitations will ultimately limit the performance improvement as the technology scales. Also the problem of long-lossy lines cannot be fixed by simply widening the metal lines and using thicker interlayer dielectric since this conventional solution will lead to a sharp increase in the number of metalization layers. Such an approach will increase the complexity, reliability, and cost, and will therefore be fundamentally incompatible with the industry trend of maximizing the number of chips per wafer, and 25% per year improvement in cost per chip function .

Furthermore, with the aggressive scaling suggested by the ITRS'99 [1], new physical and technological effects start dominating interconnect properties. It is imperative that these effects be accurately modeled, and incorporated in the wire performance and reliability analyses. The next three sub-sections provide quantitative analysis of the impact of these new effects caused by scaling on the resistivity of Cu interconnects.

It is commonly believed that material resistivity for copper would not change significantly for future interconnects [1]. However, because of an increasing dominance of electron scattering from the surface and because of a greater fraction of interconnect area being consumed by metal barrier in the future, as illustrated in Figure 9.2, the effective resistivity of copper may rise significantly. In addition, the operational temperature of wires ($\approx$ 373 K) are higher than the room temperature (300 K) and can increase further due to self-heating caused by the flow of current [16]. This increase in temperature, in turn, would also increase the wire resistivity. In this work, an attempt has been made to model all the above effects and to establish reliable future trends for resistivity as a function of operating temperature, interface quality, and barrier technology. These effects are modeled at all interconnect levels, i.e. local, semiglobal and global levels.

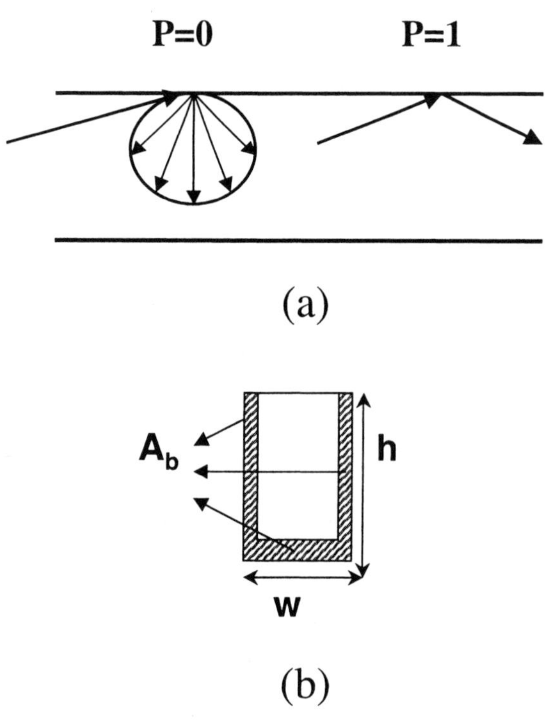

Figure 9.2. Schematic showing (a) surface scattering and (b) barrier induced effective resistivity reduction.

Effect of Interconnect Dimensions on Cu Resistivity. As dimensions shrink, the electron scattering from the surface becomes comparable to electron bulk scattering mechanisms such as phonon scattering. The dominance of this surface effect depends on the parameter, $k = d/\lambda_{mfp}$, where d is the smallest film dimension and λ_{mfp} is the bulk mean free path of electrons. Smaller k signifies a larger surface scattering effect. The surface scattering governed resistivity is given by [17].

$$\frac{\rho_s}{\rho_o} = \frac{1}{1 - \frac{3(1-P)\lambda_{mfp}}{2d} \int_1^{\infty} \left(\frac{1}{x^3} - \frac{1}{x^5}\right) \frac{1-e^{-kx}}{1-Pe^{-kx}} dx} \tag{9.1}$$

Here, ρ_s is resistivity with surface scattering effect, ρ_0 is the bulk resistivity at a given temperature, k is as defined above and x is the integration variable.

Parameter, P, is a measure of extent of specular scattering at copper/barrier interface. Its value lies between 0 and 1. $P = 0$, signifies complete diffuse scattering causing maximum decrease in mobility, hence, a maximum increase in resistivity; whereas, $P = 1$ indicates complete specular reflection leading to no change in resistivity. Values of P are influenced by technology dependent factors and have been experimentally deduced before for various materials under various conditions [18,19].

Effect of Barrier Thickness on Cu Resistivity. The second effect which contributes to increase in the effective copper resistivity results from a finite cross sectional area consumed by the higher resistivity metal barrier encapsulating copper. Barrier thickness, thus its area, depends on the deposition technology as well as the barrier material. Since barrier thickness can not scale as rapidly as the interconnect dimensions, it would occupy increasingly higher fraction of the interconnect cross section area. The effective resistivity just due to this effect is given by

$$\frac{\rho_b}{\rho_o} = \frac{1}{1 - \frac{A_b}{AR*w^2}} \tag{9.2}$$

Here, ρ_b is the effective resistivity because of barrier, ρ_0 is the bulk resistivity at a given temperature, A_b is the area occupied by the barrier, AR is the aspect ratio and w is the width of the interconnect. From the above equation it is obvious that as A_b increases, ρ_b increases.

Simulation of Surface Scattering and Barrier Thickness Effects on Cu Resistivity. The resistivities for future ITRS'99 dictated interconnects are evaluated in light of the above effects. The methodology for extracting future resistivities using various barrier deposition technologies, operating temperatures and P values is explained as follows. SPEEDIE (Stanford Profile Emulator for Etching and Deposition in IC Engineering) [20] was used to simulate the barrier profile for different deposition technologies, which was then used to extract the area consumed by the barrier. The simulations were performed on dimensions specified in the ITRS'99. The deposition time in the simulator was varied for each of the simulated geometries to obtain two conditions corresponding to a 5 nm and 10 nm minimum barrier thickness, respectively. The actual minimum barrier thickness in the future would be dictated by the quality of the barrier.

The global level copper wire resistivity as a function of future years is shown in Figure 9.3 for different P values at two different temperatures, and with an optimistic minimum barrier thickness of 10 nm, using the best available Atomic Layer Deposition (ALD) technology. From this figure it is obvious that under realistic wire temperature of 100°C and P value of 0.5 [18], resistivities as high as 2.9 $\mu\Omega$-cm will be obtained in the year 2010. This gives about 70% increase over the nominal bulk copper resistivity (1.7 $\mu\Omega$-cm) at room temperature. Under same conditions, simulations revealed resistivities of 3.45 $\mu\Omega$-cm and 3.95 $\mu\Omega$-cm, for the semiglobal and local interconnects respectively. It was also found that using any other, less conformal, barrier deposition technology such as Ionized Physical Vapor Deposition (I-PVD) or Collimated PVD (c-PVD), the resistivity values for local and semiglobal interconnects would become higher than aluminum technology for the same dimensions, in about a decade.

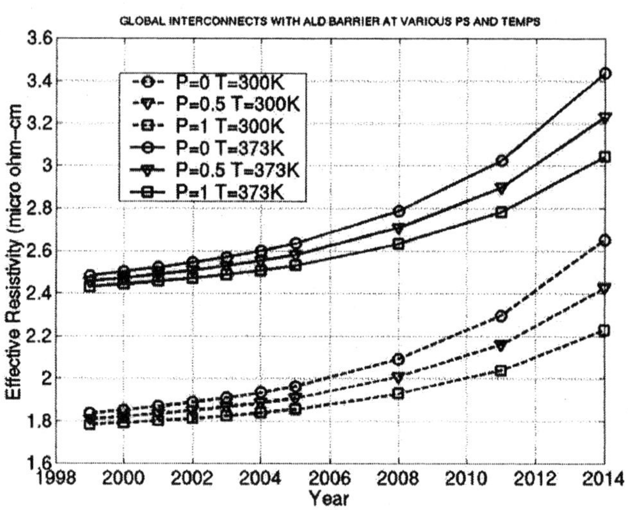

Figure 9.3. Copper resistivity of global level interconnects vs. year using most conformal technology (ALD), barrier thickness (BT) = 10 nm for various choices of P and temperatures.

Thus it has been quantitatively illustrated that in the near future the material resistivity of copper will rise to prohibitively high values even with the best available deposition and barrier technologies. At some point in future, local and semiglobal level effective resistivity of copper could become higher than corresponding resistivity for aluminum for same ITRS'99 dictated dimensions. This will make the interconnect delay even higher than depicted in Figure 9.1 where bulk resistivity was assumed. This calls for a pressing need to develop Cu technologies with smooth surfaces along the wire perimeter to maximize elastic scattering of electrons such that the value of P in (9.1) may nearly equal one. There is also an urgent need for the development of barrierless Cu technology and for lowering the operating wire temperature by going with higher thermal conductivity packaging materials and/or a radically new chip cooling mechanism.

9.2.3 DSM Interconnect Effects on VLSI Design

Aggressive scaling of interconnects in deep submicron (DSM) VLSI presents many challenges to existing computer-aided-design (CAD) methodologies and tools [21]. Apart from the increasing signal transmission delays of global signals relative to the clock period and gate delay, there are signal integrity concerns arising from electromagnetic interference such as interconnect crosstalk, wire-substrate coupling and inductance effects, as well as voltage (IR) drop effects, and pulse attenuation. Also, electromigration and thermal effects in interconnects impose severe restrictions on signal, bus, and power/ground line scaling [16].

Presently, a significant manifestation of the interconnect effects in deep submicron designs arises in the form of timing closure problems, which is caused by the inability of logic synthesis (optimization) tools to account for logic gate interconnect loading with adequate precision prior to physical design. Traditionally, logic optimization is performed using wire-load models that statistically predict the interconnect load capacitance as a function of the fanout based on technology data and design legacy information [22]. The wire-load model includes the intrinsic gate delay and an average delay due to the interconnect connecting the output of the gate to other gate inputs as well as the delay associated with the inputs of the following stage. This approach suffices if the interconnect capacitance remains a small fraction of the total capacitance. However, as illustrated in Figure 9.1, this is far from being accurate. This deficiency in the existing CAD flows causes a serious dilemma in deep submicron designs. On one hand, the increasing circuit complexity (number of gate counts) requires the CAD methodologies to adopt higher levels of abstraction (block-based and hierarchical design) to simplify and accelerate the design process, while on the other hand, the increasing interconnect delay (and other interconnect related effects) makes it difficult for them to obtain timing convergence for the design blocks within a reasonable number of iterations.

It is instructive to note that the magnitude of the interconnect problem for future deep submicron ICs with greater than 108 million transistors (269 million for both logic and memory, at the 50 nm node [1]) cannot be fully comprehended by analyzing the impact of scaling on module-level designs (with 50K gates) using standard wire-load models for average-length wires as suggested in [23]. This is due to the fact that for deep submicron designs, even if the average-length wires within small module-level blocks continue to produce wire delays such that the module-level designs can be individually handled by the traditional wire-load models, the number of such blocks required to realize the entire design would explode resulting in longer and more numerous inter-block interconnects (global wires). Unfortunately, it is these long global wires that are mainly responsible for the increasing interconnect delays as pointed out in an earlier section. Furthermore, given the various technology and material effects arising due to interconnect scaling illustrated earlier, even some of the intra-module wire delays can become unexpectedly large contrary to usual assumptions as in [24]. In order to mitigate the interconnect scaling problems some researchers have proposed combined wire planning and constant-delay synthesis [25, 26]. This methodology is also based on a block-based design where the inter-block wires are planned or constructed and the remaining wires are handled through the constant-delay synthesis [27] within the blocks. The difficulty with this method is that if the blocks are sufficiently large then the timing convergence problem persists. In contrast, if they are allowed to remain relatively small such that the constant-delay synthesis with wire-load models works, then the number of such blocks becomes so large that the majority of the wiring will be global and the physical placement of these point-like blocks becomes absolutely critical to the overall wire planning quality, which represents a daunting physical design problem. Another work proposed an interconnect fabric based on a ground-signal-ground wire grid to make wire loads more predictable [28]. However, this technique results in significant area penalty and hence not very attractive for designing high-performance VLSI circuits.

Thus it can be concluded that the interconnect problem in deep submicron VLSI design is not only going to get bigger due to ever increasing chip complexity, but will also get worse due to material and technology limitations discussed above. Hence, in the near future, existing design methodologies and CAD tools may not be adequate to deal with the wiring problem both at the modular and global levels.

Greater performance and greater complexity at lower cost are the drivers behind large scale integration. In order to maintain these driving forces it is necessary to find a way to keep increasing the number of devices on a chip, yet limit or even decrease the chip size to keep interconnect delay from affecting chip performance. A decrease in chip size will also assist in maximizing the number of chips per wafer; thus maintaining the trend of decreasing cost function. Therefore innovative solutions beyond mere materials and technology changes are required to meet future IC performance goals [29]. We need to think beyond the current paradigm of design architecture.

9.2.4 System-on-a-Chip Designs

System-on-a-chip (SoC) is a broad concept that refers to the integration of nearly all aspects of a system design on a single chip [30, 31]. These chips are often mixed-signal and/or mixed-technology designs, including such diverse combinations as embedded DRAM, high-performance and low-power logic, analog, RF, programmable platforms (software, FPGAs, Flash etc.), as schematically illustrated in Figure 9.4. They can also involve more esoteric technologies like Micro-Electromechanical Systems (MEMS), bio-electronics, micro-fluidics, and optical input/output. SoC designs are often driven by the ever-growing demand for increased system functionality and compactness at minimum cost, power consumption, and time to market. These designs form the basis for numerous novel electronic applications in the near future in areas such as wired and wireless multi-media communications including high-speed internet applications, medical applications including remote surgery, automated drug delivery, and non-invasive internal scanning and diagnosis, aircraft/automobile control and safety, fully automated industrial control systems, chemical and biological hazard detection, and home security and entertainment systems, to name a few.

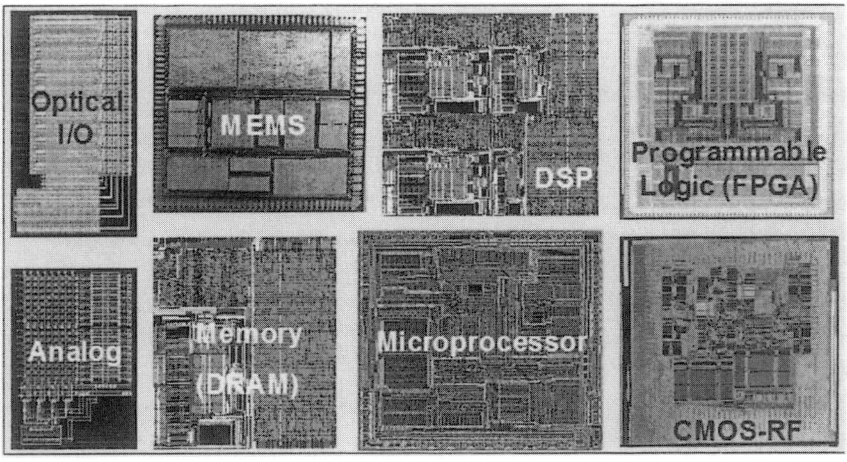

Figure 9.4. Schematic of a system-on-a-chip design.

There are several challenges to effective SoC designs. Large-scale integration of functionalities and disparate technologies on a single chip dramatically increases the chip area, which necessitates the use of numerous long global wires. These wires can lead to unacceptable signal transmission delays and increase the power consumption by increasing the total capacitance that needs to be driven by the gates. Also, integration of disparate technologies such as embedded DRAM, logic, and passive components in SoC applications introduces significant complexity in materials and process integration. Furthermore, the noise generated by the interference between different embedded circuit blocks containing digital and analog circuits becomes a challenging problem. Additionally, although SoC designs typically reduce the number of I/O pins compared to a system assembled on a printed circuit board (PCB), several high-performance SoC designs involve very high I/O pin counts, which can increase the cost/chip. Finally, integration of mixed signals and mixed-technologies on a single die requires novel design methodologies and tools, with design productivity being a key requirement.

9.2.5 3-D Architecture

3-D integration (schematically illustrated in Figure 9.5) to create multilayer Si ICs is a concept that can significantly improve deep submicron interconnect performance, increase transistor packing density, and reduce chip area and power dissipation [32]. Additionally, 3-D ICs can be very effective vehicles for large-scale on-chip integration of heterogeneous systems.

In the 3-D design architecture an entire (2-D) chip is divided into a number of logic and memory blocks which are arranged and allocated on separate layers of Si that are stacked on top of each other. Each Si layer in the 3-D structure may have its own dedicated or shared interconnect network. Each of these layers are connected together through vertical inter-layer interconnects (VILICs) and common global interconnects as shown in Figure 9.5. The 3-D architecture offers extra flexibility in system design, block placement and routing. For instance, blocks on a critical path can be placed as nearest vertical neighbors using multiple active layers. This would result in a significant reduction in *RC* delay, and can greatly enhance the performance of logic circuits. Also, the negative impact of deep submicron interconnects on VLSI design discussed earlier can be reduced significantly by replacing certain long global wires that realize the inter-block communications with short VILICs due to vertical placement of logic blocks.

Furthermore, the 3-D chip design technology offers the capability to build SoCs by placing heterogeneous circuits, such as different voltage ICs and performance requirements, in different layers. The 3-D integration would significantly alleviate many of the problems outlined in the previous section for

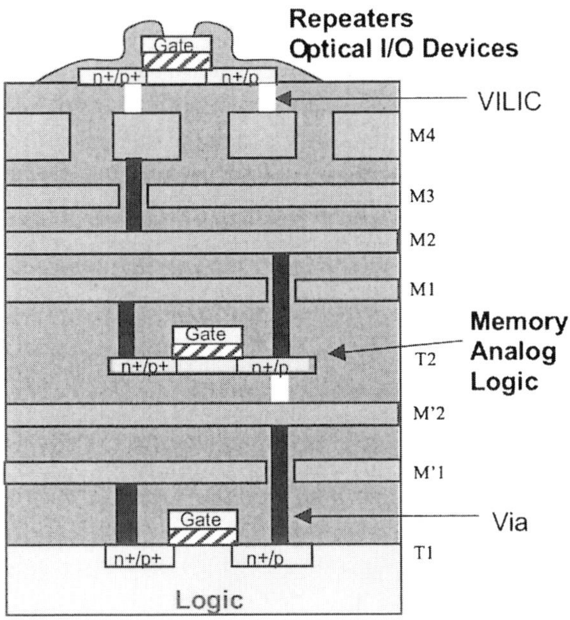

Figure 9.5. Schematic representation of 3-D integration with multilevel wiring network and VILICs. T1: first active layer device, T2: second active layer device, Optical I/O device: third active layer I/O device. M'1 and M'2 are for T1, M1 and M2 are for T2. M3 and M4 are shared by T1, T2, and the I/O device.

SoCs fabricated on a single Si layer. 3-D integration can reduce the wiring, thereby reducing the capacitance, power dissipation, and chip area and improve chip performance. It would also lower the I/O pin count, and therefore be an economically attractive option for building high-performance SoCs. Additionally, the digital and analog components in the mixed-signal systems can be placed on different Si layers thereby achieving better noise performance due to lower electromagnetic interference between such circuit blocks. From an integration point of view, mixed-technology assimilation could be made less complex and more cost effective by fabricating such technologies on separate substrates followed by physical bonding. Also, synchronous clock distribution in high-performance SoCs can be achieved by employing optical interconnects and I/Os at the topmost Si layer (as illustrated in Figure 9.5). 3-D integration of optical and CMOS circuitry have been demonstrated in the past [33]. A schematic diagram of a 3-D SoC is shown in Figure 9.6 with logic circuitry, distributed memory (SRAM) blocks (to reduce access time and enhance system performance), high-density DRAMs, analog/RF, and optical I/Os on different active layers.

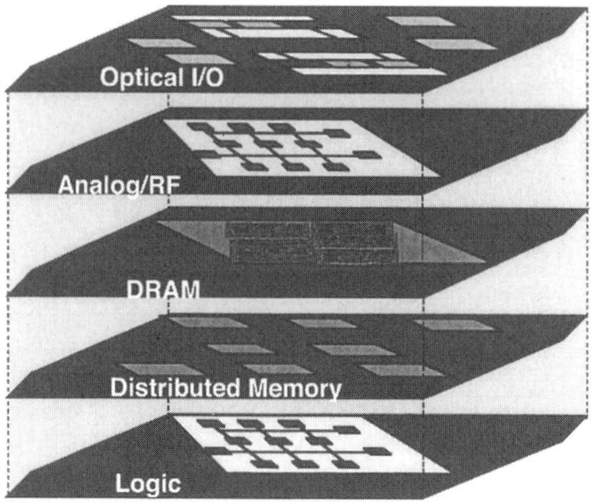

Figure 9.6. Schematic of a 3-D chip showing integrated disparate technologies.

9.3 Scope of This Study

A 3-D solution at first glance seems an obvious answer to the interconnect delay problem. Since chip size directly affects the interconnect delay, therefore by creating a second active layer, the total chip footprint can be reduced, thus shortening critical interconnects and reducing their delay. However, in today's microprocessors, the chip size is not just limited by the cell size, but also by how much metal is required to connect the cells. The transistors on the silicon surface are not actually packed to maximum density, but are spaced apart to allow metal lines above to connect one transistor or one cell to another. The metal required on a chip for interconnections is determined not only by the number of gates, but also by other factors such as architecture, average fan-out, number of I/O connections, routing complexity etc. Therefore, it is not obvious that by using a 3-D structure, the chip size will be reduced.

In this chapter the possible effects of 3-D integration of large logic circuits on key metrics such as chip area, power dissipation and performance has been quantified by modeling the optimal distribution of the metal interconnect lines. To better understand how a 3-D design will affect the amount of metal wires required for interconnections, a stochastic approach has been applied for estimating wiring requirements derived for a single layer in [34] and has been modified for 3-D ICs to quantify effects on interconnect RC delay. Unlike previous work [35], wire-pitch limited chips are considered. The study is based on Rent's Rule [36] that relates the number of I/O pins to the number of blocks in a processor. Rent's rule is suitably modified to be applicable to all the possible 3-D scenarios.

Present and future microprocessor technology trends incorporate on-chip memory alongside logic. The analysis described in this chapter includes the effect of on-chip memory on chip size and over-all interconnect performance by considering the Si real estate used by the memory components. Converse to the logic components which are assumed wire-pitch limited, memory is assumed device size limited. The modeling of chip size and interconnect performance described, therefore, is a combination of the dominating logic wiring requirement and the memory size requirement.

The results obtained in Section 9.4 indicate that when critically long metal lines that occupy lateral space are replaced with effective VILICs to connect logic blocks on different Si layers, a significant chip area reduction can be achieved. VILICs are found to be ultimately responsible for this improvement. The assumption made here is that it is possible to divide the microprocessor into different blocks such that they can be placed on different levels of active silicon.

In Section 9.5 important concerns in 3-D ICs such as power dissipation, have been analyzed. It is demonstrated that advancement in IC cooling technology will be necessary for maximizing 3-D circuit performance.

Throughout this work no differences were assumed in the performance or the properties of the individual devices on any layer. Also the treatment is independent of the 3-D technology used. However, even if the properties of the devices on the upper Si layers are different, these layers can be used for memory devices or repeaters. Some of these applications are discussed in Section 9.6. Finally, in Section 9.7, various technology options for fabricating 3-D ICs have been outlined.

The analysis described in this chapter represents a significant improvement over the work presented by Banerjee, et al.[117]. The effects of memory are incorporated in the performance modeling and an analytical thermal model is introduced which exhibits significant agreement with finite element analysis. These have the effect of providing much more realistic insights into the future of 3D ICs.

9.4 Area and Performance Estimation of 3-D ICs

We now present a methodology, which can be used to provide an initial estimate of the area and performance of high-speed logic circuits fabricated using multiple silicon layer IC technology. This approach is inspired by [37], where a delay and area estimation scheme for conventional (2-D) IC technologies was presented. The approach is primarily based on the empirical relationship known as Rent's Rule [36]. Rent's Rule correlates the number of signal input and output (I/O) pins, T, to the number of gates, N, in a random logic network and is given by the following expression:

$$T = kN^p \tag{9.3}$$

Here k and p denote the average number of fanout per gate and the degree of wiring complexity (with $p = 1$ representing the most complex wiring network) respectively, and are empirically derived as constants for a given generation of ICs. The underlying assumption of this methodology is based upon the recursive application of Rent's Rule throughout an entire logic system.

9.4.1 2-D And 3-D Wire-Length Distributions

The wire-length distribution, $I(l)$ (or the cumulative interconnect distribution function) gives the total number of interconnects that have length less than or equal to l, and is defined as,

$$I(l) = \int_1^l i(x)dx \tag{9.4}$$

where x is a variable of integration and l is the length of the interconnect in gate pitches. The derivation of the wire-length distribution in an IC is based on Rent's Rule. To derive the wire-length distribution, $I(l)$ of an integrated circuit, the latter is divided up into N logic gates, where N is related to the total number of transistors, N_t, in an integrated circuit by $N = N_t/\phi$, where ϕ is a function of the average fan-in (f.i.) and fan-out (f.o.) in the system [38]. The gate pitch is defined as the average separation between the logic gates and is equal to $\sqrt{A_c/N}$ where A_c is the logic area of the chip.

We first review the stochastic approach used for estimating the wire-length distribution of a 2-D chip and then modify it for 3-D chips. In order to derive the complete wire-length distribution for a chip, the stochastic wire-length distribution of a single gate must be calculated. The methodology is illustrated in Figure 9.7. The number of connections from the single logic gate in Block A to all other gates that are located at a distance of l gate pitches is determined using Rent's Rule. The gates shown in Figure 9.7 are grouped into three distinct but adjacent blocks (A, B, and C), such that a closed single path can encircle one, two, or three of these blocks. The number of connections between Block A and Block C is calculated by conserving all I/O terminals for blocks, A, B, and C, which states that terminals for blocks A, B, and C are either inter-block connections or external system connections.

Hence, applying the principle of conservation of I/O pins to this system of three logic blocks shown in Figure 9.7 gives,

$$T_A + T_B + T_C = T_{A-to-C} + T_{A-to-B} + T_{B-to-C} + T_{ABC} \tag{9.5}$$

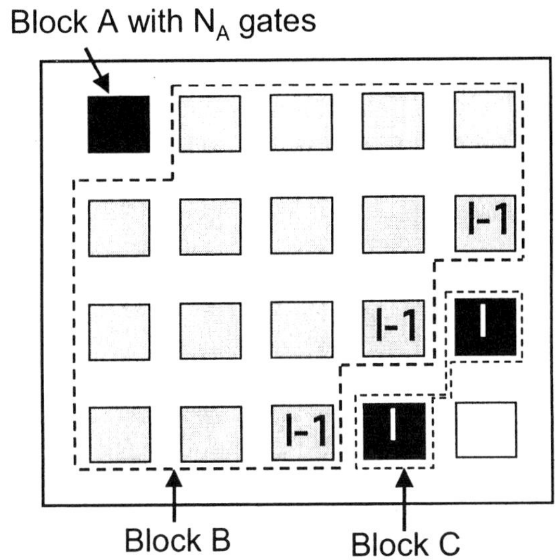

Figure 9.7. Schematic view of logic blocks used for determining wire-length distribution (adopted from [34])

where T_A, T_B, and T_C are the number of I/Os for blocks A, B, and C respectively. T_{A-to-C}, T_{A-to-B}, and T_{B-to-C} are the numbers of I/Os from block A to C, block A to B, and from block B to C respectively. T_{ABC} represents the number of I/Os for the entire system comprising of all the three blocks. From conservation of I/Os, the number of I/Os between adjacent blocks A and B, and between adjacent blocks B and C can be expressed as,

$$T_{A-to-B} = T_A + T_B - T_{AB} \tag{9.6}$$

$$T_{B-to-C} = T_B + T_C - T_{BC} \tag{9.7}$$

Substituting (9.6) and (9.7) in (9.5) gives,

$$T_{A-to-C} = T_{AB} + T_{BC} - T_B - T_{ABC} \tag{9.8}$$

Now the number of I/O pins for any single block or a group of blocks can be calculated using Rent's Rule. If we assume that N_A, N_B, and N_C are the number of gates in blocks A, B, and C respectively, then it follows that,

$$T_B = k\,(N_B)^p \tag{9.9}$$

$$T_{AB} = k\,(N_A + N_B)^p \tag{9.10}$$

$$T_{BC} = k\,(N_B + N_C)^p \tag{9.11}$$

$$T_{ABC} = k\,(N_A + N_B + N_C)^p \tag{9.12}$$

where $N = N_A + N_B + N_C$. Substituting (9.9)-(9.12) in (9.8) gives,

$$T_{A\text{-}to\text{-}C} = k\left[(N_A + N_B)^p - (N_B)^p + (N_B + N_C)^p - (N_A + N_B + N_C)^p\right] \tag{9.13}$$

The number of interconnects between Block A and Block C (I_{A-to-C}) is determined using the relation,

$$I_{A-to-C} = \alpha k\,(T_{A-to-C}) \tag{9.14}$$

Here α is related to the average fan-out (f.o.) by,

$$\alpha = \frac{f.o}{1 + f.o} \tag{9.15}$$

Equation (9.14) can be used to calculate the number of interconnects for each length l in Figure 9.7 in the range from one gate pitch to $2\sqrt{N}$ gate pitches, to generate the complete stochastic wire-length distribution for the logic gate in Block A. In the following step Block A is removed from the system of gates for calculating the remaining wiring distribution in order to prevent multiplicity in interconnect counting. The same process is repeated for all gates in the system. Finally, the wire-length distributions for the individual gates are superimposed to generate the total wire-length distribution of the chip with N gates. J. Davis et al. developed a closed form analytical expression of the wire-length distribution for a 2-D IC [34], which can be expressed as,

$$I\,(l) = I_{total}P\,(l) \tag{9.16}$$

where I_{total} is the total number of interconnects in a system derived form Rent's Rule as,

$$I_{total} = \alpha k N\left(1 - N^{p-1}\right) \tag{9.17}$$

Here $P(l)$ is the cumulative distribution function that describes the total probability that a given interconnect length is less than or equal to l, and is given by the following expressions,

$$P\,(l) = \frac{1}{2N\,(1 - N^{p-1})}\Gamma\left(\frac{l^{2p} - 1}{6p} + 2\sqrt{N}\frac{-l^{2p-1} + 1}{(2p-1)} - N\frac{-l^{2p-2} + 1}{(p-1)}\right) \tag{9.18}$$

for $I \leq l \leq \sqrt{N}$, and

$$P(l) = \frac{\Gamma}{2N\left(1 - N^{p-1}\right)} \left(\begin{array}{l} \frac{N^{2p-1}}{6p} + 2\sqrt{N}\frac{-N^{2p-1}+1}{2p-1} - N\frac{-N^{2p-2}+1}{p-1} \\ +\frac{1}{3}\left(\begin{array}{l} -8N^{3/2}\frac{-l^{2p-3}+N^{p-(3/2)}}{2p-3} + 6N\frac{-l^{2p-2}+N^{p-1}}{p-1} \\ -6\sqrt{N}\frac{-l^{2p-1}+N^{p-(1/2)}}{2p-1} + \frac{-l^{2p}+N^{p}}{2p} \end{array} \right) \end{array} \right) \tag{9.19}$$

for $\sqrt{N} \leq l \leq 2\sqrt{N}$. The factor Γ is defined by,

$$\Gamma = \frac{2N\left(1 - N^{p-1}\right)}{\left(-N^{p}\frac{1+2p-2^{2p-1}}{p(2p-1)(p-1)(2p-3)} - \frac{1}{6p} + \frac{2\sqrt{N}}{2p-1} - \frac{N}{p-1}\right)} \tag{9.20}$$

Substituting (9.17) - (9.20) in (9.16) gives the closed form expressions for the total wire length distribution as follows,

$$I(l) = \frac{\alpha k}{2}\Gamma\left(\frac{l^{2p}-1}{6p} + 2\sqrt{N}\frac{-l^{2p-1}+1}{(2p-1)} - N\frac{-l^{2p-2}+1}{(p-1)}\right) \tag{9.21}$$

for $I \leq l \leq \sqrt{N}$, and

$$I(l) = \frac{\alpha k}{2}\Gamma \left(\begin{array}{l} \frac{N^{2p-1}}{6p} + 2\sqrt{N}\frac{-N^{2p-1}+1}{2p-1} - N\frac{-N^{2p-2}+1}{p-1} \\ +\frac{1}{3}\left(\begin{array}{l} -8N^{3/2}\frac{-l^{2p-3}+N^{p-(3/2)}}{2p-3} + 6N\frac{-l^{2p-2}+N^{p-1}}{p-1} \\ -6\sqrt{N}\frac{-l^{2p-1}+N^{p-(1/2)}}{2p-1} + \frac{-l^{2p}+N^{p}}{2p} \end{array} \right) \end{array} \right) \tag{9.22}$$

The simple use of Rent's Rule above applies to 2-D IC's and requires adaptation for a valid application to 3-D IC's. For the case of 3-D ICs, different blocks can be physically placed on different silicon layers and connected to each other using VILICs. The area saving by using VILICs can be computed by modifying Rent's rule suitably. For simplicity, we illustrate the case where two silicon layers are available. The extension to more than two-layer case is straightforward. An N gate IC design is divided into two $N/2$ gate blocks (Figure 9.8a). It is assumed that the routing algorithm and overall logic style is the same for both layers. This ensures that Rent's constant, k, and Rent's exponent, p, are the same for both layers. Applying Rent's rule to both layers, we have,

$$T_1 = T_2 = k\left(\frac{N}{2}\right)^{p} \tag{9.23}$$

Here T_1 and T_2 are I/Os of each layer, p is Rent's exponent and k is the average number of I/Os per gate. Moreover, for the overall design,

$$T = kN_p = T_1 + T_2 - T_{\text{int}} \tag{9.24}$$

where T is the number of I/Os of the entire design and T_{int} represents the total number of I/O ports connecting the two layers as shown in Figure 9.8a. Each

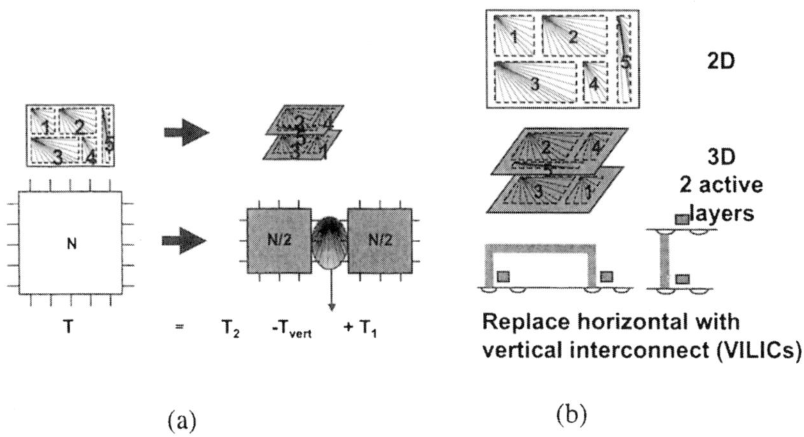

Figure 9.8. Schematic to illustrate a) conservation of total number of external I/O ports for maintaining constant functionality of chip, and b) two-layer 3-D chip with long horizontal interconnects replaced by short and vertical (VILICs) interconnects.

layer will have $(T_{int}/2)$ dedicated I/O ports for connection to the other layer. Extending this analysis to n-layer 3-D IC's, the number of external I/O ports per layer, i, is given by

$$T_{ext,i} = T_i - \frac{T_{\text{int}}}{2} = k2^{p-1}\left(\frac{N}{2}\right)^p \quad i = 1, 2 \tag{9.25}$$

Comparing this with Rent's equation, (9.3), we find that for each layer, the Rent's exponent is unchanged, the number of gates is halved and the effective number of I/Os per gate used for connecting other gates on the same layer is given by

$$k_{eff,\text{int}} = k\left(1 - 2^{p-1}\right), \tag{9.26}$$

and similarly the effective number of I/Os per gate used to connect to gates on the other active layer is given by,

$$k_{eff,ext} = k2^{p-1} \tag{9.27}$$

Now the wire-length distribution analysis discussed above can be extended to 3-D IC's using the modified values of k for each layer. Figure 9.9 shows the wire-length distributions for 2-D and 3-D ICs. It can be observed that the wiring requirement is significantly reduced for the global wires in 3-D ICs. This is due to the fact that these long wires have been converted to short VILICs as schematically illustrated in Figure 9.8b.

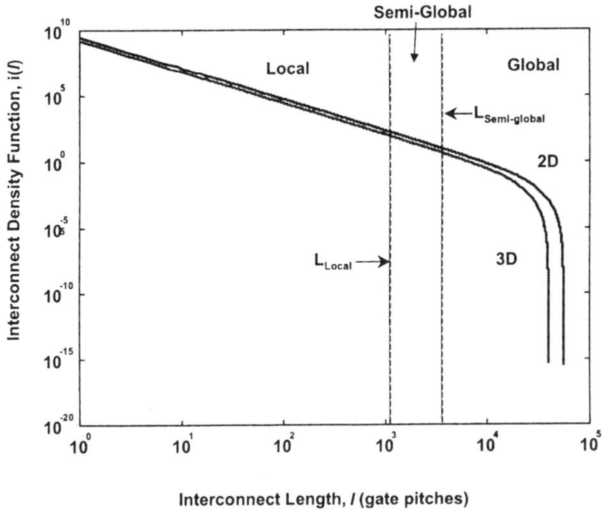

Figure 9.9. Wire-length distributions for the 2-D and 3-D ICs shown in Figure 9.8. 3-D significantly reduces requirement for longest wires. Metal tiers determined by L_{Loc} and $L_{Semi-global}$ boundaries.

9.4.2 Estimating 2-D and 3-D Chip Area

The analyses described in this work are performed on integrated circuits that are wire-pitch limited in size. Using a three tier interconnection structure (local, semi-global and global, illustrated in Figure 9.10), the semi-global tier pitch that minimizes the wire limited chip area was determined. The maximum interconnect length on any given tier was determined by the interconnect delay criterion (It is assumed that $t_{delay\_max} = 0.9T$ for global wires, and $t_{delay\_max} = 0.25T$ for semi-global and local wires, with T as the clock period).

The logic area of the chip is determined by the total wiring requirement. In terms of gate pitch, the total area required by the interconnect wiring can be expressed as:

$$A_{required} = \sqrt{\frac{A_c}{N}} \left(p_{loc} L_{total\_loc} + p_{semi} L_{total\_semi} + p_{glob} L_{total\_glob}\right) \tag{9.28}$$

where A_c is the logic chip area, N is the number of gates, p_{loc} is the local pitch, p_{semi} is the semiglobal pitch, p_{global} is the global pitch, L_{total_loc} is the total length of the local interconnects, L_{total_semi} is the total length of the semiglobal interconnects and L_{total_glob} is the total length of the global interconnects. The total interconnect length for any tier can be found by integrating

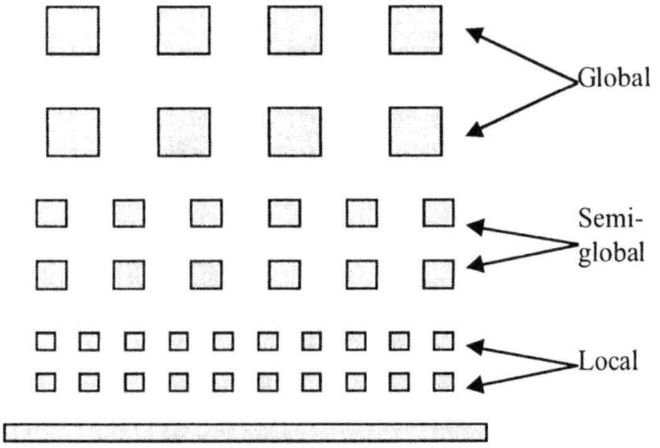

Figure 9.10. Schematic of a three-tier interconnection structure.

the wire-length distribution within the boundaries that define the tier [see Figure 9.9 where broken vertical lines define the boundaries]. Hence it follows that,

$$L_{total\_loc} = \chi \int_{1}^{L_{loc}} li(l)dl \tag{9.29}$$

$$L_{total\_semi} = \chi \int_{L_{loc}}^{L_{semi}} li(l)dl \tag{9.30}$$

$$L_{total\_glob} = \chi \int_{L_{semi}}^{2\sqrt{N}} li(l)dl \tag{9.31}$$

where χ is a correction factor that converts the point-to-point interconnect length to wiring net length (using a linear net model,$\chi = 4/(fo + 3)$. The boundaries shown in Figure 9.9 represent the length of the longest wire for each tier, L_{loc} for the local, L_{semi} for the semiglobal and L_{global} for the global tier. The length of the longest wire, L, and the pitch, p_w, for an arbitrary tier are related by the following expression [37]:

$$\frac{\beta}{f_c} = 4\frac{1.1\rho\varepsilon_r\varepsilon_o 6.08}{p_w^2}\frac{A_c}{N}L^2 + L\sqrt{\frac{A_c}{N}}\frac{\sqrt{\varepsilon_r}}{c_o} \tag{9.32}$$

where β is the maximum delay fraction of clock period, f_c is the clock frequency, ρ is the resistivity of the metal, c_0 is the speed of light in free space and ε_r is the relative permittivity of the dielectric material. Equation (9.32) can be re-arranged to solve for wire pitch or the length of the longest interconnect.

The expressions for p_{global}, L_{semi} (which is a function of p_{semi}) and L_{loc} are given by [37],

$$p_{glob} = 2\sqrt{\frac{A_c}{N}} L_{glob} \sqrt{\frac{1.1\rho\varepsilon_r\varepsilon_0 6.08}{\left(\frac{\beta_{glob}}{f_c} - \sqrt{\frac{A_c}{N}} L_{glob} \frac{\sqrt{\varepsilon_r}}{c_0}\right)}} \tag{9.33}$$

$$L_{semi} = \frac{p_{semi}^2}{4\,(2.2)\,\rho\varepsilon_r\varepsilon_0 6.08} \sqrt{\frac{N}{A_c}} \left[-\frac{\sqrt{\varepsilon_r}}{c_0} + \sqrt{\frac{\varepsilon_r}{c_0^2} + 4\beta_{semi} \frac{4.4\rho\varepsilon_r\varepsilon_0 6.08}{p_{semi}^2 f_c}} \right] \tag{9.34}$$

$$L_{loc} = \frac{p_{loc}^2}{4\,(2.2)\,\rho\varepsilon_r\varepsilon_0 6.08} \sqrt{\frac{N}{A_c}} \left[-\frac{\sqrt{\varepsilon_r}}{c_0} + \sqrt{\frac{\varepsilon_r}{c_0^2} + 4\beta_{loc} \frac{4.4\rho\varepsilon_r\varepsilon_0 6.08}{p_{loc}^2 f_c}} \right] \tag{9.35}$$

Here p_{loc} is assumed constant and equal to twice the minimum feature size. L_{global} is also assumed constant and equal to the chip die edge. Equation (9.34) for L_{semi} results in a non-unique set of possible solutions for A_c. The wire limited chip area (A_c) is calculated based on the condition that the total required wiring area ($A_{required}$) is equal to the total available area ($A_{available}$) in a multilevel network, hence it follows that,

$$A_{available} = A_c e_w n_{levels} = A_{required} \tag{9.36}$$

where e_w is the wiring efficiency factor that accounts for router efficiency and additional space needed for power and clock lines, and n_{levels} is the number of metal levels available for the multilevel network. For each possible solution of (9.34), new boundaries representing L_{loc} and L_{semi} are used with the wire-length distribution to find the new total area required by the interconnect wiring. From the total area required by the wiring, the chip area is estimated by dividing interconnects among the required number of metal layers. The resulting chip areas are then plotted as a function of p_{semi} normalized to the constant local pitch.

Effect of Memory on Chip Area. The process of quantifying the effect of on-chip memory on chip size and performance is fairly straightforward. From ITRS'99 [1] the fraction of the total number of transistors dedicated to memory is provided. Assuming on-chip memory remains device size limited across all technology nodes, the total area occupied by memory can be determined. In this analysis, it is assumed that memory is partitioned in such a manner that it appears distributed uniformly across the entire microprocessor chip. The distribution of logic and memory devices is therefore assumed homogeneous. If, instead, on-chip memory is assumed physically separate from logic, this will have minimal effect on the performance since the logic wiring would remain confined to the logic portion of the IC. Furthermore, our assumption is fairly reasonable given the futuristic trend towards distributed microprocessor designs. The average effect of such an assumption would be to effectively increase the apparent gate pitch and hence the total interconnect length.

9.4.3 Two Active Layer 3-D Circuit Performance

The above analysis is used to compare area and delay values for 2-D and 3-D ICs. The availability of additional silicon layers gives the designer extra flexibility in trading off area with delay. It is assumed that through technological advances resistivity of Cu will be maintained at the bulk value. A number of different cases are discussed below.

Chip Area Minimization with Fixed Wire Delay. The model is applied to the microprocessor example shown in Table 9.1 for the 50 nm technology node [1] for the two cases where all gates are in a single layer (2-D) and where the gates are equally divided between two layers (3-D). In this calculation VILICs are assumed to consume negligible area, interconnect line width is assumed to equal half the metal pitch at all times, and the total number of metal layers for 2-D and 3-D case was conserved.

Table 9.1. Microprocessor example (ITRS'99 for 50 nm technology node).

PHYSICAL PARAMETER	VALUE
Number of Logic Transistors, N	93.6 million
Number of Memory Transistors	86.4 million
Rent's Exponent, p	0.6
Minimum Feature Size, F	50 nm
Max number of wiring levels, n_{max}	9
Metal Resistivity, Copper	$1.673 \times 10^{-6}\Omega$-cm
Dielectric Constant, Polymer	$\epsilon_r = 2.5$
Wiring Efficiency Factor	0.4

An optimization of the chip area with respect to the wiring pitches and the total interconnect length at each tier was carried out. Figure 9.11 shows the variation in chip area with the normalized semi-global tier pitch for a fixed operating frequency of 3 GHz. The curve for the 3-D case has a minimum similar to the one obtained for the 2-D case. It can be observed that the optimum chip area for the 3-D case is 35% smaller than that of the 2-D case. Moreover, since the total wiring requirement is reduced (as shown in Figure 9.9), the semi-global tier pitch is reduced for the 3-D chip. This reduction in the semi-global pitch increases the line resistance and the line-to-line capacitance per unit length. Hence the same clock frequency, i.e., the same interconnect delay, is maintained by reducing the chip size.

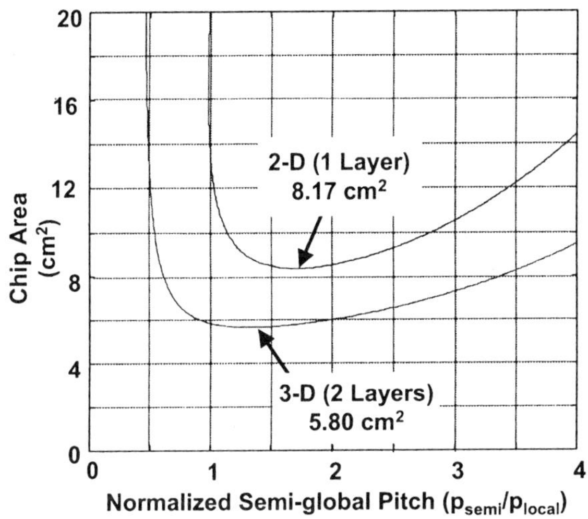

Figure 9.11. Wire-limited chip area versus normalized semi-global pitch (semi-global pitch/local tier pitch) for 2-D and 3-D ICs at a fixed operating frequency of 3 GHz. As the normalized semi-global-pitch reduces, wires are rerouted to the global tiers, which have bigger pitch, and hence the chip area increases. Note that the estimated 2-D chip area of 8.17 cm^2 is also projected by ITRS'99 for the 50 nm node. The number of metal layers for 2-D and 3-D ICs is nine (3 per tier).

Optimization of Both Chip Area and Interconnect Delay. As a second alternative, interconnect delay can be improved by increasing the wiring pitch, which causes a reduction in resistance and line-to-line capacitance per unit length. Two scenarios are considered: (a) global pitch is increased to match the global pitch for the 2-D case (b) global pitch is increased to match the chip area (footprint) for the 2-D case. Table 9.2 shows that performance can be increased by 63% for case (b) where vertical inter-layer interconnects are assumed to use up minimal area. Note that the delay requirement sets a maximum value of interconnect length on any given tier. Therefore, as interconnect lengths are increased, lines which exceeds this maximum length criterion for that particular tier need to be rerouted on upper tiers.

Moreover, an optimization similar to Figure 9.11 can be carried out for different operating frequencies. Figure 9.12 illustrates how the optimum semi-global pitch increases for higher operating frequencies. Also, as the semi-global tier pitch increases, chip area and therefore, interconnect length also

Table 9.2. Summary of delay performance improvement for 3-D ICs. The horizonal ILICs differ from the vertical ILICs in that they consume lateral area.

2-Layer Description of Inter-layer Interconnects(ILICs)	Delay Performance Improvement
Horizonal ILICs, equal global pitch	10 %
Horizonal ILICs, equal chip area	17 %
Vertical ILICs, equal global pitch	33 %
Vertical ILICs, equal chip area	63 %

increases. However, it can be observed from Figure 9.12 that the increase in chip area still remains well below the area required for the 2-D case. Figure 9.12 also helps define a maximum performance 3-D chip, which is a chip with the same (footprint) area (8.17 cm^2) as the corresponding 2-D chip, which can be obtained by increasing the semi-global pitch beyond that for the 4 GHz case.

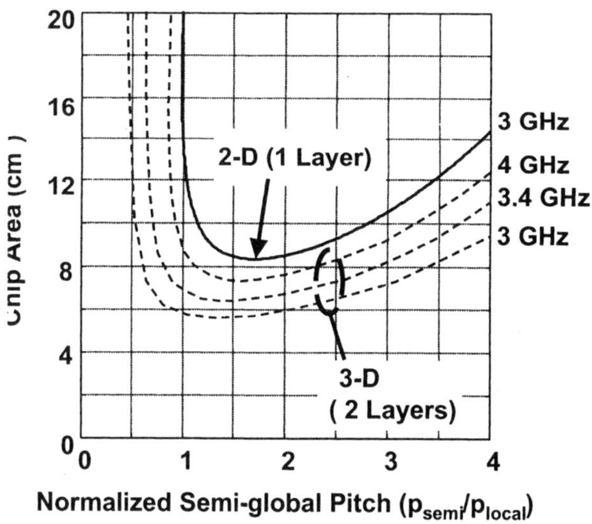

Figure 9.12. 3-D chip operating frequency(performance) increases with increases in semi-global wiring pitch. Chip area also increases but remains below the 2-D chip area.

Beyond the maximum performance point for the 3-D chip in Figure 9.12 (normalized semiglobal pitch = 1.75), the performance gain becomes increasingly smaller in comparison to the decrease in performance resulting from the increase in chip area or interconnect delay. This eventually saturates the reduction in the overall interconnect delay, and therefore, as shown in Figure

9.13, the clock frequency saturates. Furthermore, as the semi-global pitch is increased beyond the maximum performance point, semiglobal wires need to be rerouted on the global tiers, which eventually leads to overcrowding of the global tier. Any further increases in the wiring density in the global tier forces a reduction in the global pitch as shown in Figure 9.14.

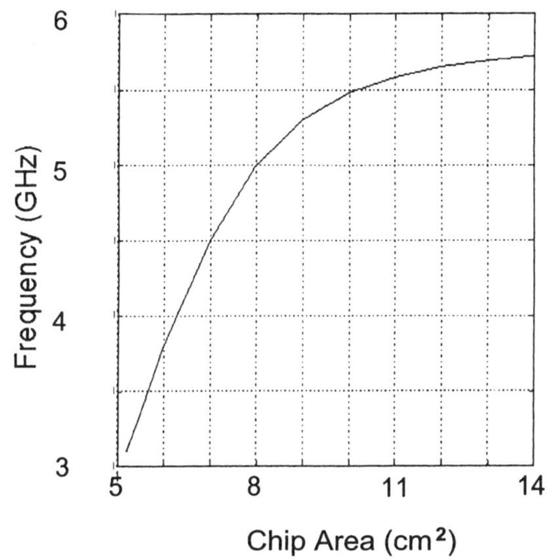

Figure 9.13. Performance improvement with increasing chip area for a two-layer 3-D IC. Chip area is increased due to increasing wire pitch.

The analysis presented so far was for a 50 nm two Si layer 3-D technology where the number of metal layers was preserved (in comparison to the 2-D case). In the next two sections, we extend this analysis to study the effect of more than two Si layers and also the effect of increasing the number of available metal layers.

9.4.4 Effect of Increasing Number of Silicon Layers

3-D technologies providing more than two active layers have also been considered. As the number of silicon layers increases beyond two, the assumption that all inter-layer interconnects (ILICs) are vertical and consume negligible area becomes less tenable. For this particular example it is assumed that 90% of all ILICs are horizontal (c.f. Table 9.2). The area used up by these horizontal ILICs can be estimated from the their total length and pitch. As shown in Figure 9.15, the decrease in interconnect delay becomes progressively smaller as the number of active layers increases. This is due to the fact that area re-

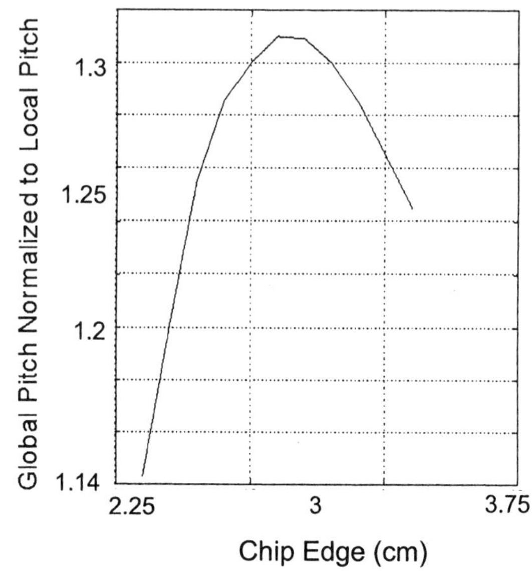

Figure 9.14. As the chip size increases due to increasing wire pitch, interconnects are re-routed to higher tiers. The global tier becomes over-crowded for large chip are as and global pitch starts to decrease.

quired by ILICs begins to offset any area saving due to increasing the number of active layers.

9.4.5 Effect of Increasing the Number of Metal Layers

In the above analysis, the total number of metal layers for 2-D and 3-D case was conserved. However, it is likely that there are local and semi-global tiers associated with every active layer, and a common global tier is used. This would result in an increase in the total number of metal layers for the 3-D case. The effect of using 3-D ICs with constant metal layers discussed earlier and the effect of employing twice the number of metal layers as in 2-D are summarized in Figure 9.16 for various technology nodes as per [1]. It can be observed that by using twice the number of metal layers the performance of the 3-D chip can be improved by an additional 35% (for the 50 nm node) as compared to the 3-D chip with same total number of metal layers as in 2-D. Figure 9.16 also shows the impact of moving only the repeaters to the second Si layer. It can be observed that a performance gain of 9% is achieved for the 50 nm node. The gate delay the interconnect delay (with repeaters) for the 2-D chip are identical to that shown in Figure 9.1 and have been included in this figure for convenience of comparison. Finally, it can also be observed that for

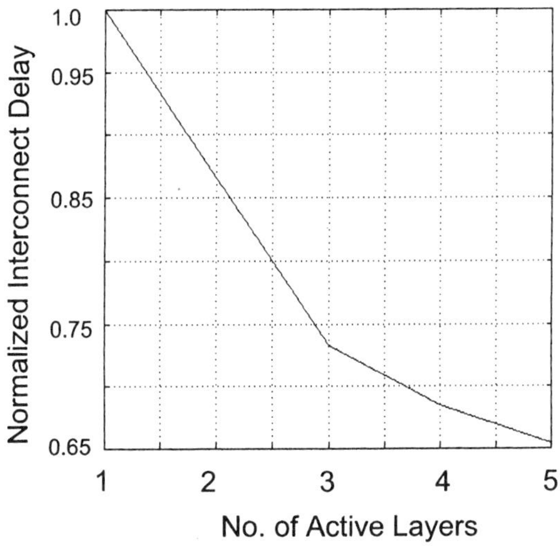

Figure 9.15. Interconnect delay normalized to single layer delay as a function of the number of active Si layers shown for 50 nm node. The VILICs are assumed to consume lateral area.

more aggressive technologies, the decrease in interconnect delay from 2-D to 3-D case is even more impressive.

9.4.6 Optimization of Interconnect Distribution

In estimating chip area, the metal requirement is calculated from the obtained wire-length distribution. The total metallization requirement is appropriately divided among the available metal layers in the corresponding technology. Thus in the example shown in Figure 9.11, each tier, the local, the semi-global and the global has three metal layers. The resulting area of the most densely packed tier, the local tier in this example, determines the chip area.

Consequently, higher tiers are routed within a larger than required area. An optimization for this scenario is possible by re-routing some of the local wires on the semi-global tier and the latter on the global, without violating the maximum allowable length (or delay) per tier. This is achieved by reducing the maximum allowed interconnect length for the local and semi-global tiers (L_{Local} and $L_{Semi-global}$ in Figure 9.9) with varying fractions, w_1 and w_2, respectively. This is implicitly achieved by suitably reducing the parameter β in (9.34) and (9.35). Minimum chip area will be achieved when all the tiers are almost equally congested. The resulting calculations for chip area with optimized interconnect distribution for the 2-D IC analyzed in Figure 9.11 are

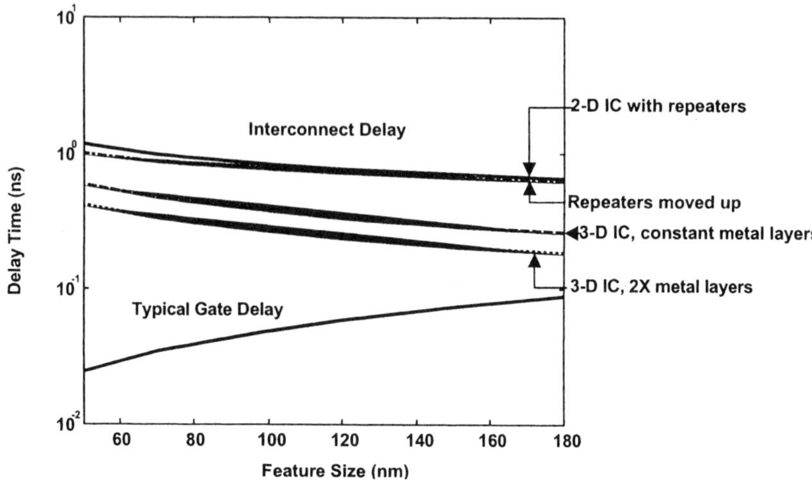

Figure 9.16. Comparison of interconnect delay as a function of technology nodes for 2-D and two-layer 3-D ICs. Moving repeaters to the upper active layer reduces interconnect delay by 9%. For the 50 nm node, 3-D (2 active layers with same number of interconnects as the 2-D chip) shows significant delay reduction (63%). Increasing the number of metal levels in 3-D reduces interconnect delay by a further 35%. This figure is based on the assumption that 3-D chip (footprint) area equals 2-D chip area.

shown in Figure 9.17. The 2-D chip area is seen to reduce by 9% as a result of this optimization. This wiring network optimization is also applied to 3-D ICs. The results are shown in Figure 9.18 where the 3-D chip area is reduced by 11%.

9.5 Challenges for 3-D ICs

9.5.1 Thermal Issues in 3-D ICs

An extremely important issue in 3-D ICs is heat dissipation [39, 40]. Thermal effects are already known to significantly impact interconnect/device reliability and performance in high-performance 2-D ICs [41, 42]. The problem is expected to be exacerbated by the reduction in chip size, assuming that same power generated in a 2-D chip will now be generated in a smaller 3-D chip, resulting in a sharp increase in the power density. Analysis of thermal problems in 3-D circuits is therefore necessary to comprehend the limitations of this technology, and also to evaluate the thermal robustness of different 3-D technology and design options.

It is well known that most of the heat energy generated in integrated circuits arises due to transistor switching. This heat is typically conducted through the silicon substrate to the package and then to the ambient by a heat sink.

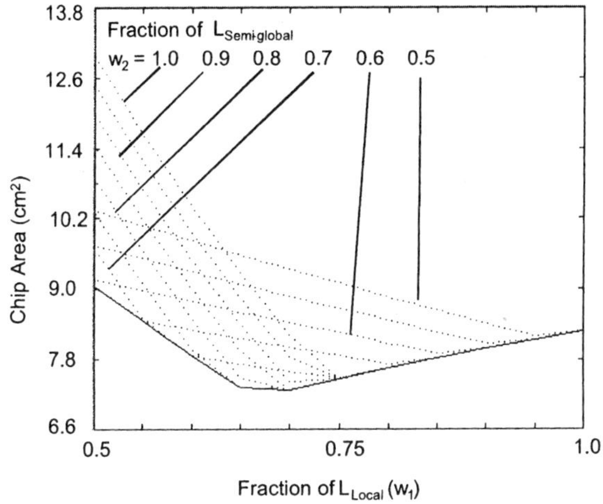

Figure 9.17. Chip area for 2-D IC with wiring network optimization. Solid line represents points of minimum area. (Based on ITRS'99 data for 50 nm node).

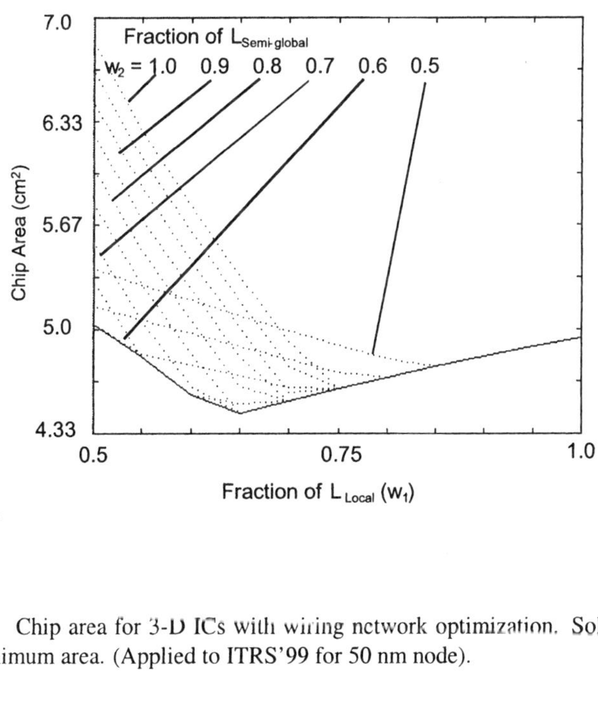

Figure 9.18. Chip area for 3-D ICs with wiring network optimization. Solid line represents points of minimum area. (Applied to ITRS'99 for 50 nm node).

With multi-layer device designs, devices in the upper layers will also generate a significant fraction of the heat. Furthermore, all the active layers will be insulated from each other by layers of dielectrics (LTO, HSQ, polyimide etc.) which typically have much lower thermal conductivity than Si [43, 44]. Hence, the heat dissipation issue can become even more acute for 3-D ICs and can cause degradation in device performance, and reduction in chip reliability due to increased junction leakage, electromigration failures, and by accelerating other failure mechanisms [41].

In this section, a detailed thermal analysis of high performance three dimensional (3-D) ICs is presented under various integration schemes. It is shown that under certain scenarios, 3-D ICs can actually lead to better thermal performance than planar (2-D) ICs. A complete thermal model including power consumption due to both transistors and interconnect joule heating from multiple strata is presented. With the effect of vias, as efficient heat dissipation paths, taken into account, this model provides more realistic temperature rise estimation for 3-D ICs. These vertical links and vias have much higher thermal conductivity and hence can effectively reduce the thermal resistance caused by the ILD layers. Ignoring the effect of these structures can result in overly pessimistic estimations predicting unacceptably high 3-D chip temperatures. Recently, a model has been developed to quantify the via thermal effect in 2-D structures [113, 114]. Here, this compact analytical model is applied to evaluate temperature rise in 3-D structures, incorporating via effect and power consumption due to both devices in active layers and interconnect joule heating. The results show excellent agreement with the 3-D finite element simulations using ANSYS. With the effect of vias, as efficient heat dissipation paths, taken into account, this model provides more realistic temperature rise estimation for 3-D ICs. It is shown that under certain scenarios, 3-D ICs can actually lead to better thermal performance than planar (2-D) ICs. Furthermore, tradeoffs between power, performance, chip area and thermal impact are evaluated.

Thermal Modeling Incorporating Via Effect. According to ITRS [1], although the average power density for high performance microprocessor will remain relatively constant throughout the technology nodes, current density in the wires will rise significantly (Figure 9.19). Furthermore, Cu resistivity will increase due to barriers, surface scattering and skin effect. Thus interconnect joule heating will become significant. In addition, low-k dielectrics with poor thermal conductivity (Figure 9.20), will not only lead to higher interconnect temperature in 2-D ICs but also impact the device temperature in various active layers in 3-D ICs (Figure 9.21). As seen in Figure 9.22, even for the case of two active layers, the ratio of thermal resistance caused by ILD layers (R_{ILD}) to required package (including glue layers, heat sink) thermal resistance (R_{pkg}) increases rapidly for future technology nodes. The required R_{pkg} is the maximum allowed value which gives the maximum junction temperature specified in ITRS. With multiple active layers, R_{ILD} will become the dominant factor to determine temperature rise in 3-D ICs. The analytical expression derived based on first principles, to evaluate temperature rise in 3-D structure is given below.

$$\overline{T_{Si\_N}} = T_{amb} + \sum_{m=1}^{N} \{ \underbrace{[\sum_{n=1}^{N_m} \frac{t_{ILD,mn}}{k_{ILD,mn} s_{mn}} \eta_{mn} (\sum_{i=n}^{N_m} j_{rms,mn}^{2} \rho H_{mn} + \sum_{j=m+1}^{M} 2\Phi_j)]}_{\text{Temperature rise caused by ILDs}} + \underbrace{R_m (\sum_{k=m}^{M} Q_k)}_{\text{Temp. Rise caused by PKG, glue layer, Si sub.}} \}$$

T_{amb}: ambient temperature,

M: number of strata,

N_m: number of metal levels in the mth stratum,

mn: the nth interconnect level in the mth stratum,

t_{ILD}: thickness of ILD,

k_{ILD}: thermal conductivity of ILD materials,

s: heat spreading factor [3],

η: via correction factor, $0 \leq \eta \leq 1$ [3],

j_{rms}: root-mean-square value of current density flowing in the wires,

ρ: electrical resistivity of metal wires,

H: thickness of metal wires,

Φ power density of active device layer,

Q_m: total power consumption of mth stratum, including power consumed by active layer and interconnect joule heating,

R: thermal resistance of glue layer and Si layer for each of the stratum, with R_1 representing the total thermal resistance of package, heat sink and Si substrate.

Via effect is incorporated in the expression by the via correction factor η ($0 \leq \eta \leq 1$), with $k_{ILD,eff} = \frac{k_{ILD}}{\eta}$, where $k_{ILD,eff}$ is the effective thermal conductivity of ILD with the help of via effect and k_{ILD} is the nominal thermal conductivity with via effect ignored. Power consumption due to both active (device) layers and interconnect joule heating are included. This expression can be better understood by comparing it with the Elmore-delay model

Elmore-Delay Analogy

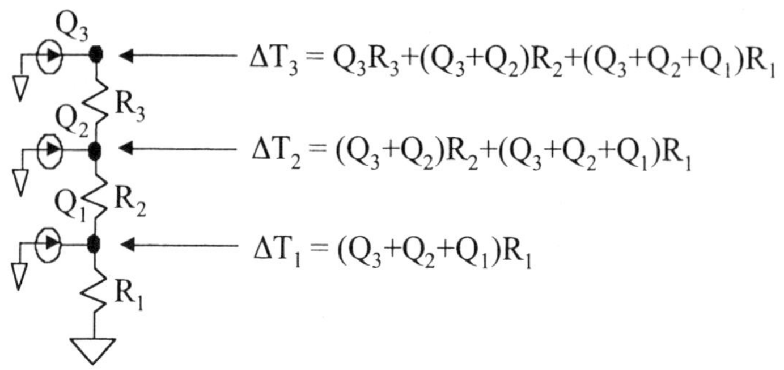

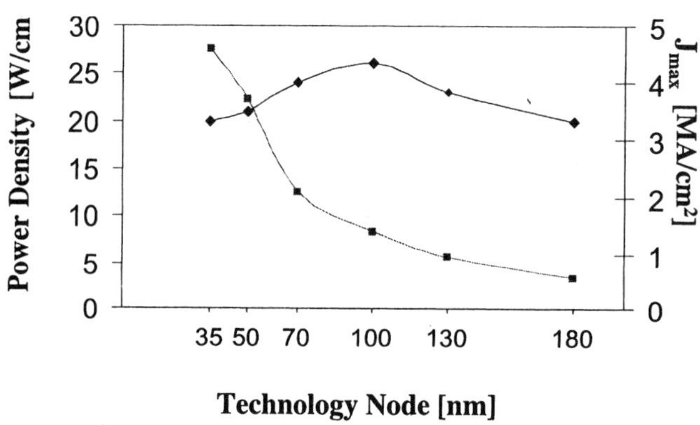

Figure 9.19. Trends of chip power density and interconnect J_{max} along technology nodes suggested by ITRS [1]. Chip power density is calculated by total power of the chip divided by chip size.

following an electrical-thermal analogy. The model is validated by comparing with full chip thermal simulation done using ANSYS in [115]. The two-layer 3-D structure with wafer bonding technique used for the validation is shown in Figure 9.23. Since it is too complicated to construct the 3-D structure with thousands of vias in ANSYS, the validation is done for the case where via effect is ignored ($\eta = 1$). The result from analytical expression shows excellent agreement with ANSYS (upper curve in Figure 9.24). However, the analytical model takes much less computation time and provides better insight. Furthermore, as shown in the lower curve of Figure 9.24, via effect greatly helps heat dissipation and the resultant temperatures are much lower. Therefore, it is crucial to include via effect for thermal analysis of 3-D ICs.

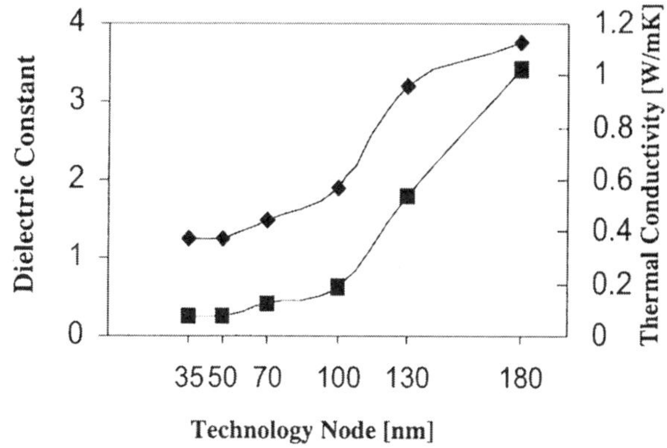

Figure 9.20. Both dielectric constant and thermal conductivity of ILD materials decrease with advanced technology nodes.

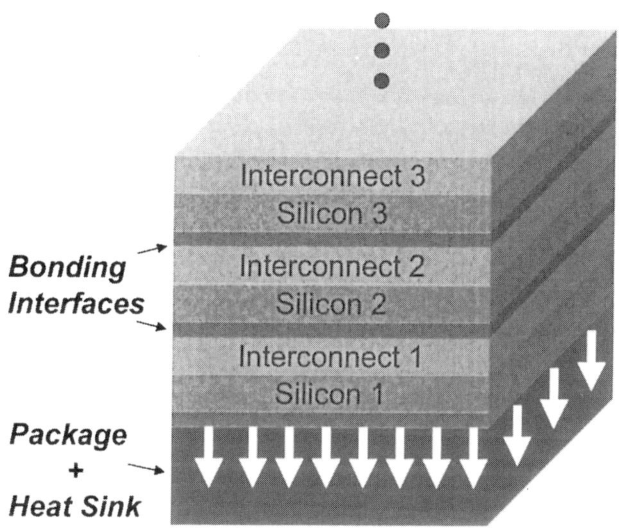

Figure 9.21. Schematic of multi-level 3-D IC with a heat sink attached to Si substrate.

Power Analysis of 3-D ICs. A power dissipation comparison between wire-pitch limited 2-D and 3-D ICs is presented in Table 9.3. All the calculations in this table are based on the 50 nm technology node and the thermal resistance of package is assumed to be 2.15 cm^2C/W from ITRS for 2-D ICs at the 50 nm node. The dynamic power dissipation components considered are due to logic; local, semi-global and global interconnects; clock distribution and repeaters and are calculated using $P_{Dynamic} = 1/2\alpha C V_{dd}^2 f_c$ where

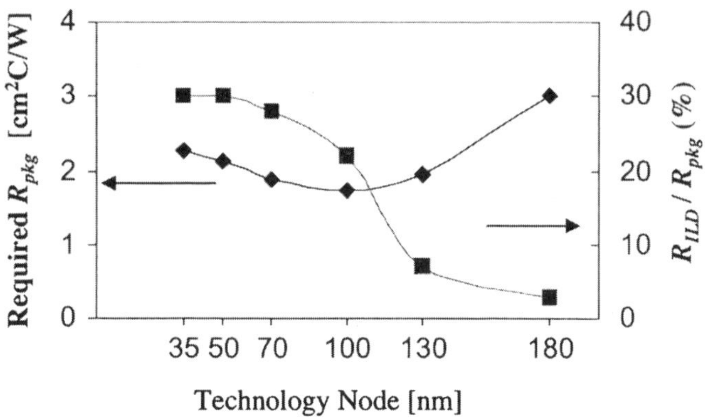

Figure 9.22. The required package thermal resistance, R_{pkg}, to achieve the maximum junction temperature specified in ITRS and the ratio of R_{ILD} and R_{pkg} vs. technology nodes.

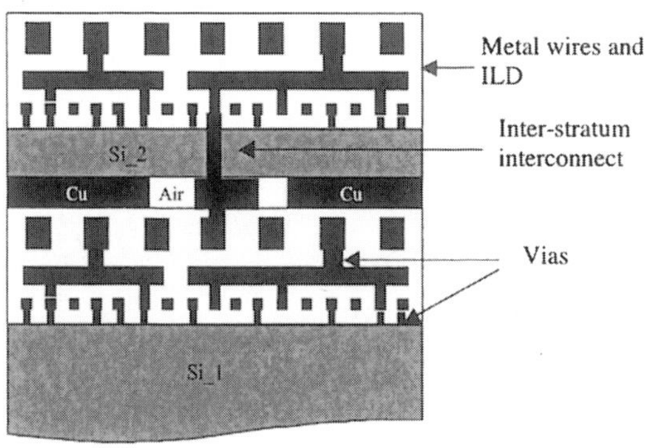

Figure 9.23. Schematic of 3-D structure fabricated by wafer bonding using Cu pad thermo-compression

α is the activity factor (assumed to be 0.1), V_{dd} is the supply voltage obtained from ITRS, f_c is the operating frequency and C is capacitance. Other power dissipating components include memory, I/O pads and static components such as leakage and short-circuit currents (which are considered negligible), are all combined under P_{Other}.

The capacitance, C, is calculated for each component to determine the associated power dissipated. For P_{Logic}, the device capacitance is calculated by considering gate oxide capacitance, overlap capacitance, and junction capacitance all of which can be easily calculated from ITRS data. Interconnect capacitances for the local, semi-global and global tiers are found from the wire-length distribution and the dimensions of the wire pitches for each tier. Clock

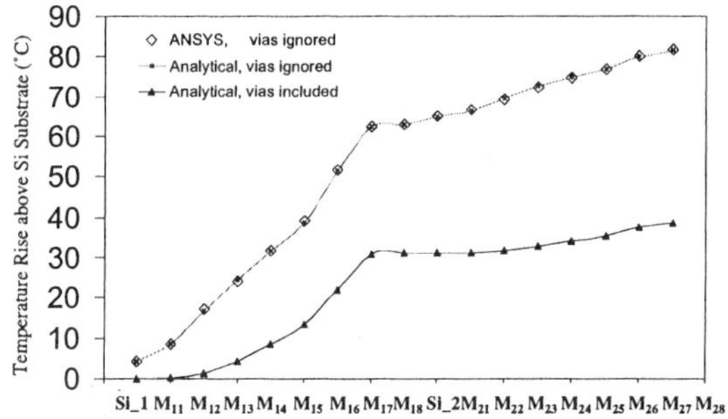

Figure 9.24. Temperature distribution along the vertical layers from the Si substrate surface (Si_1) to top metal level of the second stratum. The 3-D structure is shown in Figure 9.25. M11 represents the first metal level in stratum 1, etc.. J_{rms} is 4.85 $\times$ 10^5 A/cm^2 and active power density is $F = 0.615$ W/cm^2. Interconnect parameters are quoted from ITRS 0.1 μm technology node. For the case of via effect included, the via separations assigned to the metal layers, from 1st to 8th levels, are 1, 5, 15, 30, 50, 80, 200 and 500 μm, respectively.

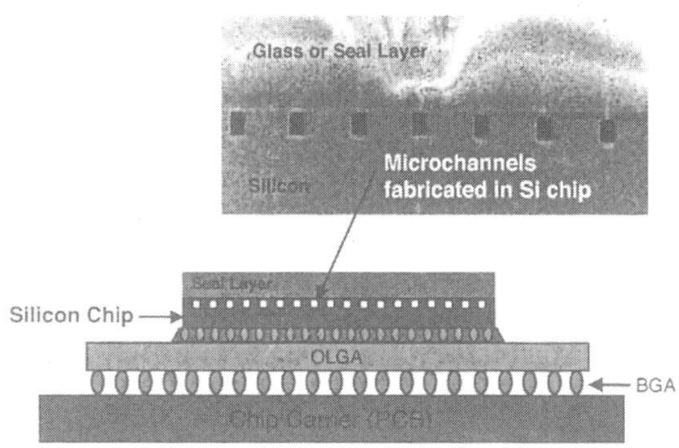

Figure 9.25. Schematic of a packaged Si chip with integrated microchannels etched in the substrate for pumping coolant to lower the package thermal resistance. BGA and OL GA denote ball grid array and organic layer ball grid array respectively. (Courtesy of Kenneth E. Goodson, Stanford University).

distribution capacitances are calculated using the BACPAC model proposed in [116] by considering a buffered H-Tree model. Power dissipated by repeaters is calculated based on the driver capacitances and the number of repeaters which is modeled earlier in this chapter. P_{Other} is determined in the 2-D case to be the sum of remaining components to achieve the ITRS projected total power dissipation for this generation. Since this component is assumed dominated by dynamic dissipation it is considered linearly dependent on the operating frequency for all 3-D cases.

Table 9.3. Comparison of power dissipation due to logic, interconnect, clock distribution and repeaters for 2-D and 3-D ICs with 2 active layers for ITRS 1999 50 nm technology node. 3-D IC cases are presented for comparison by varying the chip area, A_c, and operating frequency, f_c, and represent the same 2-D IC (conserving feature size, number of transistors and functionality) converted to 3-D with 2 active layers. 3-D Case 1 is special in that memory and logic are each dedicated to separate active layers without any modifications to wiring. Resulting A_c is determined by larger logic area and power dissipation is unchanged relative to 2-D. The following 4 3-D cases are obtained, and compared to 2-D, by modifying the wiring to achieve, respectively: equal f_c and decreased A_c; equal f_c and A_c; $2f_c$ and equal A_c; equal A_c with f_c determined by maintaining 2-D P_{Total}.

	2-D	3-D(C1)	3.D,(C2)	3-D,(C3)	3-D,(C4)	3-D,(C5)
Active Layers	1	2	2	2	2	2
f_c (MHz)	3000	3000	3000	3000	6000	3559
Feature Size (nm)	50	50	50	50	50	50
Chip Area (cm^2)	8.17	4.25	2.51	8.17	8.17	8.17
Memory Area (cm^2)	3.92	3.92	3.92	3.92	3.92	3.92
Logic Area (cm^2)	4.25	4.25	5.1	12.42	12.42	12.42
P_{Logic}(W)	34.8	34.8	34.8	34.8	69.6	41.28
P_{local} (W)	17.4	17.4	17.44	20.66	10.44	6.19
$P_{Semi-Global}$(W)	14.63	14.63	6.89	8.16	30.68	18.2
P_{Global}(W)	6.96	6.96	4.18	5.63	11.78	6.99
P_{clock}(W)	34.8	34.8	22.97	27.21	56.93	33.76
$P_{Repaters}$ (W)	45.24	45.24	29.7	35.19	73.6	43.65
P_{Other}(W)	20.17	20.17	20.17	20.17	40.34	23.93
P_{Total}(W)	174	174	136.15	151.82	293.37	174
$P^*_{density}(W/cm^2)$	21.30	20.47	15.09	9.29	17.95	10.65

* Power Density per active level

In 3-D case 2, the total power dissipation is seen to decrease primarily due to the reduction in the wiring requirement thus reducing the interconnect power dissipation, reducing number of required repeaters and reducing the clock distribution network. 3-D case 3 is associated with a larger chip area which requires longer interconnect lines, a larger number of repeaters and clock-distribution network all of which increase the power dissipation. 3-D case 4 shows a dramatic increase in the power dissipated primarily due to the significant increase in operating frequency. 3-D case 5 illustrates the increase in the operating frequency if the chip area and the power dissipation requirements are maintained constant to 2-D.

Although the total power consumption as shown in Table 9.3 is reduced by going from 2-D to 3-D ICs due to the reduction in the interconnect and the clock network related capacitance, the heat removal capability could deteriorate as the upper active layers experience longer heat dissipation path to heat sink. Therefore, the several scenarios of 3-D ICs presented in Table 9.3 are used to explore the opportunities for 3-D integration. Figure 9.26 compares the temperature rise for different 3-D integration scenarios. It shows that, in most of the cases, 3-D ICs have similar temperature rise but have the advantage of either reduced chip area (case 1) or increasing operating frequency (case 4). In the case of equal chip area and operating frequency (case 2), lower temperature than 2-D ICs can be achieved. Note that to double the operating frequency (case 3), temperature will invariably increase. Nevertheless, the temperature is still much lower than that estimated with via effect ignored [115].

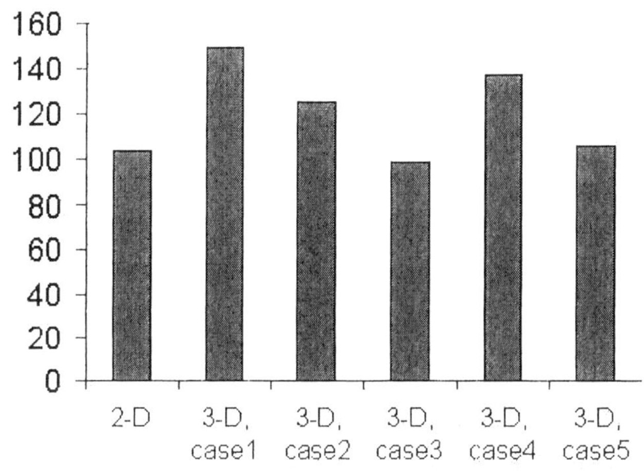

Figure 9.26. Comparison of temperature performance among 2-D ICs and four different two-active-layer 3-D ICs scenarios.

Furthermore, it is desirable to put memory in close proximity to logic circuitry to reduce latency in high performance microprocessors. 3-D ICs provides an excellent opportunity to stack memory and logic. The power consumption in on-chip memory is generally less than 10% of total power consumption and the area occupied by memory and logic are comparable. With these assumptions, Figure 9.27 shows four 3-D stack schemes along with their temperature performance. It can be observed that with logic in the bottom active layer and memory in the upper layer (scheme 1), the resultant temperature rise is the lowest. On the other hand, stacking logic parts back-to-back will experience much higher temperature rise.

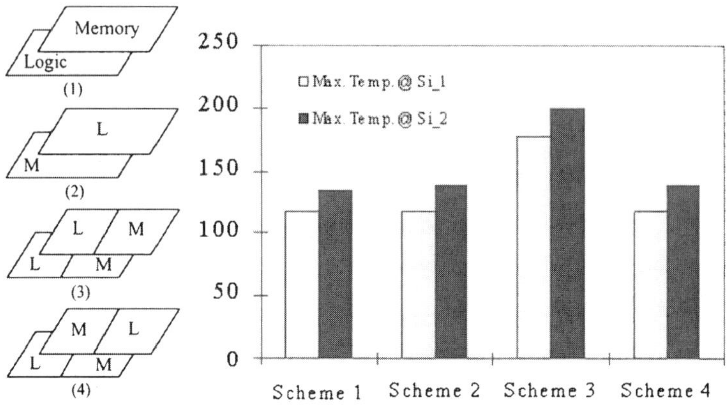

Figure 9.27. Integration Schemes and thermal capability of high performance 3-D ICs (case4) for four different 3-D logic-memory integration schemes

9.5.2 Electromagnetic Interactions (EMI) in 3-D ICs

Interconnect Coupling Capacitance and Crosstalk. In 3-D ICs an additional coupling between the top layer metal of the first active layer and the devices on the second active layer is expected to be present. This needs to be addressed at the circuit design stage. However, for deep submicron technologies, the aspect ratio of global tier interconnects is ≥ 2.5 [1]. Therefore line-to-line capacitance is the dominant portion of the overall capacitance. Hence, the presence of an additional silicon layer on top of a global metal line may not have an appreciable effect on the line capacitance per unit length. For technologies with very small aspect ratio, the change in interconnect capacitance due to the presence of an additional silicon layer could be significant, as reported in [47].

Interconnect Inductance Effects. For deep submicron interconnects on-chip inductive effects arising due to increasing clock speeds and decreasing rise times are a concern for signal integrity and overall interconnect performance [48]. Inductance causes ringing in the signal waveforms, which can adversely affect signal integrity. For global wires inductance effects are more severe due to transmission line effects and also due to the lower resistance of these lines, which makes the wire impedance due to inductance comparable to that due to the resistance, and also due to the presence of significant mutual inductive coupling between wires resulting from longer current return paths [49]. In 3-D ICs, the presence of a second substrate close to the global wires might help lowering the inductance by providing shorter return paths, provided the substrate resistance is sufficiently low or if the wafers are bonded through metal pads.

9.5.3 Reliability Issues in 3-D ICs

3-D ICs will possibly introduce some new reliability problems. These reliability issues may arise due to the electro-thermal and thermo-mechanical effects between various active layers and at the interfaces (glue layers) between the active layers, which can also influence existing IC reliability hazards such as electromigration and chip performance [40]. There will be an increasing need to understand mechanical and thermal behavior of new material interfaces, thin-film material thermal and mechanical properties, and barrier layer integrity. Additionally, from a manufacturing point of view, there might be yield issues arising due to the mismatch between the individual die-yield maps of different active layers, which will affect the net yield of 3-D chips.

9.6 Implications for Circuit Design and System-on-a-Chip Applications

9.6.1 Repeater Insertion

For deep submicron technologies, interconnect delay is the dominant component of the overall delay, especially for circuits with very long interconnects where the delay can become quadratic with line lengths. To overcome this problem, long interconnects are typically broken into shorter buffered segments. In [25] it was shown that for point-to-point interconnects, there exists an optimum interconnect length and an optimum repeater size for which the overall delay is minimum. Repeater sizes for various metal layers for different technologies have been presented in [16, 25]. For top layer interconnect, the corresponding inverter sizes were approximately 450 times the minimum inverter size available in the relevant technology. These large repeaters present a problem since they take up a lot of active silicon and routing area. The vias that connect such a repeater from the top global interconnect layers block all the metal layers present underneath them, hence taking up substantial routing area. It has been predicted [50] that the number of such repeaters can reach 10,000 for high performance designs in 100 nm technology. Inserting these repeaters will cause unacceptable area increase as shown in Figure 9.28.

However, this problem can be easily tackled using 3-D technology with just two silicon layers. The repeaters can be placed on the second silicon layer thereby saving area on the first silicon layer. Furthermore, if the second silicon layer is placed close to the common global metal layers, the vias connecting the global metal layers to the repeaters will not block the lower metal layers thereby freeing up additional routing area.

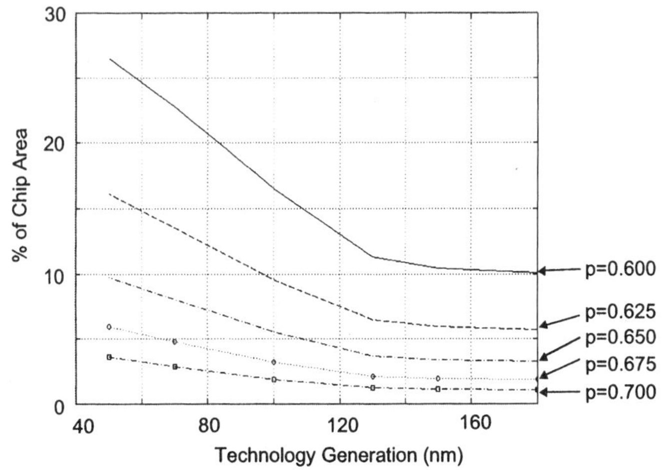

Figure 9.28. Fraction of chip area used by repeaters for different technology nodes and different Rent's exponents. As much as 27% of the chip area at 50 nm node is likely to be occupied by repeaters.

Previously Figure 9.16 had also included delay simulation results for an otherwise single active layer IC except that the repeaters had been moved to a second active layer. A conservative value of Rent's exponent ($p = 0.65$) was used to estimate the reduction in chip area and therefore reduction in overall interconnect delay. At 50 nm node, an additional reduction of 9% in the overall interconnect delay results from the resulting area reduction.

9.6.2 Layout of Critical Paths

In typical high performance ASIC and microprocessor designs, interconnect delay is a significant portion of the overall path delay [51]. Logic blocks on a critical path need to communicate to other logic blocks which, due to placement and other design constraints, may be placed far away from each other. The delay in the long interconnects between such blocks usually causes timing violations. With the availability of a second active layer, these logic blocks can be placed on different silicon layers and hence can be very close to each other, thereby minimizing interconnect delay. Even if highest quality devices are not made on the second active layer, the decrease in interconnect delay can be more that the increase in gate delay due to sub-optimal transistor characteristics.

9.6.3 Microprocessor Design

In microprocessors and DSP processors, most of the critical paths involve on-chip caches [52]. The primary reason for this is that on-chip cache is (physically) located in one corner of the die whereas the logic and computational blocks, which access this memory, are distributed all over the die. By using a technology with two silicon layers, the caches can be placed on the second active layer and the logic and computational blocks on the first layer. This arrangement ensures that logic blocks are in closer proximity to on-chip caches.

Consider a microprocessor of dimensions $L \times L$. In typical current generation microprocessors, about half the physical area is taken up by on-chip caches. Hence the worst case interconnect length in a critical path is 2L (typically the data transfer from cache takes more than one clock cycles but we assume single clock cycle transfers for simplicity). If on-chip caches are placed on the second active layer and the chip is resized accordingly to have dimensions $\frac{L}{\sqrt{2}} \times \frac{L}{\sqrt{2}}$, then the worst case interconnect length is $\sqrt{2}L$, a reduction of about 30%. Even though this analysis is very simplistic compared to the more elaborate one presented in Section 9.4, and does not perform any optimization of the interconnect pitch, it demonstrates that going from single silicon layer to two layers results in nontrivial improvement in performance. Recent studies [53] have shown that by integrating level one and level two cache and the main memory on the same silicon using 3-D technology, access times for level 2 cache and main memory can be decreased. This coupled with an increase in bandwidth between the memory, level 2 cache and level 1 cache, reduces the level 2 cache/memory miss penalty and therefore reduces average time per instruction and increases system performance.

9.6.4 Mixed-Signal Integrated Circuits

With greater emphasis on increasing the functionality that can be implemented on a single die in the system-on-a-chip paradigm, more and more analog, mixed-signal and RF components of the system are being integrated on the same piece of silicon (as illustrated in Figure 9.4). However, this presents serious design issues since switching signals from the digital portions of the chip couple into the sensitive analog and RF circuit nodes from the substrate and degrade the fidelity (or equivalently, increase the noise) of the signals present in these blocks [54]. Furthermore, different fabrication technologies are required for the two applications. However, with the availability of multiple silicon layers, RF and mixed-signal portions of the system can be realized on a separate layer (using different technologies) thereby providing substrate isolation from the digital portion. A preliminary analysis shows a 30 dB improvement in isolation by moving the RF portions of the circuit to a separate substrate [32]. Moreover, since the second Si layer is not continuous, good isolation between different analog and RF components (such as the low- noise amplifier (LNA) and power amplifier) can also be achieved.

9.6.5 Optical Interconnects for System Clocking and I/O Connections

For high performance microprocessors with operating frequencies greater than a few GHz and large die sizes (on-chip frequency= 10 GHz, and die area = 8.17 cm^2 at the 50 nm technology node [1]), interconnects responsible for global communications, including the interconnect network used for the clock distribution, can contribute significantly to the key performance metrics (area, power dissipation, and delay) and to the overall cost of the chip. As the complexity (size) of the microprocessors increases, synchronization of various blocks in the chip becomes increasingly difficult [55]. This occurs mainly due to the variation in the placement of different blocks (or clock line lengths) and due to differences in their operating temperature that affects the clock skew and the net signal delay. Additionally, data input and output (I/O) requirements drive up the number of I/O pads and the corresponding size of the I/O circuitry (or chip area). Furthermore, in high performance designs around 40-70% of the total power consumption could be due to the clock distribution network [56, 57], and as the total chip capacitance (dominated by interconnects) and the chip operating frequency increases with scaling, the power dissipation increases according to (40).

On-chip optical interconnects can eliminate most of the problems associated with clock distribution and I/O connections in large multi-GHz chips [58, 59]. They are attractive for high-density and high- bandwidth interconnections, and optical signal propagation loss is almost distance-independent. Also, the delays on optical clock and signal paths are not strongly dependent on temperature. Additionally, optical signals are immune to electromagnetic interactions discussed earlier with regards to metal interconnects. Hence optical interconnects are very attractive for large-scale synchronization of systems within multi-GHz ICs. Furthermore, optical interconnects employing short optical (laser) pulses, can reduce its optical power requirement [60]. They can also reduce the electrical power consumption since no photocurrent is generated during transition periods since optical power is incident on the transmitters and receivers only during valid output states [61]. The short duration of ultrafast laser pulses also results in large spectral bandwidth, which enables system concepts such as a single-source implementation of wavelength-division multiplexed optical interconnects [62, 63], a technique that allows multiple channels to be transmitted down a single waveguide.

Optical interconnect devices and networks integrated in a 3-D system-on-a-chip IC (schematically illustrated in Figure 9.6) can be employed to attain system synchronization and to enhance system performance. Furthermore, use of optical interconnects for clock distribution can significantly alleviate the power dissipation problem in 3-D ICs, and hence reduce the cost per chip. Integrated 3-D optical devices have been demonstrated directly on top of active silicon CMOS circuits [33], [64-66]. Also, polysilicon based optical waveguides of submicron dimensions have been demonstrated for low loss optical signal propagation and power distribution [67].

9.6.6 Implications on VLSI Design and Synthesis

VLSI design and synthesis (both logic and physical) for large digital circuits and high-performance system-on-a-chip type applications based on 3-D ICs will necessitate some new design methodologies, design and layout tools, and test strategies. At an abstract level, physical design (placement and routing) can be viewed as a graph embedding problem. The circuit graph (synthesized and mapped circuit) is embedded on a target graph which is planar (which corresponds to the physical substrate of the conventional single silicon substrate technology). However, with more than one silicon layer available, the target graph is no longer planar, and therefore placement and routing algorithms need to be suitably modified. Moreover, since placement and routing information also affects synthesis algorithms, which in turn can affect the choice of architectures, this modification needs to be propagated all the way to synthesis and architectural level. Additionally, since 3-D ICs would likely involve SOI (silicon-on-insulator) type upper active layers, the design process will need to address issues specific to SOI technology to realize significant performance improvements [68, 69].

9.7 Overview of 3-D IC Technology

9.7.1 Technology Options

Although the concept of 3-D integration was demonstrated as early as in 1979 [70], and was followed by a number of reports on its fabrication process and device characteristics [71-80], it largely remained a research technology, since microprocessor performance was device limited. However, with the growing menace of RC delay in recent times, this technology is being viewed as a potential alternative that can not only maintain chip performance well beyond the 130 nm node, but also inspire a new generation of circuit design concepts. Hence, there has been a renewed spur in research activities in 3-D technology [81-85] and their performance modeling [32, 53], [86-89].

Presently, there are several possible fabrication technologies that can be used to realize multiple layers of active area (single crystal Si or recrystallized poly-Si) separated by inter-layer dielectrics (ILDs) for 3-D circuit processing.

A brief description of these alternatives is given below. The choice of a particular technology for fabricating 3-D circuits will depend on the requirements of the circuit system, since the circuit performance is strongly influenced by the electrical characteristics of the fabricated devices as well as on the manufacturability and process compatibility with the relevant 2-D technology.

Beam Recrystallization. A very popular method of fabricating a second active (Si) layer on top of an existing substrate (oxidized Si wafer) is to deposit polysilicon and fabricate thin film transistors (TFT) (see Figure 9.29). MOS transistors fabricated on polysilicon exhibit very low surface mobility values (of the order of 10 [$cm^2/(V{\cdot}s)$]), and also have high threshold voltages (several volts) due to the high density of surface states (several 10^{12} cm^{-2}) present at the grain boundaries. To enhance the performance of such transistors, an intense laser or electron beam is used to induce re-crystallization of the polysilicon film [70-80], to reduce or even eliminate most of the grain boundaries. This technique however may not be very practical for 3-D devices because of the high temperature involved during melting of the polysilicon and also due to difficulty in controlling the grain size variations [90, 91]. Beam recrystallized polysilicon films can also suffer from lower carrier mobility (compared to single crystal Si) and unintentional impurity doping. However, high-performance TFTs fabricated using low temperature processing [92], and even low-temperature single-crystal Si TFTs have been demonstrated [93] that can be employed to fabricate advanced 3-D circuits.

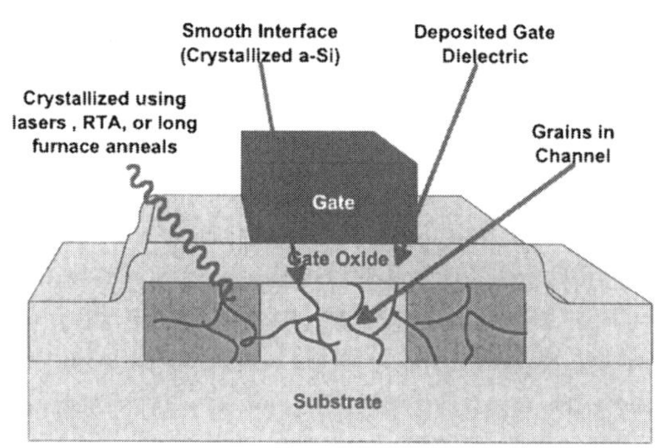

Figure 9.29. Schematic of a thin film transistor (TFT) fabricated on polysilicon depicting several grain boundaries in the active region.

Silicon Epitaxial Growth. Another technique for forming additional Si layers is to etch a hole in a passivated wafer and epitaxially grow a single crystal Si seeded from open window in the ILD. The silicon crystal grows vertically and then laterally, to cover the ILD (Figure 9.30) [84]. In principle, the quality of devices fabricated on these epitaxial layers can be as good as those fabricated underneath on the seed wafer surface, since the grown layer is single crystal with few defects. However, the high temperatures (1000°C) involved in this process cause significant degradation in the quality of devices on lower layers. Also this technique cannot be used over metallization layers. Low temperature silicon epitaxy using ultra-high-vacuum chemical vapor deposition (UHV-CVD) has been recently developed [94]. However, this process is not yet manufacturable.

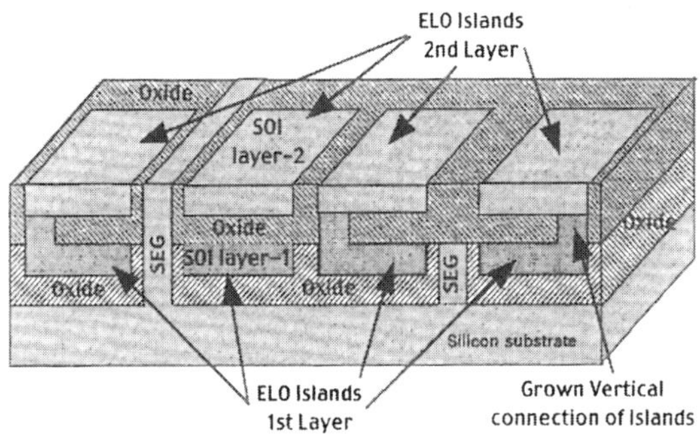

Figure 9.30. Schematic of an epitaxially grown second active layer. ELO denotes epitaxial layer overgrowth. (Courtesy of Gerold W. Neudeck, Purdue University, West Lafayette, IN).

Processed Wafer Bonding. An attractive alternative is to bond two fully processed wafers, on which devices are fabricated on the surface including some interconnects, such that the wafers completely overlap (Figure 9.31) [82, 95]. Interchip vias are etched to electrically connect both wafers after metallization and prior to the bonding process at 400 0C (discussed in section 6.2 below). This technique is very suitable for further processing or the bonding of more pairs in this vertical fashion. Other advantages of this technology lie in the similar electrical properties of devices on all active levels and the independence of processing temperature since all chips can be fabricated separately and later bonded. One limitation of this technique is its lack of precision (best-case alignment ± 2 μm) which restricts the inter-chip communication to global metal lines. However, for applications where each chip is required to perform independent processing before communicating with its neighbor, this technology can prove attractive. Additionally, bonding techniques based on the thermocompression of metal pads [95] offer low thermal- resistance interfaces between bonded wafers, which can help in heat dissipation.

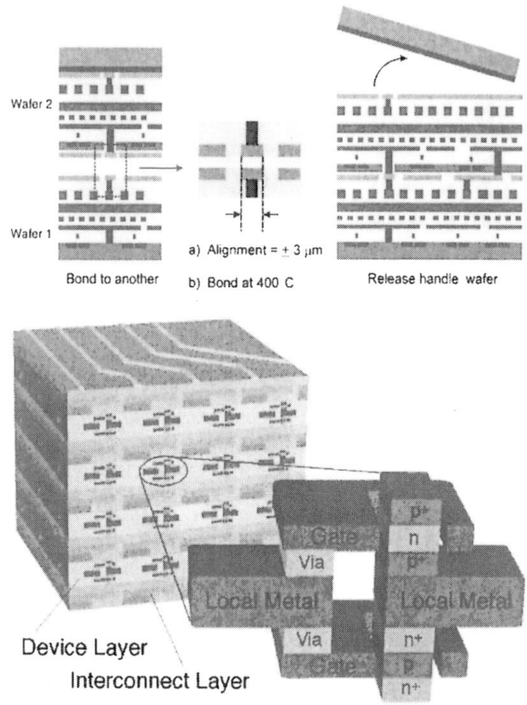

Figure 9.31. Schematic of steps used in one of the wafer bonding technologies based on metal thermocompression and a finished 3-D chip (bottom). (Courtesy of Rafael Reif and Dimitri Antoniadis, Massachusetts Institute of Technology, Cambridge, MA).

Solid Phase Crystallization (SPC). As an alternative to high temperature epitaxial growth discussed above, low temperature deposition and crystallization of amorphous silicon (a-Si), on top of the lower active layer devices, can be employed. The amorphous film can be randomly crystallized to form a polysilicon film [96-99]. Device performance can be enhanced by eliminating the grain boundaries in the polysilicon film. For this purpose, local crystallization can be induced using low temperature processes ($< 600°C$) such as using patterned seeding of Germanium (Figure 9.32) [83, 100]. In this method Ge seeds implanted in narrow patterns made on a-Si can be used to induce lateral crystallization and inhibit additional nucleation. This results in the formation of small islands, which are nearly single crystal. CMOS transistors can then be fabricated within these islands to give SOI like performance. Another approach based on the seeding technique employs metal (Ni) seeding to induce simultaneous lateral recrystallization and dopant activation after the fabrication of the entire transistor on an a-Si layer. This technique known as the Metal Induced Lateral Crystallization (MILC) (see Figure 9.33) [101, 102] offers even lower thermal budget ($< 500^* C$) and can be employed to fabricate high-performance devices (MOSFETS or optical devices) on upper active layers even with metallization layers below.

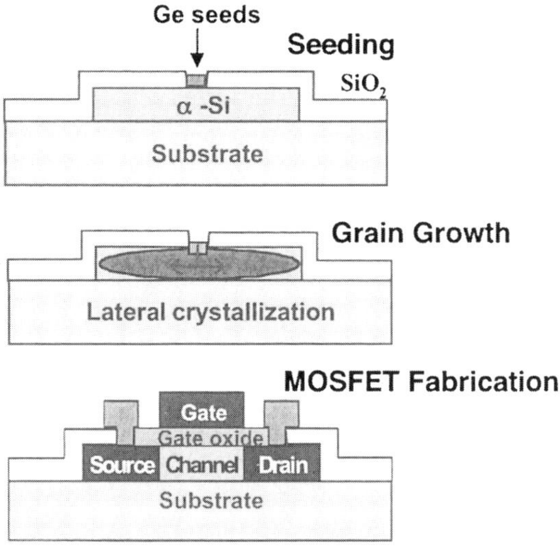

Figure 9.32. Schematic of the Ge seeded SPC fabrication steps.

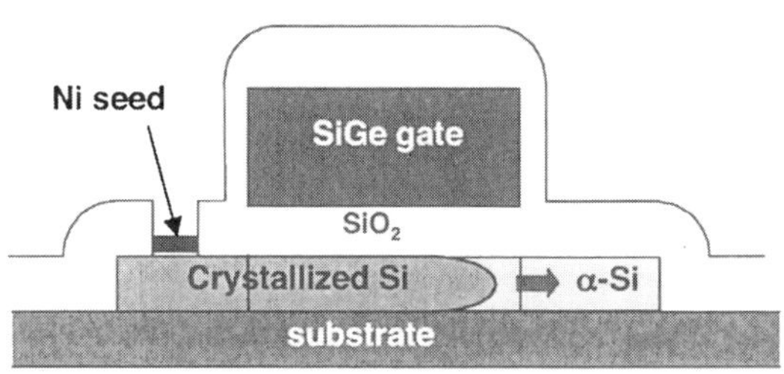

Figure 9.33. Schematic of the MILC process using Ni seeding.

The SPC technique offers the flexibility of creating multiple active layers and is compatible with current CMOS processing environments. Recent results using the MILC technique prove the feasibility of building high performance devices at low processing temperatures, which can be compatible with lower level metallization [103]. It is found that the electrical characteristics of these devices (although superior among their peers) are still inferior to single crystal devices. However, technological advances to overcome the thermal budget problem have been made to allow fabrication of high-performance devices using SPC [104-106].

It is possible to conceive of several 3-D circuits for which SPC will be a suitable technology, such as in upper-level non-volatile memory, or by simply sizing up the upper level transistors to match their single crystal CMOS counterparts. For example, deep sub-micron polysilicon TFTs [107], stacked SRAM cells [108, 109], and EEPROM cells [110] have already been demonstrated. With technological improvements, the MILC (Ni seeding) process can be used to fabricate islands of single-grain-devices to maximize circuit performance.

9.7.2 Vertical Inter-Layer Interconnect Technology

The performance modeling presented in this study directly relates improved chip performance with increased utility of VILICs. It is therefore important to understand how to connect different active layers with a reliable and compatible process. Upper-layer processing needs to be compatible with metal lines underneath connecting lower layer devices and metal layers. With Cu technologies, this limits the processing temperatures to $< 450°C$ for upper layers. Otherwise, Cu diffusion through barrier layers, and the reliability and thermal stability of material interfaces can degrade significantly. Tungsten is a refractory metal that can be used to withstand higher processing temperatures, but it has higher resistivity. Current via technology can also be employed to achieve VILIC functionality. The underlying assumption here requires that intra-layer gates are interconnected using regular horizontal metal wires and inter-layer interconnects can be vias connecting the wiring network for each layer, as schematically illustrated in Figure 9.5.

Recently, inter-layer (VILIC) metallization schemes for 3-D ICs have been demonstrated using direct wafer bonding. These techniques are based on the bonding of two wafers with their active layers connected through high aspect ratio vias, which serve as VILICs. One method is based on the optically adjusted bonding of a thinned (10 μm) top wafer to a bottom wafer with an organic adhesive layer of polyimide (2 μm) in between [111]. Interchip vias are etched through the ILD (inter level dielectric), the thinned top Si wafer and through the cured adhesive layer, with an approximate depth of 20 μm prior to the bonding process (see Figure 9.34a). The interchip via made of chemical vapor deposited (CVD) TiN liner and CVD W plug provides a vertical interconnect (VILIC) between the uppermost metallization levels of both layers. The bonding between the two wafers (misalignment ($\leq 1\mu m$) is done using a flip-chip bonder with split beam optics at a temperature of $400°C$.

A second technique relies on the thermocompression bonding between metal pads in each wafer [95]. In this method Cu/Ta pads on both wafers (illustrated in Figure 9.34b) serve as electrical contacts between the interchip via on the top thinned Si wafer and the uppermost interconnects on the bottom Si wafer. The

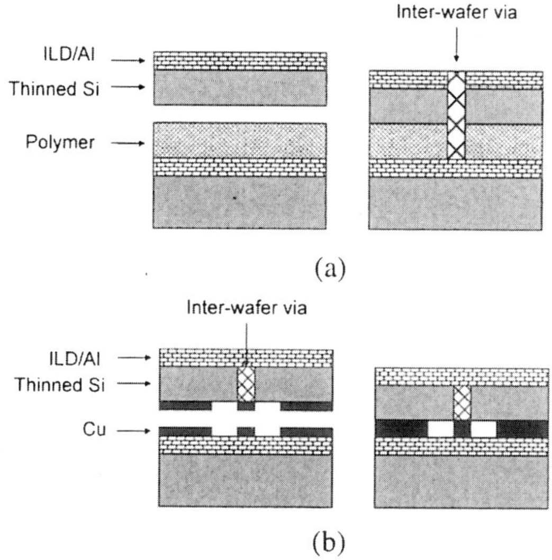

Figure 9.34. Schematic of the wafer bonding techniques a) with adhesive layer of polymer in between, and b) through thermocompression of Cu metal. (Courtesy of Rafael Reif, Massachusetts Institute of Technology, Cambridge, MA).

Cu/Ta pads can also function as small bond pads for wafer bonding. Additionally, dummy metal patterns can be made to increase the surface area for wafer bonding. The Cu/Ta bilayer pads with a combined thickness of 700 nm are fused together by applying a compressive force at $400C$. This technique offers the advantage of a metal-metal interface that will lower the interface thermal resistance between the two wafers (hence provide better heat conduction and can be beneficial as a partial ground plane for lowering the electromagnetic effects discussed in section 4.2.

Table 9.4. Summary of ITRS 1999 data for high-performance 2-D microprocessor ICs showing projected chip area,memory and logic device allocation

	1999	2001	2003	2005	2008	2011
Technology Node	180 nm	150 nm	120 nm	100 nm	70 nm	50 nm
A_c(cm^2)	4.5	4.5	5.76	6.22	7.13	8.17
N_{logic}(M)	12	24	48	96	272	769
N_{Memory}(M)	98	196	393	786	2222	6284

9.8 Conclusions

In this chapter we have motivated the need for 3-D IC technologies with multiple active layers, as a promising alternative to the present single Si layer IC technologies, to alleviate the ever increasing problem of interconnect delays in high-performance logic circuits and in future high-performance systems, and to realize large scale integration of disparate systems and technologies in a single die.

The interconnect delay problem associated with Cu/low-k technologies was discussed using estimated RC values based on the data from ITRS'99. The implications of material effects arising at deep submicron dimensions such as increasing metal resistivity due to increased electron scattering and finite metal barrier layer thickness were quantified. The increasing impact of interconnect delays on VLSI design was also discussed and the limitations of various proposed solutions to overcome the interconnect problem were highlighted, especially in light of ITRS based interconnect trends and their associated effects. It was concluded that Cu/low-k interconnects alone will not be able to solve the interconnect problem and that the design based solutions have not been adequate to deal with the wiring problem.

A methodology was presented to obtain the wire-length distributions for 3-D ICs which can be used to accurately predict area, delay, and power dissipation, and provide examples of some of these trade-offs which result in area and/or delay reduction over the 2-D case. A scheme to optimize the interconnect distribution among different interconnect tiers was also presented and the effect of transferring the repeaters to upper Si layers was quantified in this analysis for a two-layer 3-D chip. Our analysis predicts significant performance improvements of up to 145% over the 2-D case. The primary target technology for this analysis has been the ITRS'99 based 50 nm node with two active layers of silicon. Other technology nodes with two active layers have also been considered. It was shown that the availability of additional silicon layers gives extra flexibility to designers which can be exploited to minimize area, improve performance and power dissipation or any combinations of these.

Additionally, we addressed some of the concerns associated with 3-D circuits including that of thermal problem. A compact analytical thermal model for temperature rise in 3-D ICs was presented, which can be used to estimate the temperature of individual active layers. It was demonstrated that for circuits with two silicon layers running at maximum performance, maintenance of acceptable die temperatures may require advanced heat-sinking technologies. Implications on reliability and electromagnetic interactions (such as capacitance and inductance effects) arising in 3-D ICs were also briefly discussed.

Furthermore, we highlighted some scenarios in current and future VLSI and systems-on-chip type applications involving mixed signals and technologies, where the use of 3-D circuits will have an immediate and beneficial impact on performance. We also briefly discussed the implications of using this technology on the design process, as conventional VLSI design methodologies and tools, gate level and architecture level synthesis algorithms need to be suitably adapted. Finally, an overview of some of the manufacturing technologies under investigation, which can be used to fabricate these circuits, was provided.

References

[1] The International Technology Roadmap for Semiconductors (ITRS), 1999.

[2] C. R. Barrett, "Microprocessor evolution and technology impact," *Symp. VLSI Technol., Digest*, 1993, pp. 7-10.

[3] C. Hu, "MOSFET scaling in the next decade and beyond," *Semiconductor International*, pp. 105-114, 1994.

[4] B. Davari, R. H. Dennard, and G. G. Shahidi, "CMOS scaling for high performance and low power-The next ten years," Proc. of the IEEE, vol. 83, no. 4, pp. 595-606, 1995.

[5] G. A. Sai-Halasz, "Performance trends in high-end processors," Proc. of the IEEE, vol. 83, no. 1, pp. 20-36, 1995.

[6] K. C. Saraswat and F. Mohammadi, "Effect of interconnection scaling on time delay of VLSI circuits," IEEE Trans. Electron Devices, vol. ED-29, pp. 645-650, 1982.

[7] M. T. Bohr, "Interconnect scaling-the real limiter to high performance ULSI," IEDM Tech. Dig., 1995, pp. 241-244.

[8] J. D. Meindl, "Low power microelectronics: retrospect and prospect," Proc. of the IEEE, vol. 83, no. 4, pp. 619-635, 1995.

[9] S-Y Oh and K-J Chang, "2001 needs for multi-level interconnect technology," Circuits and Devices, pp. 16-21, 1995.

[10] M. T. Bohr and Y. A. El-Mansy, "Technology for advanced high-performance microprocessors," IEEE Trans. Electron Devices, vol 45, no. 3, pp. 620-625, 1998.

[11] D. Edelstein et al., "Full copper wiring in a sub-0.25 ?m CMOS ULSI technology," IEDM Tech. Dig., 1997, pp. 773-776.

[12] S. Venkatesan et al., "A high performance 1.8V, 0.20 ?m CMOS technology with copper metallization," IEDM Tech. Dig., 1997, pp. 769-772.

[13] E. M. Ziclinski et al., "Damascene integration of copper and ultra-low-k xerogel for high performance interconnects," IEDM Tech. Dig., 1997, pp. 936-938.

[14] N. Rohrer et al., "A 480MHz RISC microprocessor in a 0.12 ?m Leff CMOS technology with copper interconnects," Int. Solid-State Circuits Conf., Tech. Digest, 1998, pp. 240-241.

[15] B. Zhao et al., " A Cu/low-k dual damascene interconnect for high performance and low cost integrated circuits," Symp. VLSI Technology, Tech. Digest, 1998, pp. 28-29.

[16] K. Banerjee, A. Mehrotra, A. Sangiovanni-Vincentelli, and C. Hu, "On thermal effects in deep sub-micron VLSI interconnects," Proc. 36th ACM Design Automation Conference, 1999, pp. 885-891.

[17] The Use of Thin Films in Physical Investigation; Ed. J.C Anderson, Academic Press, London, 1966.

[18] F. Chen and D. Gardner, "Influence of line dimensions on the resistance of Cu interconnections," IEEE Electron Device Letters, vol. 19, no.12, pp. 508-510, 1998.

[19] Handbook of Thin Film Technology, Edited by L.I. Maissel and R. Glang, Chapter 13, McGraw Hill Book Company, 1970.

[20] J.P. McVittie et. al., SPEEDIE 3.5 Manual, Stanford University, 1998.

[21] Private Communications, Lukas P.P.P. Van Ginneken, Magma Design Automation, Cupertino, CA.

[22] S. Devadas, A. Ghosh, and K. Keutzer, Logic Synthesis, McGraw Hill, 1994.

[23] D. Sylvester and K. Keutzer, "Getting to the bottom of deep submicron," Proc. Int. Conf. on Computer Aided Design, 1998, pp. 203-211.

[24] R. Ho, K. Mai, H. Kapadia, and M Horowitz, "Interconnect scaling implications for CAD," Proc. Int. Conf. on Computer Aided Design, 1999.

[25] R. H.J.M. Otten and R. K. Brayton, "Planning for performance," Proc. 35th Annual Design Automation Conference, 1998, pp. 122-127.

[26] W. Gosti, A. Narayan, R. K. Brayton and A. L. Sangiovanni-Vincentelli, "Wireplanning in logic synthesis," Proc. Int. Conf. on Computer Aided Design, 1998, pp. 26-33.

[27] J. Grodstein, E. Lehman, H. Harkness, B. Grundmann, and Y. Watanabe, "A delay model for logic synthesis of continuously sized networks," Proc. Int. Conf. on Computer Aided Design, 1995.

[28] S. P. Khatri, A. Mehrotra, R. K. Brayton, A. Sangiovanni-Vincentelli, and R. H.J.M Otten, "A Novel VLSI Layout Fabric for Deep Sub-Micron Applications," Proc. 36th ACM Design Automation Conference, 1999, pp. 491-496.

[29] W. J. Dally, "Interconnect-limited VLSI architecture," Int. Interconnect Technology Conf. Proceedings, 1999, pp. 15-17.

[30] M. J. M. Pelgrom, "System-On-Chip Concepts," Chapter 11, ULSI Devices, Eds. C. Y. Chang and S. M. Sze, Wiley Inter-Science, 2000.

[31] H. De Man, "System Design Challenges in the Post PC Era," Keynote Address Presentation Slides, 37th ACM Design Automation Conf., 2000.

[32] S. J. Souri, K. Banerjee, A. Mehrotra, and K. C. Saraswat, "Multiple Si layer ICs: motivation, performance analysis, and design implications," Proc. 37th ACM Design Automation Conf., 2000, pp. 873-880.

[33] A. V. Krishnamoorthy, et al., "3-D integration of MQW modulators over active submicron CMOS circuits: 375 Mb/s transimpedance receiver-transmitter circuit," IEEE Photonics Technology Letters, vol. 7, no. 11, pp. 1288-1290, 1995.

[34] J. A. Davis, V. K. De, and J. D. Meindl, "A stochastic wire-length distribution for gigascale integration (GSI) - Part I: Derivation and validation," IEEE Trans. Electron Devices, Vol. 45, no. 3, March 1998.

[35] L. Robinson, L. A. Glasser, and D. A. Antoniadis, "A simple interconnect delay model for multilayer integrated circuits," IEEE VMIC Conf., 1986.

[36] B. S. Landman, and R. L. Russo, "On a pin versus block relationship for partitions of logic graphs," IEEE Trans. Computers, vol. C-20, no. 12, Dec. 1971.

[37] J. A. Davis, V. K. De, and J. D. Meindl, "A stochastic wire-length distribution for gigascale integration (GSI) - Part II: Applications to clock frequency, power dissipation, and chip size estimation," IEEE Trans. Electron Devices, Vol. 45, no. 3, March 1998.

[38] H. B. Bakoglu and J. D. Meindl, "Optimal interconnection circuits for VLSI," IEEE Trans. Elec. Dev. vol. 32, no. 5, pp. 903- 909, 1985.

[39] S. A. Kuhn, M. B. Kleiner, P. Ramm, and W. Weber, "Thermal analysis of vertically integrated circuits," IEDM Tech. Dig., 1995, pp. 487-490.

[40] S. Im and K. Banerjee, "Full chip thermal analysis of planar (2-D) and vertically integrated (3-D) high performance ICs," IEDM Tech. Dig., 2000 (To appear).

[41] K. Banerjee, Ph.D. Thesis, University of California at Berkeley, 1999.

[42] K. Banerjee, "Thermal effects in deep submicron VLSI interconnects," Tutorial Notes, IEEE International Symposium on Quality Electronic Design, March 20-22, 2000.

[43] K. Banerjee, A. Amerasekera, G. Dixit, and C. Hu, "The effect of interconnect scaling and low-k dielectric on the thermal characteristics of the IC metal," IEDM Tech. Dig., 1996, pp. 65-68.

[44] K. E. Goodson and Y. S. Ju, "Heat conduction in novel electronic films," Annu. Rev. Mater. Sci., 29: pp. 261-293, 1999.

[45] D. B. Tuckerman, R. F. W. Pease, "High-performance heat sinking for VLSI," IEEE Electron Device Lett., vol. EDL-2, no.5, pp.126-129, 1981.

[46] Private Communication, K. E. Goodson, Stanford University.

[47] S. A. Kuhn, M. B. Kleiner, P. Ramm, and W. Weber, "Interconnect capacitances, crosstalk, and signal delay in vertically integrated circuits," IEDM Tech. Dig., 1995, pp. 487-490.

[48] E. E. Davidson, B. D. McCredie, and W. V. Vilkelis, "Long lossy lines (L3) and their impact upon chip performance," IEEE Trans. Components, Packaging, and Manufacturing Technology-Part B, vol. 20, no. 4, pp. 361-375, 1997.

[49] A. Deutsch et al., "Frequency-dependent crosstalk simulation for on-chip interconnections," IEEE Trans. Advanced Packaging, vol. 22, no. 3, pp. 292-308, 1999.

[50] J. Cong and L. He, "An efficient technique for device and interconnect optimization in deep submicron designs," Int. Symp. on Physical Design, 1998, pp. 45-51.

[51] P. D. Fisher, "Clock cycle estimation for future microprocessor generations," Technical Report, SEMATECH 1997.

[52] D. Greenhill et al, "A 330 MHz 4-way superscalar microprocessor," ISSCC, Digest of Tech. Papers, 1997, pp. 166-167.

[53] M. B. Kleiner, S. A. Kuhn, P. Ramm, and W. Weber, "Performance improvement of the memory hierarchy of RISC-systems by application of 3-D technology," IEEE Trans. Components, Packaging, and Manufacturing Technology-Part B, vol. 19, no. 4, pp. 709-718, 1996.

[54] B. Razavi, "Challenges and trends in RF design," Proc. 9th Annual IEEE Int. ASIC Conf. and Exhibit, 1996, pp. 81-86.

[55] J. M. Rabaey, "Digital Integrated Circuits: A Design Perspective," Prentice Hall Inc., 1996.

[56] H. Kawaguchi and T. Sakurai, "A reduced clock-swing flip-flop (RCSFF) for 63% power reduction," IEEE Journal of Solid-State Circuits, vol. 33, no. 5, pp. 807-811, 1998.

[57] T. Sakurai, "Design challenges for 0.1 um and beyond," Proceedings of the ASP DAC, 2000, pp. 553-558.

[58] J. W. Goodwin, F. J. Leonberger, S. C. Kung, R. A. Athale, "Optical interconnections for VLSI systems," Proceedings of the IEEE, vol. 72, pp. 850-866, 1984.

[59] D. A. B. Miller, "Rationale and challenges for optical interconnects to electronic chips," To appear in Special Issue of Proceedings of the IEEE.

[60] A. L. Lentine, L. M. F. Chirovsky, and T. K. Woodward, "Optical energy considerations for diode-clamped smart pixel optical receivers," IEEE J. Quantum Electron., 30, pp. 1167-1171, 1994.

[61] G. A. Keeler, B. E. Nelson, D. Agarwal, and D. A. B. Miller, "Skew and jitter removal using optical pulses for optical interconnection," IEEE Photonics Letters, 12, pp. 714-716, 2000.

[62] E. A. De Souza, M. C. Nuss, W. H. Knox, and D. A. B. Miller, "Wavelength-division multiplexing with femtosecond pulses," Optics Letters, 20, pp. 1166-1168, 1995.

[63] D. Agarwal, G. A. Keeler, B. E. Nelson, and D. A. B. Miller, "Wavelength division multiplexed optical interconnects using femtosecond optical pulses," Proc. IEEE LEOS Annual Meeting, 2, pp. 828-829, 1999.

[64] K. W. Goossen, et al., "GaAs MQW modulators integrated with silicon CMOS," IEEE Photonics Letter, vol. 7, no. 4, pp. 360-362, 1995.

[65] A. V. Krishnamoorthy and Keith W. Goossen, "Optoelectronic-VLSI: photonics integrated with VLSI circuits," IEEE Journal of Selected Topics in Quantum Electronics, vol. 4, no. 6, pp. 899-912, 1998.

[66] T. K. Woodward and A. V. Krishnamoorthy, "1-Gb/s integrated optical detectors and receivers in commercial CMOS technologies," IEEE Journal of Selected Topics in Quantum Electronics, vol. 5, no. 2, pp. 146-156, 1999.

[67] L. C. Kimerling, "Photons to the rescue: microelectronics becomes microphotonics," The Electrochemical Society Interface, pp. 28-31, Summer 2000.

[68] C. T. Chuang, P. F. Lu, and C. J. Anderson, "SOI for digital CMOS VLSI: design considerations and advances," Proc. IEEE, vol. 86, no. 4, pp. 689-720, 1998.

[69] D. Allen, D. Behrends, and B. Stanisic, "Converting a 64b PowerPC processor from CMOS bulk to SOI technology," Proc. 36th ACM Design Automation Conference, 1999, pp. 892-897.

[70] M. W. Geis, D. C. Flanders, D. A. Antoniadis, and H. I. Smith, "Crystalline silicon on insulators by graphoepitaxy," IEDM Tech. Dig., 1979, pp. 210-212.

[71] J. P. Colinge and E. Demoulin, "ST-CMOS (Stacked Transistor CMOS): a double-poly-NMOS-compatible CMOS technology," IEDM Tech. Dig., 1981, pp. 557-560.

[72] G. T. Goeloc, E. W. Maby, D. J. Silversmith, R. W. Mountain, and D. A. Antoniadis, "Vertical single-gate CMOS inverters on laser-processed multilayer substrates," IEDM Tech. Dig., 1981, pp. 554-556.

[73] S. Kawamura, N. Sasaki, T. Iwai, M. Nakano, and M. Takagi, "Three-dimensional CMOS IC's fabricated by using beam recrystallization," IEEE Electron Device Lett., vol. EDL-4, no. 10, pp. 366-368, 1983.

[74] S. Akiyama, S. Ogawa, M. Yoneda, N. Yoshii, and Y. Terui, "Multilayer CMOS device fabricated on laser recrystallized silicon islands," IEDM Tech. Dig., 1983, pp. 352-355.

[75] M. Nakano, "3-D SOI/CMOS," IEDM Tech. Dig., 1984, pp. 792-795.

[76] K. Sugahara, T. Nishimura, S. Kusunoki, Y. Akasaka, and H. Nakata, "SOI/SOI/Bulk-Si triple level structure for three-dimensional devices," IEEE Electron Device Lett., vol. EDL-7, no. 3, pp. 193-195, 1986.

[77] Y. Akasaka and T. Nishimura, "Concept and basic technologies for 3-D IC structure," IEDM Tech. Dig., 1986, pp. 488-491.

[78] S. Tatsuno, "Japan's push into creative semiconductor research: 3-dimension IC's," Solid State Technology, March 30, pp. 29-30, 1987.

[79] T. Nishimura, Y. Inoue, K. Sugahara, S. Kusunoki, T. Kumamoto, S. Nakagawa, M. Nakaya, Y. Horiba, and Y. Akasaka, "Three dimensional IC for high performance image signal processor," IEDM Tech. Dig., 1987, pp. 111-114.

[80] T. Kunio, K. Oyama, Y. Hayashi, and M. Morimoto, "Three dimensional ICs, having four stacked active device layers," IEDM Tech. Dig., 1989, pp. 837-840.

[81] S. Strickland, et al., "VLSI design in the 3rd dimension," INTEGRATION, Elsevier Science, pp. 1-16, 1998.

[82] D. Antoniadis, "3-dimensional 25 nm - scale CMOS technology," Advanced Microelectronics Program Review Proceedings Book, Sept. 1-2, Lexington, MA, 1998.

[83] V. Subramanian and K. C. Saraswat, "High-performance germanium-seeded laterally crystallized TFT's for vertical device integration," IEEE Trans. Electron Devices, vol. 45, no. 9, pp. 1934-1939, 1998.

[84] G. W. Neudeck, S. Pae, J. P. Denton, and T. Su, "Multiple layers of silicon-on-insulator for nanostructure devices," J. Vac. Sci. Technol. B 17(3), pp. 994-998, 1999.

[85] K. C. Saraswat, S. J. Souri, V. Subramanian, A. R. Joshi, and A. W. Wang, "Novel 3-D Structures," IEEE Int. SOI Conf., 1999, pp. 54-55.

[86] S. A. Kuhn, M. B. Kleiner, P. Ramm, and W. Weber, "Performance modeling of the interconnect structure of a three-dimensional integrated RISC processor/cache system," IEEE Trans. Components, Packaging, and Manufacturing Technology-Part B, vol. 19, no. 4, pp. 719-727, 1996.

[87] S. J. Souri and K. C. Saraswat, "Interconnect performance modeling for 3D integrated circuits with multiple Si layers," Int. Interconnect Technology Conf. Proceedings, 1999, pp. 24-26.

[88] A. Rahman, A. Fan, J. Chung, and R. Reif, "Wire-length distribution of three-dimensional integrated circuits," Int. Interconnect Technology Conf. Proceedings, 1999, pp. 233-235.

[89] R. Zhang, K. Roy, and D. B. Jones, "Architecture and performance of 3-dimensional SOI circuits," IEEE Int. SOI Conf., 1999, pp. 44-45.

[90] A. W. Wang and K. C. Saraswat, "A strategy for modeling of variations due to grain size in polycrystalline thin film transistors," IEEE Trans. Electron Dev. vol. 47, pp. 1035-1043, 2000.

[91] V. Subramanian, M. Toita, N. R. Ibrahim, S. J. Souri and K. C. Saraswat," Low-leakage Germanium-seeded Laterally-crystallized Single-grain 100 nm TFTs for Vertical Integration Applications," IEEE Electron Dev. Lett., vol. 20, no. 7, pp. 341-343, 1999.

[92] A. Kohno, T. Sameshima, N. Sano, M. Sekiya, and M. Hara, "High performance poly-Si TFTs fabricated using pulsed laser annealing and remote plasma CVD with low temperature processing," IEEE Trans. Electron Devices, vol 42, no. 2, pp. 251-257, 1995.

[93] M. A. Crowder, P. G. Carey, P. M. Smith, R. S. Sposili, H. S. Cho, and J. S. Im, "Low-temperature single crystal Si TFT's fabricated on Si-films processed via sequential lateral solidification," IEEE Electron Device Lett., vol. 19, no. 8, pp. 306-308, 1986.

[94] H-Y. Lin, C-Y. Chang, T. F. Lei, J-Y. Cheng, H-C. Tseng, and L-P. Chen, "Characterization of polycrystalline silicon thin film transistors fabricated by ultrahigh-vacuum chemical vapor deposition and chemical mechanical polishing," Jpn. J. Appl. Phys., Part 1, vol.36, (no.7A), pp. 4278-4282, July 1997.

[95] A. Fan, A. Rahman, and R. Reif, "Copper wafer bonding," Electrochemical and Solid State Letters, vol. 2(10), pp. 534-536, 1999.

[96] T. Noguchi, "Appearance of single-crystalline properties in fine-patterned Si thin film transistors (TFTs) by solid phase crystallization (SPC)," Jpn. J. Appl. Phys., Part 2, no.11A, vol.32, pp. 1584-1587, Nov. 1993.

[97] T. W. Little, H. Koike, K. Takahara, T. Nakazawa, and H. Oshima, "A 9.5-in. 1.3-Mpixel low-temperature poly-Si TFT-LCD fabricated by solid-phase crystallization of very thin films and an ECR-CVD gate insulator," J. Society for Information Display, 1/2, pp. 203-209, 1993.

[98] N. Yamauchi, "Polycrystalline silicon thin films processed with silicon ion implantation and subsequent solid-phase crystallization: theory, exper-

iments, and thin-film transistor applications," J. Appl. Phys., 75(7), pp. 3235-3257, 1994.

[99] D. N. Kouvatsos, A. T. Voutsas, and M. K. Hatalis, "Polycrystalline silicon thin film transistors fabricated in various solid phase crystallized films deposited on glass substrates," J. Electronic Materials, vol. 28, no. 1, pp. 19-25, 1999.

[100] J. A. Tsai, A. J. Tang, T. Noguchi, and R. Reif, "Effects of Ge on material and electrical properties of polycrystalline Si1-xGex for thin film transistors," J. Electrochem. Soc., vol. 142, no. 9, pp. 3220-3225, 1995.

[101] S-W. Lee and S-K. Joo, "Low temperature poly-Si thin film transistor fabrication by metal-induced lateral crystallization," IEEE Electron Device Lett., vol. 17, no. 4, pp. 160-162, 1983.

[102] S. Y. Yoon, S. K. Kim, J. Y. Oh, Y. J. Choi, W. S. Shon, C. O. Kim, and J. Jang, "A high-performance polycrystalline silicon thin-film transistor using metal-induced crystallization with Ni solution," Jpn. J. Appl. Phys., Part 1, pp. 7193-7197, Dec. 1998.

[103] A. R. Joshi and K. C. Saraswat, "Sub-micron thin film transistors with metal induced lateral crystallization," Abstract no. 1358, Proc. 196th Meeting of the Electrochemical Society, Honolulu, HI, 1999.

[104] J. Nakata and K. Kajiyama, "Novel low-temperature recrystalization of amorphous silicon by high energy beam," Appl. Phys. Lett., pp. 686-688, 1982.

[105] Y. W. Choi, J. N. Lee, T. W. Jang, and B. T. Ahn, "Thin-film transistors fabricated with poly-Si films crystallized at low temperature by microwave annealing," IEEE Electron Device Lett., vol. 20, no. 1, pp. 2-4, 1999.

[106] A. Heya, A. Masuda, and H. Matsumura, "Low-temperature crystallization of amorphous silicon using atomic hydrogen generated by catalytic reaction on heated tungsten," Appl. Phys. Lett., vol. 74, no. 15, pp. 2143-2145, 1999.

[107] R K. Watts and J. T. C. Lee, "Tenth-micron polysilicon thin-film transistors," IEEE Electron Device Lett., vol. 14, no. 11, pp. 515-517, 1993.

[108] M. Rodder and S. Aur, "Utilization of plasma hydrogenation in stacked SRAMs with poly-Si PMOSFETs and bulk Si NMOSFETs," IEEE Electron Device Lett., vol. 12, no. 5, pp. 233-235, 1991.

[109] T. Yamanaka et al., "Advanced TFT SRAM cell technology using a phase-shift lithography," IEEE Trans. Electron Devices, vol. 42, no. 7, pp. 1305-1312, 1995.

[110] M. Cao, T. Zhao, K. C. Saraswat, and J. D. Plummer, "A simple EEPROM cell using twin polysilicon thin film transistor," IEEE Electron Device Lett., vol. 15, no. 8, pp. 304-306, 1994.

[111] P. Ramm et al., "Three dimensional metallization for vertically integrated circuits," Microelectronic Engineering, 37/38 pp. 39-47, 1997.

[112] Bakoglu, H.B., Circuits, Interconnections, and Packaging for VLSI, Addison-Wesley Company: Reading, Mass., 1990.

[113] T-Y, Chiang et al., IITC, 2001.

[114] T-Y, Chiang et al., VLSI Tech. Symp., 2001. pp. 141-142

[115] S. Im et al., IEDM Tech. Dig., 2000, pp. 727-730.

[116] D. Sylvester et al., Proc. IEEE, pp. 634-664, May 2001.

[117] K. Banerjee, S. Souri, P. Kapur, K. Saraswat, "3-D ICs: A Novel Chip Design for Improving Deep Submicron Interconnect Performance and Systems-on-Chip Integration". Proceedings of the IEEE, May 2001.

Chapter 10

SILICON MICROPHOTONICS

L.C. Kimerling
MIT Microphotonics Center

10.1 Introduction

It is now one-half century since the advent of solid state electronics with the invention of the transistor. Through unparalleled gains in functionality at relatively constant cost, integrated circuits have enabled telecommunications, computation and manufacturing to move to the leading edge of societal change. This revolution has been conducted with "the turn of a single knob": the shrinking of device dimensions. During the last two decades a new "killer technology" has emerged in the telecommunications field. This photonic technology uses optical fibers for interconnection, and has delivered an exponential increase with time of information carrying capacity to the industry. Photonics has provided unparalleled bandwidth to the backbone of the discrete, point-to-point long distance (Wide Area Network) telephone line technology. However, the circuit architecture (central switching office) that worked so well for voice communications now limits the universal access that is required for Internet and data communications. A single optical fiber, with several hundred gigabits/second of capacity, is limited by electronic processing at each circuit node. To avoid this problem optical signal processors are required: e.g., optical cross-connects and optical add/drop multiplexers. To provide full functionality, these components must be integrated at densities compatible with microelectronic integration. This microphotonics platform represents not only a solution to information access, but it can also solve the problems of bandwidth, pin-out density, reliability and complexity that threaten to end the advance of the silicon integrated circuit technology. A good figure of merit for the performance of microphotonic integrated circuits is $speed/(power \times area)$.

The Information Age was ushered into existence by microelectronics. The future will depend on the networking of databases for universal accessibility. This new Age of Connectivity will require a mating of microelectronic and fiber optic technology through integrated of microphotonic functionality.

The technology pull for integrated optical interconnects in electronic systems is based on cost and performance trends. The cost-per-interconnect increases exponentially with interconnect length: from the shortest interconnects on a chip to the longest interconnects in global telecommunications networks. This economic pressure is a driving force to condense functions locally at the chip, board or system level. Lightwave telecommunications systems utilize silicon transistors to drive gigahertz rate modulation of optical sources. These transistors, together with compound semiconductor lasers and detectors, and silica optical fibers, have increased the system information carrying capacity to the extent that voice and image transmission have enabled the explosive growth of the Internet. This lightwave technology systems driver is potentially the key enabler for extension of chip performance beyond the limits of dimensional scaling.

By 2010 microprocessors will have clock frequencies greater than 6 GHz with chip dimensions greater than 3 cm across the diagonal [1]. One key constraint in the conventional scaling of silicon ULSI is the high interconnect-related power dissipation per unit area. Additionally, in future electronic systems, utilization of chip performance will be limited by the crosstalk introduced by high-densities of chip-to-chip electronic interconnects operating at frequencies near 500 MHz. The number of I/O pins per chip, determined by word size and power distribution, will drive cost and will, hence, dictate the future of interconnect architecture. Multiple levels of interconnection are employed at the chip level to minimize area and crosstalk, but long metal interconnects, 1-100 μm in width, operating at frequencies of 6 GHz, present process yield, power dissipation and thermal management problems. Universal implementation of copper wiring offers an intermediate solution. Optical interconnects at the board level and for the longest interconnection lines on chip offer a scalable, low cost interconnection architecture with the capability of reducing the number of I/O pins by wavelength division multiplexing. Additionally, these optical links can feature reduced power dissipation and crosstalk at gigahertz frequencies.

10.2 Optical Interconnection

Several approaches to optical interconnection have been proposed in the literature. Challenges in materials, optical device design, CMOS receiver circuit design and overall system architecture have been recognized, and some fundamental limits have been presented. Krishnamoorthy, et al. [2], [3] have summarized the limitations of optoelectronic-VLSI (OE-VLSI) using flip-chip bonded MQW modulators coupled with standard Si-CMOS transimpedance-amplifier based receivers. They concluded that the main barriers to continued scaling of OE-VLSI circuits based on modulator technology can be expected to come from the yield of the modulators, the finite laser source power, and the on-chip power dissipation of the I/O circuits. Additionally, they show that the receiver circuits will be a significant source of electrical power dissipation for any OE-VLSI technology that attempts to provide a large number of low-energy, high-speed surface-normal optical links. According to their calculations, an aggregate I/O bandwidth of 0.5 Tb/s with 0.7 μm CMOS to 25 Tb/s with future 0.1 μm CMOS technologies appears to be possible, despite some relatively conservative assumptions on optoelectronic device performance.

Iwata et al., [4] compare the signal propagation delay time of optical interconnection to that of electrical interconnection as a function of length and design rule. Their calculations, based on modeling electrical interconnects as a distributed RC line, show that for interconnection lengths greater than 3 mm, optical interconnects have much smaller delay times for all of the 0.2 μm, 0.5 μm or 1 μm design rules considered. For example, the delay time of a 20 mm long electrical interconnect with a 0.5 μm design rule is over 5 ns, and it exceeds 20 ns with a 0.2 μm design rule. Both values are much larger than the 0.1 ns delay of a 20 mm long optical interconnect. They conclude that higher frequency(i.e. higher speed) systems are possible only with optical interconnection, because there is no intrinsic coupling between optical interconnection lines, or between an optical line and an electrical line.

In a laboratory report, Miller et al., [5] convincingly argue that there is a limit on the bit-rate capacity, B, of electrical interconnects. B depends only on the aspect ratio of the interconnect, i.e. the ratio of interconnect length l to the square root of the total cross-sectional area of the interconnect lines, according to the relation $B = 10^{15} A/l^2$ bits/s. They arrive at this limit by calculating the bit-rate capacity of electrical lines using the L and C of a coaxial cable as an example to obtain signal propagation velocity. Their conclusions remain unchanged even with the inclusion of skin-effect limited lines with resistive losses. Therefore, miniaturization of electrical interconnects will not help in improving the bit-rate capacity. Optical interconnects have no such aspect ratio limit because loss in optical media is essentially independent of the modulation bit rate, even into the terahertz regime. Long, thin optical interconnects work especially well. For example, a single-mode optical fiber, 125 μm in diameter and 15 km long, is able to carry 100 Mb/s of data well within expected dispersion limits. However, an equivalent electrical interconnect with similar dimensions can carry less than 0.1 b/s of data. This comparison makes a case for using optical signal links for chip packaging operating in the tens of GHz regime.

Pin-out density, cost, power dissipation and crosstalk are ultimate limiters to IC chip architectures [6], [7]. In current microprocessors, the clock signal is the master controller for all activities and the highest bandwidth interconnect function on the chip. The signal must synchronize every subsystem on the chip, and it determines the number of instructions executed per second. Current electronic designs trade area (capacitance) and power (speed) to provide a clock skew of less than 1% of the clock period. Data input and output to a chip present a slightly different problem. As word sizes increase, 64, then 128, etc. separate inputs and outputs must exist on every chip. This layout problem lies much deeper, though, as the capacitive loads at the end of the pins require larger and larger devices, limiting speed and taking up precious real estate. Additionally, when gate dielectrics reach fundamental scaling limits, electronic circuits will reach a speed plateau determined by transistor gate delay. At that point we envision a paradigm shift to parallelism facilitated by optical interconnection as the favored option for maintaining chip performance roadmap trends.

Optical interconnection is a solution to both the clock distribution and data I/O problems. An optical signal can be split without adding delay or skew. Many signals can be multiplexed into one waveguide; even clock and data signals can be combined, reducing the number of required inputs and outputs. Global optical bus networks promise to enable high density interconnection and signal multiplexing while featuring the low crosstalk of photons (as opposed to electrons).

10.3 Monolithic Silicon Microphotonics

Silicon microphotonics is the optical equivalent of microelectronics for integrated circuits. Photonics enables unlimited interconnection transmission bandwidth. In addition photonic switching promises ultrafast routing and encoding of information for telecommunications and data transmission. The key issue is the integration of photonic components with integrated circuit electronics. Silicon microphotonics utilizes silicon-based materials that are process compatible with standard integrated circuit fabrication methods [7], [8].

Advances in materials and processing technology have established a new class of optical media. High dielectric contrast confinement (e.g., the Si/SiO_2 system) shrinks the wavelength of the light to dimensions of λ/η, where η is the refractive index of the core material. Photons of the telecommunications standard wavelength, 1.55 μm, can propagate in single mode transmission within silicon waveguides of less than 0.5 μm in cross section dimension. Smaller sized devices enable faster optoelectronic transduction, higher local fields to drive nonlinear interactions and high levels of integration to provide new functionality at lower cost.

Silicon as a photonic medium has unique advantages. Silicon's process maturity and thermal conductivity are superior to the III-V compounds; its high index of refraction, $\eta(Si) = 3.5$, allows for submicron-dimensioned optical devices and waveguide structures. We envision an entry of silicon microphotonic interconnection at the hybrid, multichip module level with migration to monolithic on-chip architectures early in the next decade.

10.3.1 Photon Propagation

Several SOI platforms have been explored. High performance strip waveguides are dimensioned at 0.5 μm $\times$ 0.2 μm for single mode transmission in the λ=1.3-1.5 μm range with a 7000 A SiO_2 cladding layer required to prevent tunneling losses to the underlying silicon substrate. The single crystal silicon waveguide by the 'Smart-Cut' technology offers high flexibility in layer thicknesses with a high-quality, silicon transmission medium. Polycrystalline silicon offers the maximum flexibility in layer positioning as well as thickness. The defect engineering challenges are reduction of optical loss in polycrystalline waveguides, and high perfection for multilayer wafer bonding.

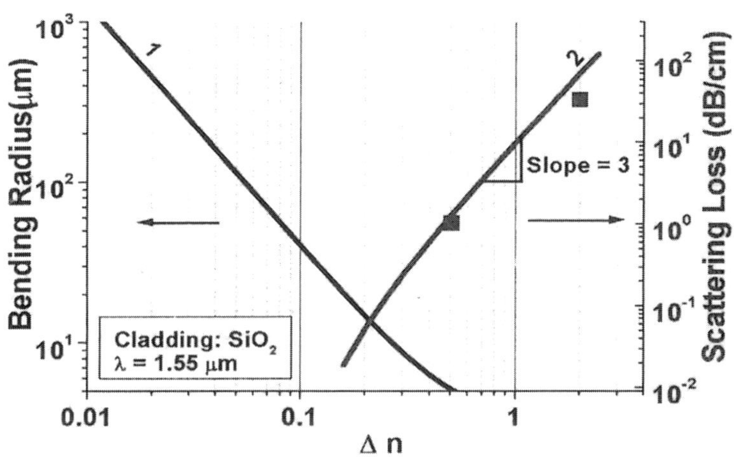

Figure 10.1. Data and model calculations for scattering loss as a function of waveguide refractive index for λ=1.55 μm. The 2001 data employed Ultrabond, single crystalline silicon waveguides. The data points employ silicon nitride and polySi waveguides that have been subjected to a high temperature (1100^{circ}C) anneal with an additional hydrogen passivation step.

We find that the main sources of transmission loss are light scattering and absorption at deep levels. Figure 10.1 shows the measured and calculated loss for single mode strip waveguides as a function of the refractive index difference between the waveguide and its cladding. As the index difference decreases, scattering loss is reduced. However, circuit size must increase, because the bending radius of the guide must decrease with the reduced optical confinement to prevent radiation loss. We have shown that propagation, splitting and bending losses may be reduced with smaller guides by applying these design rules. When these small, single-mode polySi waveguides with non-perfect surfaces and edges are used, scattering losses from edge roughness (as opposed to bulk losses) becomes the dominant source of loss in light transmission [9]. Therefore, using single-crystalline silicon waveguides is no longer a significant advantage because transmission losses in the crystalline and polycrystalline materials are almost equivalent. Post process smoothing can reduce loss below 1 dB/cm [10]. A related source of loss is heterogeneous strain. This strain modifies the core/cladding refractive index locally in a way similar to roughness [11].

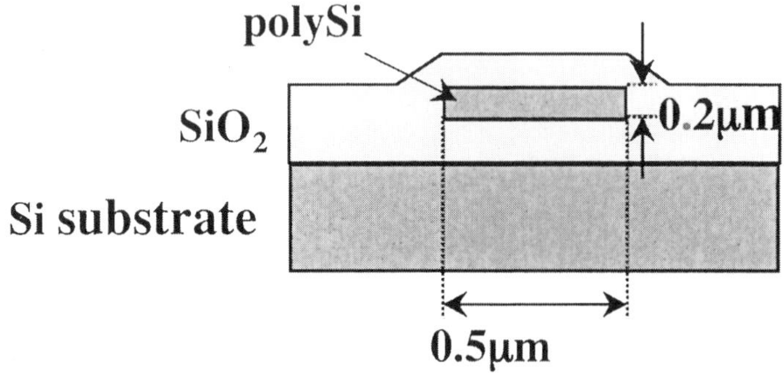

Figure 10.2. Planar polysilicon optical interconnect on SiO_2 cladding. The dimensions of 0.5 μm × 0.2 μm is for single mode transmission at λ= 1.55 μm.

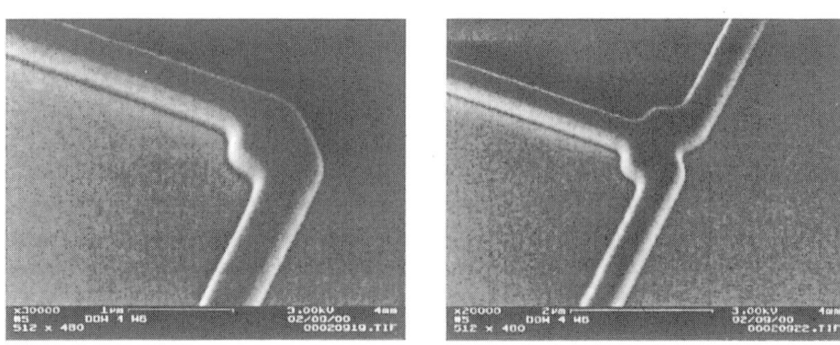

Figure 10.3. Measured losses are less than 0.3 dB for polySi waveguides with bending radii between 1 μm and 100 μm. The 180-degree emergent angle splitters (right) show losses of less than 1 dB.

We have shown [11], [12], that Si/SiO_2 or $polySi/SiO_2$ is an ideal materials system for monolithically integrated optical signal transmission (Figure 10.2) that is capable of high confinement, submicron-dimensioned waveguides and micron-dimensioned radii of curvature for flexible routing [13] (See Figure 10.3).

This capability suggests new degrees of freedom in 3-D architectural design. The materials system is robust, and can be introduced prior to high temperature front-end processing as an interlevel, smart interconnect between layers of electronics, or by low temperature deposition or wafer bonding at the back end. Conventionally one could replace several layers of metal interconnects with a single top layer of optical waveguide circuits. Base level and interlevel structures could act as configurable interconnection for chip sources and loads.

10.3.2 Photon Creation

One of the most intriguing silicon photonic devices is the Si:Er light emitting diode (LED) [14]. Energy is coupled from injected minority carriers to excite the erbium by an intracenter Auger process. Rare earth elements have been employed in photon generation (e.g., Nd:YAG laser) and amplification (e.g., SiO_2:Er optical fiber amplifier) routinely. In particular, the Er^{3+} ion exhibits a 4f intrashell, atomic-like electronic transition that generates a nearly monochromatic photon beam of λ=1.54 μm. The spectral width of the emission is less than 0.6 nm at 4K and less than 1.3 nm below temperatures of 200K in CZ Si [15]. The local crystal field of the ligands adds four spectral features at longer wavelengths in a cubic crystalline matrix. In disordered hosts, the spectral width broadens significantly. The performance goals for optical amplification and laser devices are narrow linewidth, high quantum efficiency and high erbium doping levels.

Figure 10.4 shows the process window for MeV implanted erbium in silicon. An electronegative ligand field surrounding the Er^{3+} in silicon is required for optical activity. [16] Additional oxygen implants promote optical activity; 30 min anneals at 900°C remove the implant damage; and higher temperature treatments deactivate the erbium. The process window is defined by ligand dissociation. Optically active erbium is a donor. The residual optical activity of erbium correlates with the outdiffusion of donors [17]. The diffusivity for this process is given by D(Er-O)=10^3exp(-4 eV/kT), where the activation energy is the Er-trap limited outdiffusion of oxygen. The Er-O association equilibrium constant is given by K=10^{22}exp(-1.5 eV/kT). Therefore, metastable concentrations of implanted erbium can be maintained as long as the thermal budget of the heat treatment remains below the level required for ligand dissociation, but above the level for implant damage removal.

An alternative to implantation is in-situ doping with Er-ligand complexes during silane CVD layer growth. We have explored a variety of metallorganic precursors for this purpose. The main process constraint is the temperature

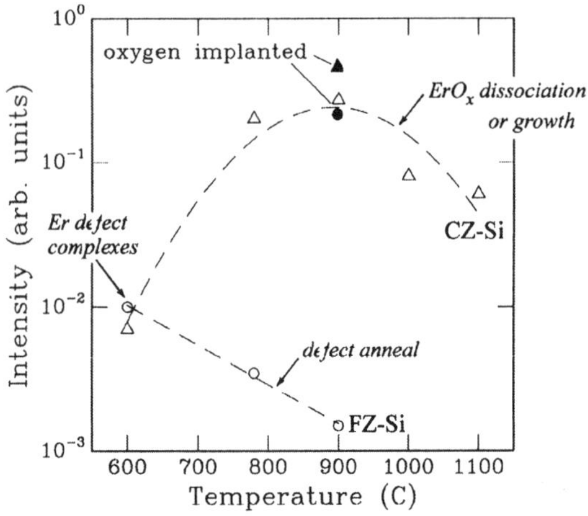

Figure 10.4. The process window of implanted oxygen ligands for Si:Er light emission. Cz-Si and implanted oxygen exhibit stronger light emission.

difference between volatilization and transport of the precursor, and the deposition temperature where both the precursor and the silane dissociate. We have used $Er(tmhd)_3$, $Er(fod)_3$ and $Er(tmsa)_3$ as precursors to provide oxygen, fluorine and nitrogen ligands, respectively [18]. All yield strong Er emission lines. Metastable erbium concentrations as high as $2 \times 10^{21} cm^{-3}$ have been obtained at deposition temperatures near 600°C. The films have a higher quantum efficiency than the best implanted samples. The $Er(tmsa)_3$ has provided the best control of the precursor fragmentation process to give a high yield of Er-ligand complexes in the films.

CMOS process integration of Si:Er LEDs has been demonstrated. Er has a low diffusivity in Si, and it is readily oxidized as an external phase. Hence, it does not cross contaminate during processing [19]. Optical waveguides have been integrated with Si:Er LEDs and drive electronics as shown in Figures 10.5 and 10.6. Modulation of the MOSFET gate voltage provides direct modulation of the LED light output. The circuit exhibits 20 kHz roll-off of the direct modulation frequency due to the long excited state lifetime of Er^{3+}. The mesa design with top contacts allows the Er/O implant and anneal to be independent of the MOSFET thermal budget.

A most valuable application of Si:Er devices would be optical amplification. High photon fluxes at local electronic signal distribution nodes reduce the latency associated with amplification and signal conditioning. An Si:Er optical amplifier with direct injection pumping should be a factor of 10^6 more efficient in excitation than the 980 nm optical pump for conventional fiber amplifiers. This property can reduce amplifier size to levels compatible with component

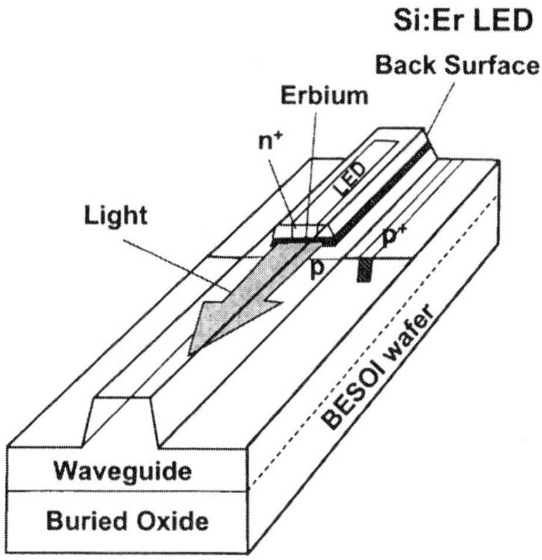

Figure 10.5. Integration of Si:Er LED and Si waveguide.

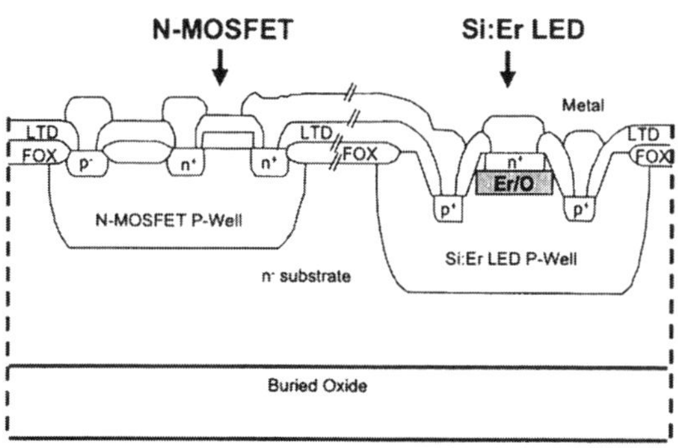

Figure 10.6. Si:Er LED integration with MOSFET driver.

integration. Optical amplification is a key function in the optical interconnection paradigm.

10.3.3 Photon Detection

Transduction of photons to electrons is required for the integration of microphotonics with microelectronics. Since photons of energy greater than the silicon band gap can be absorbed throughout the integrated circuit and inject spurious signals, 'sub-gap' energies are preferred. Materials with bandgaps less than the photon energy must be used for detection. The primary candidate for monolithic integration with silicon is the SiGe alloy. Figure 10.7 shows the design window for these alloys [8].

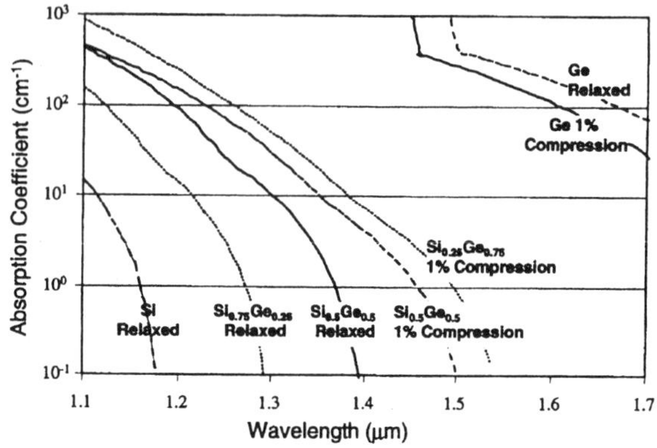

Figure 10.7. Calculated values of absorption coefficient of SiGe alloys in the 1.1 μm$<\lambda<$ 1.7 μm range.

For high speed and high levels of integration small detector sizes and short absorption lengths are desired. As shown in Figure 10.7, Ge can provide high absorption coefficients for λ=1.3-1.5 μm photons that are comparable to compound semiconductors. Threading dislocations are sources of leakage current that introduces noise to the detector performance. Leakage current is introduced in SiGe alloy pn junctions at a rate of 200 pA/μm of dislocation length in the depletion region [7]. The defect engineering challenge is to accommodate the 4% Si/Ge lattice misfit without threading dislocations in the detector device.

This requirement has recently been met using a novel direct growth process combined with post-growth annealing [20] as shown in Figure 10.8. By using a unique thermal cycling procedure that drives dislocations out of the layer, pure Ge layers were grown on Si, with threading dislocation densities below 10^7 cm^{-2} on a wafer, and with zero dislocation on mesa structures within 100 μm $\times$ 100 μm oxide windows. Ge p-I-n photodetectors have been fabricated using these structures with measured responsivities of 770 mA/W at λ=1.3 μm with bit rate capabilities greater than 1 Gb/s.

10.3.4 Photon confinement

Photon confining structures represent a new silicon process challenge. Photonic crystals represent the ultimate in photon confinement. These materials are periodic composites of high and low refractive index materials. The periodicity of the microstructure is the major process and performance constraint.

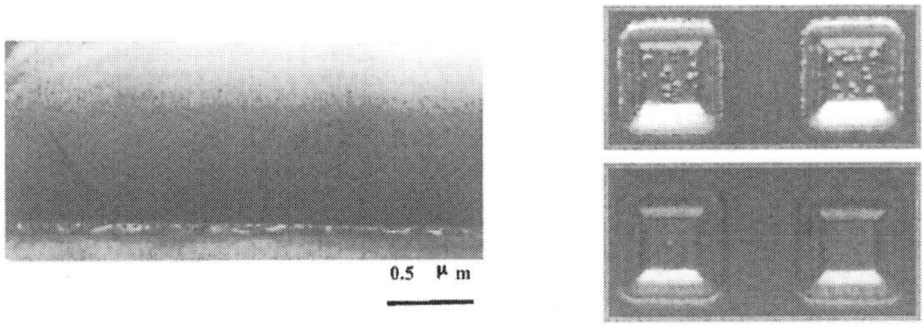

Figure 10.8. SEM picture of an epitaxial layer of Ge directly-grown on Si; as-grown and annealed Ge-on-Si mesas (10×10 μm).

Air holes and Si/SiO_2 multilayers in one dimension, post arrays in two dimensions and displaced checkerboards in three dimensions have been addressed by a variety of approaches. The multilayer structure research closely parallels the development of 3-D SOI integration of electronic circuits. Wafer bonding, CVD and sputter deposition have been successful, because the high index contrast of the materials system requires only four layer pairs (GaAs/AlGaAs requires many tens of layer pairs.) for a sufficiently high cavity Q. The critical microstructural dimension for light of λ=1 μm is of the order of 0.1 μm for silicon. This size range is highly desirable for microphotonics, and the capability to achieve these dimensions with routine fabline processes is likely within the next five years.

A one dimensional photon confining structure is shown in Figure 10.9. The structure is composed of a 0.5 μm wide silicon waveguide with a series of periodically spaced air holes to create the photonic band gap: a frequency range of perfect reflection. The spectral range of the photonic band gap was designed to span the amplification spectrum of the industry standard SiO_2:Er fiber optical amplifier. A missing air hole in the center is the defect. A defect in the photonic crystal constitutes a deep level that traps photons. The device performance closely follows the photonic crystal design (Figure 10.9), and the measured Q of 250 is capable of fitting 128 different wavelength channels within the amplifier spectrum. The optical mode volume of the device, 0.055 μm^3, is the smallest ever created [11]. Optical microcavity resonators can enhance the intensity of Si:Er light emission by as much as 1000$\times$. High Q cavities comprised of Er_2O_3 exhibit strong photon coupling with cavity/Er coupled resonances [21].

Optical resonator devices coupled to high index contrast silicon waveguides have applications as add/drop filters for wavelength division multiplexing, for modulators and for cavity enhanced light emitters. We have designed and fabricated microdisk cavities consisting of Si disks on SiO_2 and observed whis-

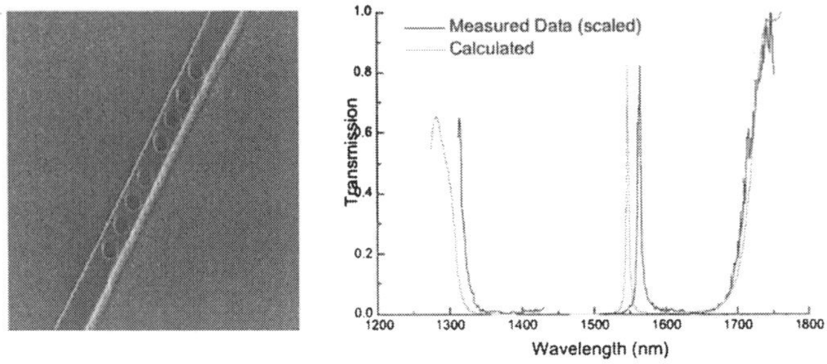

Figure 10.9. Structure and performance of a 1D photonic bandgap resonator.

pering gallery mode resonances near λ= 1.5 μm in these structures. Since the resonant mode travels at the disk periphery, a microring behaves similarly. Microrings with 3 μm diameter show a high Q of 250 and a free spectral range selectivity of 25 nm between channels near λ= 1.5 μm [11], [22]. A silicon microring resonator is shown in Figure 10.10. This racetrack configuration features an enhanced coupling length and an external Q of 2000 [23].

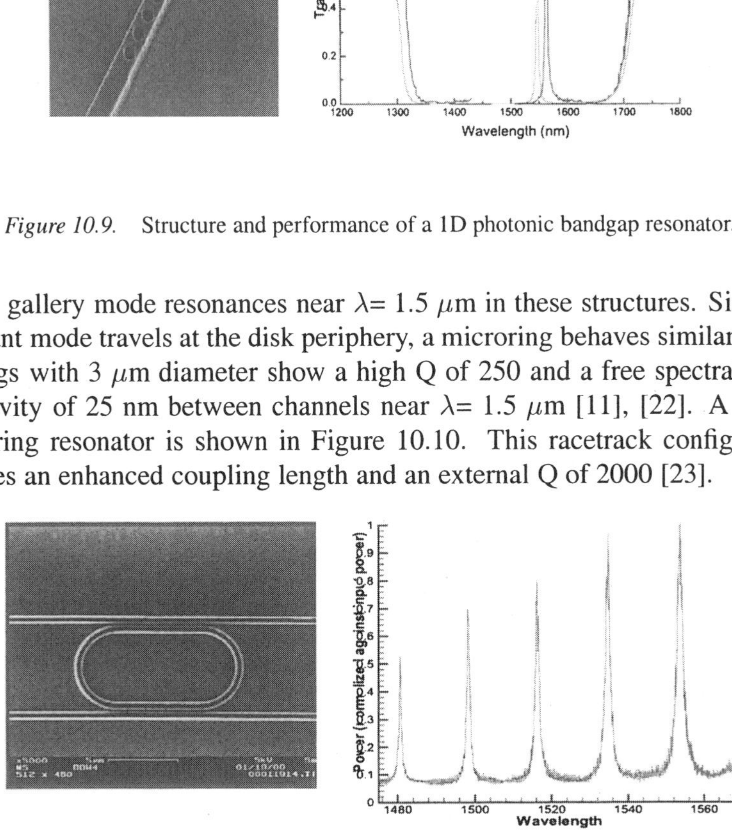

Figure 10.10. Racetrack configuration of ring resonator using silicon waveguides and resonances observed in the drop port. The lower guide is the through port and the upper guide is the drop port.

10.4 Optical Clock Distribution and Data I/O

Electrical clock distribution and data input/output to a chip are two areas facing physical limits as circuit speeds increase and dimensions shrink. Optoelectronic integration can potentially relieve this constraint on microprocessor design in future generations. Optical signal transmission is not limited by the resistance and capacitance of transmission lines. Independent signal carrier wavelengths can be multiplexed on a single transmission line; and the signals are immune from crosstalk interference. The main challenge of silicon microphotonics is reduction of device size to dimensions comparable to devices in integrated CMOS circuits. This size reduction scales in linear dimension with the refractive index (n), difference across the core (high index) and cladding (low index) boundary of the component. Typical optical fiber and planar waveguide structures feature Δn=0.01 by doping of the SiO_2 core. This relatively weak confinement limits not only device size, but also the ability to navigate photons around the sharp turns required for intrachip optical interconnection. The Si/SiO_2 materials system with η(Si)=3.5 and η(SiO2)=1.5 meets all of the requirements for microphotonic waveguides. An Si/SiO_2 waveguide structure has Δn=2, and the scaling factor of 200 allows turn radii of one micrometer. Silicon is transparent for the industry standard λ=1.3-1.5 μm photons, and the index match to semiconductor emitters and detectors facilitates low insertion loss. The small dimensions of the waveguides and devices enable low power optoelectronic transduction and multilevel, highly integrated functionality.

10.4.1 Circuit Architecture

A circuit architecture for global clock signal distribution with local electronic distribution is shown in Figure 10.11. Since the passive optical components (waveguides, splitters and bends) can present negligible losses, the remote photon source can reduce the processor chip power dissipation by as much as 50%. As in electronics, the signal propagation path lengths need to be matched. However, this matching tolerance is of the order of 1 mm for 5 psec skew (signal propagation skew $\approx \Delta L\eta/c$, η-index of refraction, ΔL path length difference). The major skew contribution comes from the integration of receiver electronics involving multiple amplification stages. A preliminary analysis indicates that both power dissipation and skew increase exponentially with the number of transistors in the receiver circuit.

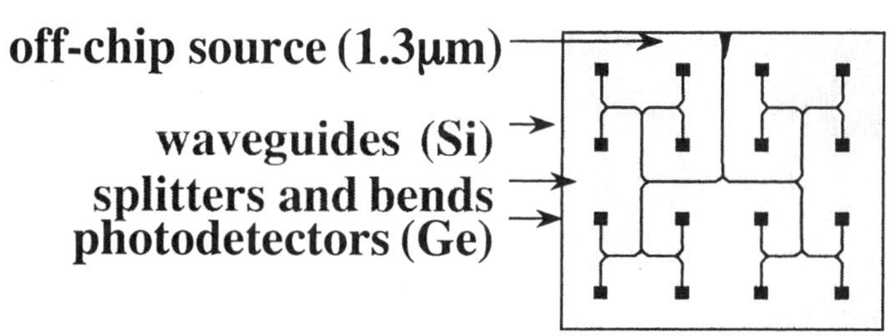

Figure 10.11. Optical bus architecture for on-chip optical signal distribution.

The optical clock and the I/O signal distribution make use of the optical elements described above. The 'photon bus' concept is directly analagous to the chip power supply. A photon flux from the light source is routed and encoded in much the same manner as current is treated in electronic circuits. For the I/O architecture, optical modulators driven by local electro-optic devices will be used. To maximize bandwidth using optics for I/O, optical signal multiplexing using the components shown in Figure 10.9 and 10.10 can be used.

Optical interconnection is a scalable solution to both the clock distribution and the data I/O problems. An optical signal can be split without adding delay or skew. The only optical delay is the difference in transmission time for lines of different lengths. This delay is less than 50 ps for a 1 cm difference in path length. It should be relatively easy, however, to control line lengths to differences less than 1 mm. More importantly, the delay does not depend on the circuitry surrounding the transmission line, nor does it depend on the switching speed of the signal being sent. Finally, many signals can be multiplexed into one waveguide; even the clock and data signals can be combined, reducing the number of required inputs and outputs.

Optical interconnects can also provide relief from the power consumption problem. Waveguides are passive devices, so the only power consumed will be in the emitter/modulator and detector. The power consumed in electrical interconnects is directly proportional to the frequency of the signal and the total capacitance. At very high frequencies and small dimensions, the capacitance between different lines, as well as between the lines and the substrate, contributes to both signal delay and power consumption.

10.4.2 System Design

The architecture for a microprocessor with optical clock distribution could be as shown in Figure 10.11. The clock signal is generated in a separate resonator circuit and converted to an optical signal with a relatively high power laser. The alternate light source configurations in order of probability of adoption are: 1) a mode locked laser with optical delay timing, 2) a laser with an electro-absorption modulator, 3) a laser directly pumped by oscillator/power supply. This signal can originate on a dedicated source chip, or die in an MCM configuration, and the clock signal can be distributed to all synchronized chips in the system. The clock signal is received on-chip optically and split optically (10-100 times) in distribution to the major sections of the chip. At each major section the signal is converted to an electrical signal through a detector and receiver circuitry to yield the appropriate voltage level, and it is distributed electrically to subsections. Measured data on existing silicon avalanche photo detectors (APD) show that a total skew of 100 ps can be achieved with a silicon APD and one stage of amplification with an incident optical power of 1 μW

[24]. Since APD detectors require high voltages to operate, which may be incompatible with high levels of integration, PIN detectors with a higher optical power requirement may be preferred.

Starting with the 1 μW required power at each detector, the specifications for the emitter and waveguide depend on the total fanout level and the coupling and distribution losses. For 16 levels of fanout with a 10dB loss budget, the coupled emitter power must be 160 μW. For a p-i-n detector a factor of 10 more power, 1.6mW, is required. The requirements for data transmission are somewhat less stringent. In order to achieve a bit error rate of 10^{-9} from a silicon APD detector, an incident photon flux of 125 photons/bit, or 1.65×10^{-8}W is required. Thus, for data transmission, two orders of magnitude less optical power is required than for clock distribution. For the clock application with current architectures, the switching of individual latches dominates the current requirements. More than 100 μA can be required to switch one transistor latch at a 1nS clock cycle time, a 100 ps skew limit and a 1V power supply level. Electronic amplification after the photodetector can only add to skew. Therefore, a major design consideration for optical clock is the depth of optical distribution. "A million points of light" represents the lowest skew configuration, but a shallower fanout with local electronic distribution could demand less total power.

Optical data transmission between chips illustrates an opportunity to take advantage of the high optical bandwidth. By using a large array of emitters and detectors operating at high modulation frequencies, data transmission in the terabaud region could be realized. Rather than requiring hundreds of fibers running between the chips, which would be difficult and expensive, free space connections with mirrors and lenses are likely to be preferable.

The remaining issues are the operating wavelength and the waveguide materials. A photon wavelength of 1.5 micron is compatible with telecommunications standard, and since its energy is below the bandgap of silicon, there is no effect of stray light on device performance. One problem this wavelength creates is detector materials. Silicon is an excellent material for APDs operating at 0.88 μm), but it cannot be used at 1.5 μm. Germanium possesses the appropriate bandgap, but its devices will have a higher dark current, raising the power required at the detector to maintain a bit error rate of 10^{-9}.

Doped silicon dioxide waveguides are compatible with optical fiber interconnection between chips. However, the low refractive index difference between the core and the cladding of these structures can lead to very high bending losses. As a result, turning mirrors must be incorporated to route the signal around sharp bends. At the 1.5 μm wavelength, silicon waveguides clad with silicon dioxide can be used. These waveguides can accommodate very small radii of curvature turns, $r < 10$ μm with low loss, due to the large index difference between the waveguide and the cladding.

An important, but often neglected, part of the silicon optoelectronic circuit is the packaging. This packaging issue involves the integration of all on-chip optical components as well as the coupling to off-chip sources. Critical tapering of waveguides is required for the coupling of optical fiber to on-chip nano-waveguides. Every internal coupling must be optimized to reduce insertion loss of the components. Options for packaging and integration need careful evaluation.

10.5 Summary

Component and circuit engineering for microelectronics is structured to meet the constraints posed by carrier transport and lifetime, and electric field design criteria. One is concerned with the control of scattering sites, recombination centers and localized sources of dielectric breakdown and leakage current. Microphotonics focuses concern on photon creation, propagation and optoelectronic detection. The figures of merit for photonics are optical power transmission loss, quantum efficiency of photonic/electronic transduction and optical microcavity Q (photon lifetime). Photon absorption and scattering are sources of loss. Localized dipole coupling determines the efficiency of photon generation and detection. Precise dimensional control at length scales of the photon wavelength in the material (wavelength divided by the refractive index) is necessary. These metrics provide the engineering design constraints. This paper has reviewed the principles and progress in component and circuit engineering for silicon microphotonic emitters, waveguides, detectors and optical resonators.

We envision an architecture for the future where global clock distribution will enable parallelism among chips and across the entire chip. The ultimate limiter to such an architecture has been the skew-compensating electronic circuitry needed to achieve uniform signal delay. To circumvent this "electronic" limitation, we need to have several minichips within the chips. Each minichip locally performs the electronic logic functions required by the device. For this vision to become a reality in the future, signals have to be communicated between minichips over long distances ($>$10 mm) without skew. Optical interconnects provide an optimal solution to this "electronic" limitation. The fundamental limit of the speed of $\lambda = 1.5\ \mu$m light in silicon waveguides is a signal propagation delay at about 100 ps/cm. By designing clock distribution interconnects to be the same length, high frequency clock distribution over long distances ($>$10mm) as well as over shorter on-chip distances, can be achieved. The fundamental quantum limit of an ideal photodetector with unity quantum efficiency and no dark current is 20 photons/bit for a bit-error-rate (BER) of 10^{-9}. The envisioned entry technology architecture, Figure 10.11, consists of off-chip light sources, on-chip optical bus signals for clock and data; on-chip modulators to encode data; on-chip detectors to transduce optical clock signals for local electronic distribution; and optical add/drop filters for MUX/DEMUX.

A complete appraisal of silicon microphotonic interconnection, requires detailed analysis of the issues of speed, area and power. Evaluation of an electrical interconnect approach involves optimization of materials, wiring designs, and systems architecture. For the evaluation of optical interconnects, one must to optimize each optical component and the integration of the components to retain the benefit of high bandwidth, minimal crosstalk and low power consumption. Optical impedance matching is critical to minimization of insertion loss and standing wave reflections. Large scale monolithic integration has the potential of solving an untenable insertion loss issue that is orders of magnitude greater than exists in current point-to-point communications. The principles for integration must be established early in the technology with simple circuit layouts. Together, this new class of devices constitutes a gateway to on-chip and interchip optical interconnection , and a potential solution to the integrated circuit, interconnection bottleneck.

It is interesting then to think of the optoelectronic system design as the driving technology of the future with current chip-size Si CMOS as the basic units of tomorrow. The optoelectronic interconnection and hybrid chip integration will be the back end of tomorrow. In other words, the interconnection hierarchy above the current chip-level ("the board") becomes the new "chip" of tomorrow. Additionally, SOI wafer bonding and solder bump die attach can be major enablers of hybrid optoelectronic to electronic integration. The industries that possess this design and manufacturing capability for integrated optical interconnects will be the leaders of semiconductor electronics of tomorrow.

Acknowledgments

I am indebted to the students and colleagues listed in the references who inspired and implemented this work. Professor Hermann Haus and Dr. Anu Agarwal, in particular, were instrumental in developing concepts for integration and in writing. This research was supported by the NSF-MRSEC under contract DMR9808941, AFOSR under contract F49620-95-1-0312, the Semiconductor Research Corporation, and the MARCO Interconnect Focus Center.

References

[1] "The National Technology Roadmap For Semiconductors 1997 Edition", SIA Semiconductor Industry Association, (San Jose, CA), 1997, pp. 1-196.

[2] Ashok V. Krishnamoorthy and David A. B. Miller, "Scaling Optoelectronic-VLSI Circuits into the 21st Century: A Technology

Roadmap", *IEEE Journal of Selected Topics in Quantum Electronics*, Vol. 2, No. 1, April 1996.

[3] Ashok V. Krishnamoorthy and D. A. B. Miller, "Firehose architectures for free-space optically-interconnected VLSI circuits", *Ginzton Laboratory Report No. 5474*, Stanford University, Stanford, CA, November 1996.

[4] Atsushi Iwata and Izuo Hayashi, "Optical Interconnections as a New LSI Technology", *IEICE Trans. Electron.*, Vol E76-C, No. 1, January 1993.

[5] D.A.B. Miller and H. M. Ozaktas, "Limit to the Bit-Rate Capacity of Electrical Interconnects from the Aspect Ratio of the System Architecture", *Ginzton Laboratory Report No. 5458*, Stanford University, Stanford, CA, October 1996.

[6] E. A. Fitzgerald and L. C. Kimerling, "Silicon-based microphotonics and integrated optoelectronics", *MRS Bulletin*, Volume 23, Number 4, Page 39, April 1998.

[7] A. Agarwal, J.S. Foresi, L.M. Giovane, L. Liao, J. Michel, K. Wada, and L. C. Kimerling, "Defect Engineering for Silicon Microphotonics", *Defects in Silicon III*, T. Abe, W. M. Bullis, S. Kobayashi, W. Lin and P. Wagner, Eds. (The Electrochemical Society, Pennington, NJ, 1999) p. 215.

[8] L.M. Giovane, D.R. Lim, S.H. Ahn, T.D. Chen, J.S. Foresi, L. Liao, E.J.Oulette, A.M. Agarwal, X. Duan, J. Michel, A. Thilderkvist, L.C. Kimerling, "Materials For Monolithic Silicon Microphotonics," *Proceedings of the MRS 1997 Fall Meeting*, (Boston, MA), December 1997, Vol.486 (1998), pp. 45-56.

[9] K.K. Lee, D.R. Lim, H-C Luan, A. Agarwal, J. Foresi and L.C. Kimerling, "Effects of Size and Roughness on Light Transmission in Si/SiO2 Waveguides: experiments and models", *Applied Phys. Lett.*, 77, 1617 (2000).

[10] K. K. Lee, D. R. Lim, and L. C. Kimerling; J. Shin, F. Cerrina, "Fabrication of Ultra-low Loss Si/SiO2 Waveguides by Roughness Reduction", *Optics Letters*, 26, 1888 (2001).

[11] J. S. Foresi, D. R. Lim, A. M. Agarwal, L. C. Kimerling, M. Tavassoli, M. Cox, M. Cao and W. Greene, "Small radius bends and large angle splitters in SOI waveguides", *Proceedings of the SPIE "Photonics West" Symposium*, (San Jose), Vol. 3007, pp 112-118, February 1997.

[12] Anuradha M. Agarwal, Ling Liao, James S. Foresi, Marcie R. Black, Xiaoman Duan and L. C. Kimerling, "Low-loss polycrystalline silicon waveguides for silicon photonics", *J. Appl. Phys.*, vol. 80, (11), Dec. 1996.

[13] C. Manolatou, Steven G. Johnson, Shanhui Fan, Pierre Villeneuve, H.A. Haus and J.D. Joannopoulos, "High-Density Integrated Optics", *Journal of Lightwave Technology*, 17, 1682 (1999).

[14] B. Zheng, J. Michel, F. Y. G. Ren, D. C. Jacobson, J. M. Poate, and L. C. Kimerling, "Room Temperature Sharp Line Electroluminescence at = 1.54 ţm from an Erbium doped, Silicon Light-emitting Diode," *Applied Physics Letters*, 64, 2842, (1994).

[15] T. Chen, A.M. Agarwal, L.M. Giovane, J.S. Foresi, L. Liao, D.R. Lim, M.T. Morse, E.J. Ouellette, S.H. Ahn, X. Duan, J. Michel, L.C. Kimerling, "Erbium-doped polycrystalline silicon light emitters for resonant microcavity devices," in "Light-Emitting Diodes: Research, Manufacturing, and Applications II," *Proceedings of the SPIE "Photonics West" Symposium*, (San Jose, CA), Jan. 1998 (Ed) Fred Schubert, Vol. 3279, pp. 136-145.

[16] Lionel C. Kimerling, K. D. Kolenbrander, J. Michel, J. Palm and L. M. Giovane, "Light Emission from Silicon", *Solid State Physics* 50, 333 (1996).

[17] S. Ahn, J. Palm, B. Zheng, X. Duan, A. Agarwal, S. Nelson, J. Michel, L.C. Kimerling, "Electrical Study of Crystalline Silicon Coimplanted with Erbium and Oxygen," *Proceedings of the SPIE "Photonics West" Symposium*, Vol. 3007, (San Jose, CA), Feb. 1997, pp 144-151.

[18] M. Morse, B. Zheng, J. Palm, X. Duan, L.C. Kimerling, "Properties of Ion-Implanted and UHV-CVD Grown Si:Er," *Proceedings of the MRS Spring Meeting*, Vol. 422 (1996) pp 41-46.

[19] F. Y. G. Ren, J. Michel, Q. Sun-Paduano, B. Zheng, H. Kitagawa, D. C. Jacobson, J. M. Poate and L. C. Kimerling, "IC Compatible Processing of Si:Er for Optoelectronics", *Proceedings of the MRS Spring Meeting*, Vol. 298 (1993) pp. 415-423.

[20] L. Colace, G. Masini, G. Assanto, H.C. Luan, K. Wada and L.C. Kimerling, "Efficient, High Speed Near Infrared Ge Photodetectors Integrated on a Si substrate," *Applied Phys. Letters*, Vol. 76, 1231, 2000.

[21] M. Lipson and L. C. Kimerling, "Er^{3+} in strong light confining microcavity," *Applied Phys. Lett.* 77, 1150 (2000).

[22] "Ultra-Compact Si-SiO2 microring Resonator Optical Channel Dropping Filters," *IEEE Photonics Technology Letters*, Vol. 10 (4), pp. 549-551, April 1998.

[23] D. Lim, B.E. Little, K. Lee, M. Morse, H. Fujimoto, H.A. Haus and L.C. Kimerling, "Micron-sized channel dropping filters using Si waveguide devices", *Proc. SPIE 3847*, 65 (1999).

[24] SPIE Vol. 1178 (1989), pp 171-176.

Index